Studien zur Interdisziplinären Anthropologie

Herausgegeben von
G. Hartung
Wuppertal, Deutschland

Unter dem Leitbegriff der Interdisziplinären Anthropologie formiert sich aktuell eine Forschungslandschaft, die dem Rätsel des Menschen angesichts seiner Riskiertheit und nicht-garantierten Überlebenschancen, aber auch seiner technologischen Gestaltungschancen nachgeht. Einerseits liefern die neueren Forschungen zur evolutionären Anthropologie in kurzen Fristen immer präzisere Daten zur Bestimmung der menschlichen Lebensform; andererseits stellen uns die neuen technologischen Möglichkeiten in den Lebenswissenschaften vor praktische Probleme der Folgenabschätzung unseres Handelns. Derzeit sind alle Wissensdisziplinen sowohl der Natur- als auch der Geistes- und Kulturwissenschaften gefragt, ihren Beitrag zur Orientierung in dieser Situation zu liefern. Es geht dabei um theoretische Durchdringung komplexer Forschungsfragen und deren ethische Reflexion. Wir können daher mit Blick auf die Interdisziplinäre Anthropologie von einem Schlüsselthema aktueller Forschung sprechen. Die vorliegenden Studien zur Interdisziplinären Anthropologie stellen – neben dem Jahrbuch Interdisziplinäre Anthropologie – einen weiteren Versuch dar, diesem weiten Forschungsfeld ein Gesprächsforum zu bieten.

Herausgegeben von:
Gerald Hartung
Bergische Universität Wuppertal
Deutschland

Editorial Board:
Jörn Ahrens, Universität Gießen, Deutschland
Cornelia Brink, Universität Freiburg, Deutschland
Dirk Evers, Universität Halle, Deutschland
Thomas Fuchs, Universität Heidelberg, Deutschland
Matthias Herrgen, Universität Münster, Deutschland
Matthias Jung, Universität Koblenz, Deutschland
Katja Liebal, Freie Universität Berlin, Deutschland
Stephan Rixen, Universität Bayreuth, Deutschland
Hartmut Rosa, Universität Jena, Deutschland

Matthias Jung · Michaela Bauks
Andreas Ackermann (Hrsg.)

Dem Körper eingeschrieben

Verkörperung zwischen
Leiberleben und kulturellem Sinn

 Springer VS

Herausgeber
Matthias Jung
Michaela Bauks
Andreas Ackermann

Universität Koblenz-Landau
Deutschland

Studien zur Interdisziplinären Anthropologie
ISBN 978-3-658-10473-3   ISBN 978-3-658-10474-0 (eBook)
DOI 10.1007/978-3-658-10474-0

Die Deutsche Nationalbibliothek verzeichnet diese Publikation in der Deutschen Nationalbi-
bliografie; detaillierte bibliografische Daten sind im Internet über http://dnb.d-nb.de abrufbar.

Springer VS

Lektorat: Frank Schindler, Katharina Gonsior

Gedruckt auf säurefreiem und chlorfrei gebleichtem Papier

Springer Fachmedien Wiesbaden ist Teil der Fachverlagsgruppe Springer Science+Business Media
(www.springer.com)

Inhalt

3   Die Leiblichkeit verkörperter Gründe

4   Zur historischen Anthropologie von Körperritualen

Einleitung

Alle kulturellen Praktiken sind verkörpert, müssen also als Spannungseinheit zwischen einer symbolischen und einer physischen Komponente verstanden werden. Kultureller Sinn kann daher weder als rein geistige Bedeutung/Struktur noch als beobachtbares Verhalten allein beschrieben werden, und der menschliche Organismus ist nie bloß ein von außen beschreibbarer Körper oder von innen erlebter Leib. Auch der normative Charakter von Kultur, ihre essentielle Prägung durch erwünschte oder geforderte Möglichkeiten, kann nur als die „symbolische Prägnanz" (Ernst Cassirer) physischer bzw. physiologischer Realitäten gedacht werden. Normativität existiert nicht freischwebend, sie verkörpert sich in Vorbildern, Narrationen, Ritualen und nicht zuletzt in Einschreibungen in den Körper selbst. Diese basale Einsicht war in den antidualistischen Unterströmungen des anthropologischen Denkens von Giambattista Vico über Johann Gottfried Herder, Wilhelm von Humboldt, den Pragmatismus und die Leibphänomenologie immer schon gegenwärtig, steht aber dennoch quer zu den etablierten Routinen des Forschungsbetriebs und hat erst in jüngster Zeit wieder die ihr gebührende Aufmerksamkeit gefunden.[1] Das Thema Verkörperung/Embodiment ist zum Zentrum einer großen Zahl neuerer Publikationen geworden, ob sie nun antidualistisch Ausdrucksphänomene analysieren[2], die Kognitions- und Neurowissenschaften anthropologisch kontextualisieren[3] oder

---

1 Vgl. etwa A. Barkhaus/M. Mayer/N. Roughley/D. Thürnau (Hg.), Identität, Leiblichkeit, Normativität. Neue Horizonte anthropologischen Denkens (stw 1247), Frankfurt a. M. 1996; ²1999.

2 Vgl. etwa N. Meuter, Anthropologie des Ausdrucks. Die Expressivität des Menschen zwischen Natur und Kultur, München 2006; M. Jung, Der bewusste Ausdruck. Anthropologie der Artikulation, Berlin/New York 2009.

3 T. Fuchs, Das Gehirn – ein Beziehungsorgan. Eine phänomenologisch-ökologische Konzeption, Stuttgart 2008.

Bausteine zu einer „Philosophie der Verkörperung" vorlegen wollen.[4] Parallel zu diesem neuen *begrifflichen* Interesse an Verkörperung und interdisziplinär mit ihm verknüpft entstehen neue Forschungsperspektiven etwa in Kunstwissenschaft[5], Ethnologie[6] und Bibelwissenschaft/antiker Religionsgeschichte.[7]

Dabei lässt sich allerdings beobachten, dass der Ausdruck „Verkörperung" häufig als eine Art *umbrella term* verwendet wird, unter dem sich höchst Unterschiedliches verbergen kann. Vor allem folgende Differenzierung ist hier von zentraler Bedeutung: Während sich das Interesse in den Kognitionswissenschaften, aber auch in der philosophischen Anthropologie oft auf *somatische* Verkörperung (also auf jene Phänomene, die durch die Leib/Körper-Unterscheidung sichtbar gemacht werden) konzentriert, richten die ethnologisch bzw. historisch-anthropologisch beheimateten Kulturwissenschaften ihren Blick eher auf *kulturelle* Verkörperungsformen. Diese lassen sich wiederum dreifach auffächern: In *sozialen Praktiken* (speziell Ritualen) internalisieren die Subjekte kulturellen Sinn und gestalten ihn durch die Art ihrer Performanz gleichzeitig mit. *Soziale Institutionen* stellen die Interaktionseffekte solcher Praktiken auf Dauer ein, und in *materiellen Artefakten* schließlich wird Naturmaterialien kultureller Sinn eingeschrieben. Durch die so etablierte Arbeitsteilung zwischen Kognitions- und Kulturwissenschaftlern entsteht jedoch die Gefahr, dass die integrative, antidualistische Kraft des Verkörperungsbegriffs verlorengeht und sich ein neuer Dualismus ausbreitet, der nun die Differenz Leib/Kultur an die Stelle des klassischen Hiatus zwischen Leib/Seele zu setzen droht.

Die Beiträge des hier vorgelegten Bandes wollen daher sichtbar machen, dass innovative Forschung und gelingende Interdisziplinarität davon lebt, somatische und kulturelle Verkörperung gerade aufeinander zu beziehen, statt sie dualistisch zu separieren. Sie wenden sich deshalb kulturellen Praktiken zu, in denen der *physische Körper* eine konstitutive Rolle spielt, als Subjekt/Objekt sozialer Semantisierungen und zugleich als Quelle von Sinn und Manifestationsort kultureller Sinndeutungen.

---

4   M. Wild/R. Hufendiek/J. Fingerhut (Hg.), Philosophie der Verkörperung (stw 2016), Berlin 2013.

5   Ein Beispiel für viele: U. Feist/M. Rath, Et in imagine ego. Facetten von Bildakt und Verkörperung (FS Horst Bredekamp), Berlin 2012.

6   F. E. Mascia-Lees (Hg.), A Companion to the Anthropology of the Body and Embodiment. Malden, Mass. 2011.

7   Vgl. etwa U. Steinert, Aspekte des Menschseins im Alten Mesopotamien (Cuneiform monographs 44), Groningen 2012; A. Berlejung/J. Dietrich/J.F. Quack (Hg.), Menschenbilder und Körperkonzepte im Alten Israel, in Ägypten und im Alten Orient (ORA 9), Tübingen 2012; zu den Antikenwissenschaften L. Thommen, Antike Körpergeschichte (utb 2899), Zürich 2007.

Der Band enthält die Beiträge einer im Rahmen des geisteswissenschaftlichen Forschungsschwerpunkts „Kulturelle Orientierung und normative Bindung" an der Universität Koblenz-Landau im Dezember 2013 durchgeführten Tagung. Die Veranstaltung sowie die Redaktionsarbeiten des Tagungsbandes wurden finanziert aus Mitteln der Forschungsinitiative des Landes Rheinland-Pfalz. Unser Dank geht an Lilli Ohliger, die den Band redaktionell betreut und die Register erstellt hat, sowie an Tina Massing und Franziska Schwan, die sämtliche Beiträge Korrektur gelesen haben. Aus der Sicht von Bibelwissenschaft, Ethnologie, Pädagogik, Philosophie, Religionsgeschichte sowie Religionswissenschaft widmet er sich aus diachroner wie synchroner Perspektive den vielfältigen Manifestationen sinnhaften Handelns und Sprechens im und am menschlichen Organismus in der Spannung zwischen Körper und Leib, Individualisierung und sozialer Normativität.[8]

Die integrative Herausforderung liegt in der transdisziplinären Bedeutung des Forschungsgegenstands. Mit der Vielfalt methodischer (z.B. phänomenologischer, strukturalistischer, hermeneutischer, semiotischer etc.) Ansätze geht nämlich die Gefahr einher, über der Theorieabhängigkeit der Gegenstandskonstitution aus dem Auge zu verlieren, wie *alle* Verkörperungsphänomene jeweils zwischen biologischem Organismus und sozialem Sinn vermitteln. Die verschiedenen Beiträge zu diesem Band gelangen hier zu durchaus unterschiedlichen Einschätzungen. Als zentrale Variable fungiert jeweils die Frage, wie tief symbolischer, jeweils kulturspezifisch geformter Sinn in die Funktionen des Organismus und das individuelle Leiberleben eindringt. Kann man z.B. sinnvoll zwischen primärem Erleben und sekundärer Semantisierung unterscheiden oder imprägniert Letztere noch die elementarsten Selbstvollzüge des Organismus? Lassen historisch weit zurückliegende Beispiele überhaupt Aussagen zu einer solchen Unterscheidung zu? Wenn ja, wie lassen sich diese methodisch dingfest machen? Kulturelle Rituale können als „Schnittstellen" begriffen werden, an denen sich solche Fragen besonders eindringlich stellen, und von einer einheitlichen Antwort kann hier keine Rede sein. Gleichwohl sind alle Beiträge zu dem vorgelegten Band im Bewusstsein dieser Problematik geschrieben und arbeiten sich aus sehr unterschiedlichen methodischen Perspektiven an ihr ab.

Die Gliederung des Bandes in vier thematische Abschnitte dient nur einer groben, vom methodischen Ausgangspunkt bestimmten Orientierung. Sie darf gerade nicht im Sinne einer Fixierung inhaltlich abgegrenzter Bereiche verstanden werden, die den transdisziplinären Gedanken konterkarieren würde. Mit diesem Vorbehalt lassen sich *phänomenologische* (1), kritisch an der *Textmetapher* orien-

---

8   Vgl. auch C. Brosius/A. Michaels/P. Schrode (Hg.), Ritual und Ritualdynamik (utb 3854), Göttingen 2013.

tierte (2), *begründungstheoretische* (3) und *historisch-anthropologische* (4) Zugänge unterscheiden.

(1) Für die *Phänomenologie* ist die theoretisch möglichst enthaltsame Beschreibung leiblicher Vollzüge entscheidend; der Schwerpunkt liegt entsprechend nicht auf den Einschreibungen kulturellen Sinns in das Erleben der Subjekte, sondern umgekehrt auf deren präsemantischen Selbstvollzügen als leibliche Wesen, die als Voraussetzungen für kulturelle Rituale im engeren Sinn begriffen werden. Alle drei Beiträge konzentrieren sich dabei auf die *soziale Leiblichkeit*. So untersucht der Beitrag von *Susan Stuart* das Zusammenspiel von Verhalten, Wahrnehmung und begrifflicher Sinnzuweisung innerhalb des Konzepts der *Enkinaesthesia*, der Beschreibung der funktionalen Integration von Bewegungsschemata und Wahrnehmungsmustern leiblich Handelnder, in denen *ego* und *alter* als getrennte Subjekte erst entstehen. Am Beispiel von Artikulation und Rezeption menschlicher Sinneswahrnehmung reflektiert Stuart auch die Unmöglichkeit, somatische und kulturelle Verkörperung zu trennen. Damit schreibt sie sich in die phänomenologische Perspektive ein. In dieser wird, ausgehend vom „Leib als Wahrnehmungsorgan, als Nullpunkt der Orientierung, als Weise des Weltzugangs"[9] die Bedeutung der *Leib-Körper-Differenz* hervorgehoben, und dies in der Absicht, die Wende zum *Embodiment* aus der drohenden Vereinseitigung von Verdinglichung des Körpers oder aber seiner Diskursivierung herauszuführen.[10]

Der phänomenologische Ansatz ist weiterhin vertreten durch *Käte Meyer-Drawe*s Untersuchung des Blicks und seiner Bedeutung für zwischenmenschliche Begegnungen. Meyer-Drawe entwickelt eine differenzierte Phänomenologie der vielfältigen Formen, in denen Blicke vorab und zusätzlich zur sprachlichen Kommunikation die Ko-Präsenz von Menschen in einem geteilten Raum – ebenso wie den Rückzug aus ihm – vermitteln können. Die Metapher des Kreuzens schließt hierbei die vielfältigen Spielräume symmetrischer wie asymmetrischer Begegnungen in besonders prägnanter Weise auf. Mit seinem ethnographisch-phänomenologischen Beitrag stellt *Thorsten Gieser* die Verkörperungs-Debatten der jüngsten Zeit auf empirisch trittsichere Füße und treibt dabei jegliche Restbestände von kartesianischem Dualismus aus. Vor dem Hintergrund seiner im besten Sinne des Wortes teilnehmenden Beobachtungen bei Druiden in England vertritt er die These, dass das Konzept der Verkörperung selbst in phänomenologischen Kreisen noch zu wenig konsequent im Hinblick auf den sinnlich wahrnehmenden Körper,

---

9   Einleitung in Wild, Hufendiek, Fingerhut, Philosophie der Verkörperung, 2.
10 E. Alloa, T. Bodorf, C. Grüny, T.N. Klass (Hg.), Leiblichkeit. Geschichte und Aktualität eines Konzepts (utb 3633), Tübingen 2012.

den Leib gedacht wird. Am Beispiel eigener körperlicher Erfahrungen während eines druidischen Rituals zeigt Gieser, dass druidische Rituale weniger eine formalisierte Abfolge symbolisch strukturierten Verhaltens darstellen, sondern vor allem dazu dienen, den Teilnehmern eine bestimmte Qualität von Erfahrung des In-der-Welt-Seins zu ermöglichen, die in das Gefühl mündet, Teil der Natur zu sein. Dieser Vorgang verweist auf das transformative Potenzial von Ritualen und wird von Gieser unter Rückgriff auf Schmitz als Dynamik von „Ausleibung" und „Einverleibung" beschrieben. Dabei kann Erfahrung zum Erlebnis werden, das erinnert, reflektiert und besprochen wird, aber im sich in seiner Umwelt bewegenden und spürenden Leib begründet ist.

(2) Mit Mary Douglas' Interpretation des Körpers als doppelt „natürlichem" Symbol, welches nicht nur Ausdruck der körperlichen Natur des Menschen ist, sondern auch soziale Strukturen ausdrückt, etabliert sich eine strukturalistische Perspektive, die Homologien bzw. Repräsentationen von Gesellschaft und Körper in den Blick nimmt und häufig Textmetaphern verwendet.[11] Den Körper als Zugang zu Machtverhältnissen bzw. umgekehrt den Körper disziplinierende *body politics* thematisieren u. a. S. Freud (Kompensation), Marcel Mauss (Körpertechniken), Michel Foucault sowie Pierre Bourdieu (Habitus).[12] Zu solchen Ansätzen nimmt der Beitrag von *Andreas Ackermann* eine so ergänzende wie kritische Position ein. Aus ethnologischer Perspektive gibt er einen Überblick über die Rolle des Körpers insbesondere im Kontext des Rituals und präsentiert in diesem Zusammenhang eine den Tagungsband orientierende Definition von Ritual. Überblickt man die sozial- und kulturwissenschaftlichen Debatten der letzten 100 Jahre zum Thema Körper bzw. Verkörperung und Ritual, so lässt sich eine mangelnde Berücksichtigung des Somatischen feststellen sowie die Tendenz, körperliche Erfahrung mit kognitiven und linguistischen Modellen von Sinn bzw. Bedeutung zu interpretieren. Dies zeigt sich an der Betrachtung von Körpern als Objekten bzw. Oberflächen, die u. a. gesellschaftliche Normen repräsentieren, symbolisieren oder widerspiegeln (z.B. Foucault, Douglas). Dasselbe gilt für Rituale, die als Handlungen menschlicher Gemeinschaften (Émile Durkheim), als Teil von Bedeutungssystemen (Victor Turner) oder als Möglichkeit der Lesbarkeit dieser kulturellen Bedeutungssysteme

11 M. Douglas, Ritual, Tabu und Körpersymbolik, Frankfurt a. M. 1998 (=1970).
12 M. Featherstone/M. Hepworth/B.S. Turner, The Body. Social Process and Cultural Theory, London u. a. 1991; R. Gugutzer, Soziologie des Körpers, Bielefeld 2004; B. Turner, Body and Society. Explorations in Social Theory, London u. a. 2008.

verstanden werden (Clifford Geertz).[13] Dass Rituale aber einschneidende Handlungen sind, die tatsächliche körperliche Bedeutungen und Effekte haben können, wurde für die Wissenschaft erst interessant, als Ethnologen und Religionswissenschaftler spätestens in den 1990er-Jahren begannen, die zugrunde liegende kartesianische Körper-Geist-Dichotomie in Frage zu stellen. Die mit dieser Dichotomie einhergehende Unterordnung der Phänomenologie unter die Semiotik bzw. der Erfahrung unter die Sprache, so die Kritik, blendet wesentliche Bereiche gelebter Erfahrung aus, wie sie gerade für das Verständnis von Ritualen und die Verkörperung von Normen von wesentlicher Bedeutung sind.

Positiv zur Textmetapher verhält sich der Beitrag von *Thomas Reinhardt*, der die problematische Unterscheidung von Körper und Bewusstsein am Beispiel des Erinnerns behandelt. Vor dem Hintergrund der Einsicht, dass Thematisierungen des Gedächtnisses nicht ohne Metaphorisierung auskommen, beschreibt er den Körper als Palimpsest, der zu keinem Zeitpunkt als unbeschriebenes Blatt gelten kann. Dabei interessiert ihn vor allem die Frage, wie wir unser Gedächtnis „lesen". Reinhardt kritisiert die herkömmlichen Gedächtnis-Metaphern (z.B. „Magazin" und „Tafel"), da sie auf die kartesianische Trennung von Körper und Bewusstsein rekurrieren und daher immer auch den Einsatz von Medien, allen voran der Schrift, voraussetzen. Mit Prousts Madeleine-Episode setzt er dagegen das somatische Erinnern, bei dem der erinnernde Körper als Erinnerung und Erinnerndes die medial vermittelte Zeichenhaftigkeit des bewussten Gedächtnisses unterläuft und sich gleichsam selbst liest. Aus dieser Perspektive wird die Frage danach, wer oder was erinnert, obsolet, denn Körper, Sprache, Schrift und Schreiben sind so vielfältig miteinander verwoben, dass es Reinhardt wenig sinnvoll erscheint, sie auseinander dividieren zu wollen. Es bleibt das Paradox des sich selbst lesenden Körpers.

(3) Philosophisch und kognitionswissenschaftlich drängt sich vor allem die Frage auf, wie die Spielräume zwischen somatischer und kultureller Verkörperung normativ beschaffen sind und wie sie mit dem Bedürfnis nach rechtfertigenden Gründen zusammenhängen, die in demokratischen Gesellschaften allen sozialen Gruppen abverlangt werden, wenn sie ihre Position in einer demokratischen Öffentlichkeit geltend machen wollen. *Matthias Jung* konzentriert sich in seinem Beitrag daher auf die Frage nach dem Zusammenhang zwischen solchen Gründen und Körperpraktiken, zumal ritueller Art. Es geht ihm darum zu zeigen, dass argumentative Gründe für Überzeugungen und Werte häufig mit Narrationen und diese mit Körperpraktiken verknüpft sind. Das wird am Beispiel der biblischen

---

13 Vgl. K. Polit, „Verkörperung" in Brosius, Michaels, Schrode (Hg.), Ritual und Ritualdynamik (utb 3854), Göttingen 2013, 215-221.

Erzählungen von Bundesschluss und gelobtem Land plausibilisiert und dann auf die normative Struktur öffentlicher Debatten in pluralistischen modernen Gesellschaften bezogen. Im Beitrag von *Magnus Schlette* steht die Verkörperung von Werten im Zentrum. Anhand zahlreicher Beispiele, zumeist im Blick auf die populäre Fernsehserie „Breaking Bad", wird herausgearbeitet, wie Werthaltungen weniger über diskursive Verständigung als über Rollenvorbilder und „Helden" tradiert und modifiziert werden. Das individualistisch-anarchistische Verständnis von Freiheit und Selbstverwirklichung, das Schlette als die implizite Wertorientierung (*social imaginary* im Sinne Charles Taylors) des Protagonisten sichtbar macht, schreibt sich sogar dessen Physiognomie ein und entfaltet gerade in diesem Durchgriff kultureller Verkörperung auf den Leib seine so unauffällige wie nachhaltige normative Wirkung.

(4) Historisch-anthropologische Fragestellungen eröffnen einen weiteren zentralen Zugang, indem hier anhand historischer Fallbeispiele Zugänge zur Körpersymbolik untersucht werden. Vorausgesetzt ist das Votum Lucien Febvres in seiner Antritts-vorlesung am Collège de France (1933)[14], dass die Rekonstruktion menschlicher Geschichte nicht auf Schriftzeugnisse reduziert werden dürfe, und widmet sich deshalb – wenn auch vorwiegend auf der Basis literarischer Zeugnisse – Aspekten des Körpers und der Körperlichkeit bezüglich ihrer anthropologischen und sozialen Bedeutung im Altertum.

Martin Meyer untersucht den antiken Blick auf das Fremde ausgehend von Herodots Beschreibung der Körperriten fremder Völker und des dieser Beschreibung inhärenten Normierungsanspruchs anhand von Detailbeobachtungen zu Sexualität, Reinlichkeit und Reinheit sowie Tod und Trauer. Die Beobachtungen münden in die Überlegung, inwieweit die Wahrnehmung des Körpers auf soziale Konstruktionen verweist, die Autoren wie Herodot zu Reflexionen über das Frem-de und nicht etwa zu Leib oder Leiblichkeit führen. *Rüdiger Schmitt* untersucht Phänomene wie Ekstase und Selbstlazeration im Kontext der westsemitischen Mantik. Sie bieten aussagekräftige Beispiele sowohl für die gemeinschaftsbildende Funktion von Körperriten als auch für ihre trennende Funktion, z.B. in Form einer Unterwerfungsgeste (*devotio*). Er stellt heraus, dass diese Praktiken nicht – im Sinne eines kulturgeschichtlichen Fortschrittsmodells – als eine historisch allmählich überwundene Form von Körperritualen im alten Israel bewertet werden können, sondern im Konzept einer *unio mystica* bis weit in die hellenistische Zeit belegbar sind. *Saul Olyan* untersucht das Verhältnis von Hinzufügung bzw. Wegnahme im Rahmen der Körperveränderung an Priestern oder geweihten Personen. Er zeigt,

---

14 L. Febvre, Das Gewissen des Historikers, Frankfurt a. M. 1990, 18.

wie z.B. die Rasur in verschiedenen Kontexten wie der Initiation, eines (temporären) Statuswechsels oder für die raum-zeitliche Markierung der Übergänge vom Profan- zum Sakralbereich sehr unterschiedliche Bedeutungen innehaben kann. Denn sie kann sowohl der Separierung als auch dem Zusammenschluss oder aber der kultischen Reinigung dienen. Die durchweg in einen besonderen literarischen Kontext eingefassten Belege lassen, wie z.B. in Levitikus 14,33, mitunter auch ein didaktisches Interesse an Körperritualen erkennen. *Judith Hartenstein* untersucht Askese als ein Beispiel „verkörperter Entkörperung" und thematisiert den Zusammenhang von Askese und christlichem Weltbild im Sinne der Gender-Anthropologie. Anhand dieses Themas lässt sich ein Körper- und Reinheitskonzept rekonstruieren, welches weibliche Körper aufgrund ihrer natürlichen Beschaffenheit ablehnend bzw. einschränkend bewertet oder aber durch asketische Praxis zu einem „würdigen Gefäß der Seele" werden lässt. *Thomas Römer* wendet sich dem Thema der Beschneidung in alttestamentlichen Texten erstens in der Absicht zu, sie in ihrer jeweiligen Funktion gemäß den seit van Gennep eingeführten religionswissenschaftlichen Kategorien zu beschreiben, um, zweitens, den allmählichen Bedeutungswandel des Ritus seit der babylonisch-persischen Zeit zu reflektieren: Als Beschneidung nicht mehr zu den allgemein in dem Kulturkreis gängigen Initiationsriten zählte, konnte der Ritus zu einem Identitätsmarker neu ausgestaltet werden. Genesis 17 ist nicht nur eine Ätiologie dieses religiösen Identitätsmarkers, sondern nimmt drittens die Umwidmung von einem pubertären *Rite de Passage* zur Säuglingsbeschneidung am achten Tag nach der Geburt vor. Parallel zeichnet sich in weiteren Belegen aber eine aufkommende Tendenz für ein metaphorisiertes Verständnis von Beschneidung ab, das eine Loslösung bzw. Abstrahierung des Körperritus vorbereitet. *Michaela Bauks* untersucht Beschneidung als ein seit Sigmund Freud sehr kontrovers diskutiertes Thema, welches durch gezielte Rekontextualisierung vom Marker äußeren Befremdens bis hin zur Verunglimpfung von Körperritualen als pathologische Zwangshandlung reicht. Freud hat die bereits innerhalb der biblischen Traditionen als besonders rätselhaft zu beurteilende Passage (Exodus 4,24-26) als ein Beschneidungsritual gedeutet, welches, eingebettet in einen dramatisch anmutenden Handlungsrahmen, Beschneidung als Ersatzhandlung eines Menschenopfers darstellt. Wenn auch die ursprüngliche Funktion der Passage umstritten ist und bis in die rabbinische Rezeption kontrovers diskutiert wurde (vgl. Ned. 32a), entwickelte sich der Text doch zu einem Zentralbeleg für den Imperativ der Beschneidung und schuf Raum für eine kulturhermeneutisch imposante Wirkungsgeschichte.

Die hier versammelten Beiträge verdeutlichen, dass und in welcher Weise alle diskursiv-sprachlichen Sinnfiguren in einen größeren Kontext kultureller Praktiken eingeordnet werden müssen, in denen sich traditionell vorgegebene Einstellungen, Handlungsmotivationen und Werte *verkörpert darstellen,* also auch dem Leib der jeweils Beteiligten eingeschrieben werden. Dies gilt nicht etwa nur für von einer aufgeklärten Vernunft gerne als überholt angesehene Rituale, es prägt auch den von Philosophen häufig beschworenen „Raum der Gründe" im Ganzen.

Aus all dem folgt, dass wir das anthropologische Selbstverständnis des Menschen als „Animal rationale" nicht so auffassen dürfen, als ob unsere Rationalität von unserer Lebendigkeit als organische Wesen getrennt werden könnte. Es muss darum gehen, beides gleichermaßen ernst zu nehmen: den verkörperten Charakter sozialen Sinns nicht weniger als die Fähigkeit zur kritischen Distanzierung von ihm. Diese Einsicht ist für das Verständnis menschlicher Praxen der Begründung von Normen entscheidend. Die methodische Orientierung an *Verkörperung* wendet sich eben gleichermaßen gegen ein rein geistiges Verständnis des Kulturellen wie gegen eine naturalistische Reduktion von Sinn auf physische Prozesse. Dabei zeigt sich immer wieder, wie sich leibliches Erleben – als vermutlich evolutionär invarianter Bestandteil unserer anthropologischen Grundausstattung – und kulturell extrem variable Körperpraktiken gegenseitig beeinflussen und durchdringen, so dass beides immer unterschieden werden muss, aber nie getrennt werden kann. Die Forschung steht in dieser Frage noch ganz am Anfang. Kognitionswissenschaftler, Ethnologen, Altertumswissenschaftler, Philosophen, Bibelwissenschaftler und Vertreter weiterer Disziplinen finden hier ein neuartiges, ergiebiges und integratives Forschungsfeld vor. Es ist ein wichtiges Ziel des vorgelegten Sammelbands, einer Dichotomisierung dieses Feldes in die Erforschung anthropologischer Invarianten einerseits und kultureller Ritual- und Körperpraktiken andererseits entgegenzuwirken. Die Vielfalt der versammelten Beiträge zeigt jedenfalls deutlich, dass die etablierten Grenzen zwischen den kultur-, sozial- und naturwissenschaftlichen Disziplinen hier eher hemmend wirken. Wenn Verkörperungsphänomene im Fokus stehen, ist Transdisziplinarität nicht nur ein wünschenswerter, aber entbehrlicher Zusatz zur Disziplinarität, sondern methodisch konstitutiv.

1
Der Körper-Leib: Phänomenologische Zugänge

The Articulation of Enkinaesthetic Entanglement

Susan Stuart

Susan Stuart

I   Introduction

In a similar way to Merleau-Ponty [1962, 1964a, 1964b 1968 & 1970] my task will be to uncover, one might even say elicit, the sensual layers of pre-theoretical living within our lifeworld or *Lebenswelt*. Yet I go further than this; my central concern is with the articulation of a world of meaningful experience, created, in part, at least, by a process of prenoetic mutually transgressive affective neuro-muscular entanglement; a great deal of the paper will be spent working towards a satisfactory account of what is meant by this mutual transgression. At first blush it will seem paradoxical to use the term 'articulation' in the context of pre-reflective felt experience, for if that experience is pre-reflective, it will be pre-conceptual, non-propositional, non-representational, and, we would imagine, incapable of being articulated in the usual way we conceive of articulation. So, let's begin by making clear what will not be our concern: (i) the more usual uses of 'articulation' to mean the articulation of clear sounds in speech, (ii) or the putting into words of a previously inchoate idea, or (iii) any notion of articulation which is centred on the individual and potentially solipsistic. Whilst the first two are interesting and have been the subject of sustained concern for centuries' worth of the writings of others, the third, I argue elsewhere [Stuart 2012], is specious and, what's more, impossible. Thus, our much more specific concern is with the articulation of meaning through and by the intentionally-saturated activity of the living body in an affective community of other living bodies and things.[1] It is a richly affective,

---

1   I have elsewhere, for example, Stuart (2010) & Stuart (2012), used the categories of 'agential' and 'non-agential', but given Latour's arguments for the action or participation of non-humans ('actants') within rhizomic-networks (see, for example, Latour 2005), I am content to construe the former as intentional, the latter as non-intentional, and

plenisentient community, characterised by the implicit intricacy of a pre-reflective neuro-muscular entanglement through which we are the unthinking co-creators, that is, co-articulators of our shared world.[2] I will argue that this affective co-articulation is made possible by a condition of our own subjectivity, and that this condition is a prenoetic somatosensory intentional transgression of the living dynamic experiencing being of the other. Thus, I will argue, our lived experience is always tempered by the direct spontaneous reception, or passive synthesis, of the experientially entangled living being of the other as they transgress our own experience and we theirs. I will refer to this as the enkinaesthetic community and reciprocity of our affective being *with* our world – our *Mitseinwelt*, and it is our folding into, enfolding with, and unfolding from this community which *is* the co-articulation of our shared meanings.

The most immediate question that arises is what is involved in this process of somatic, semantic and, by no means necessarily, conceptual articulation; so, we will start by an elaboration of the notions of articulation *of* and *through* the body, and then, through the articulated and articulating body, we will reach out through our enkinaesthetic experiential entanglement with other living bodies to the role that living, breathing, feeling bodies play in rendering our shared worlds meaningful and our actions values-realising. In doing this I will extend Husserl's notion of intentional transgression to the enkinaesthetic sphere of lived experience, and in support of this claim I will examine the theoretical and practical work of osteopathic manual listening [Gens & Roche], and the 'felt sense' in focusing [Gendlin 1966, 1992, 1997 & 2015] which emphasises the possible shift from a somatic articulation to a semantic one. Throughout the whole my position will be compatible with Merleau-Ponty's claim that

> Whenever I try to understand myself, the whole fabric of the perceptible world comes too, and with it comes the others who are caught in it. …For [others] are not fictions with which I might people my desert–offspring of my spirit and forever unactualized possibilities–but my twins or the flesh of my flesh.[3]

---

both as significant, contributing to the co-articulation of meaning, within a dynamic material-semiotic.

2   None of this is to imply that there is no cognitive activity going on, there may be, there may not; and nor is it to suggest that there is no linguistic community, there may well be; it is merely to suggest that this is not the concern of this current paper.

3   Merleau-Ponty (1964), 15.

Let's start with a skeleton, put some living, breathing, sensing flesh on its bones, and then proceed, by feeling our way towards an articulation of our twins, the flesh of our flesh.

II   Articulation

There is a character, Mr Venus, in Charles Dickens' *Our Mutual Friend*, whose employ is as an articulator of bones, someone who reconstructs the skeleton of animals, including human animals, and who occasionally extends their skills to taxidermy, that is, making the animal appear as though it still has flesh on its bones. Mr Venus, who possesses the amputated leg of his interlocutor, Mr Wegg, exclaims that:

> [I]f you was brought here loose in a bag to be articulated, I'd name your smallest bones blindfold equally with your largest, as fast as I could pick 'em out, and I'd sort 'em all, and sort your wertebræ, in a manner that would equally surprise and charm you. [Chp. VII, p.64]

It's a vivid passage, not least because of the ironic suggestion that Mr Wegg would be around to be delighted by Mr Venus's remarkable feats of reassembly when he has re-articulated his bones, but also for the ease and skill with which Mr Venus makes sense of a skeletal world so familiar to his touch, to his vision, to his life-world. Each of the bones, even the smallest, has significance for, that is, has meaning and matters to, Mr Venus. Through his affective acquaintance he would be able to re-articulate and make sense of Mr Wegg's frame, with each of the individual bones articulating where they meet, in the microcosm of their intertwinings. A single bone is inarticulate, yet in Mr Venus's hands it has meaning, it is a value-object, and with Mr Venus's intervention and in conjunction with other bones, all value-objects, its complex social meanings are articulated. Their articulation can be described in functional terms as permitting this or that extent of movement, this or that orientation, these or those degrees of freedom, as having single or multiple axes of movement, and as having flat, concave or convex surfaces, but more importantly their articulation is hermeneutic, making sense and giving voice to the individually unintelligible or incoherent. Mr Venus, as an articulator of bones, is a re-creator of worlds of meaning, where the elements act not as isolated individuals free from any impingement on others, and where the sense of affecting change and being affected brings forth, that is, articulates worlds for the multiplicity of unions. Such unions are not simplified dyadic interactions, and not only for Mr Venus. The whole, including Mr Wegg and beyond, is massively polyadic, consisting not just of other living beings and things, but also memories, utterances, and events, past, present and future, all of which matter or have significance for us because they affect us and alter us, and in this universal dialogue we affect and alter them. It is within this affective community and reciprocity we feel our way in a co-articulated values-realising co-constituting dynamics.

Yet, for all this, the world of meaning that Mr Venus re-creates cannot yet be our 'twin' in the way Merleau-Ponty implies. To be our 'twin' Mr Wegg's skeleton must have feeling and sensing flesh on its bones, and this would have to comprise some part of its neurodynamical enkinaesthetic ability to feel the givenness and ownership of its own experience as entwined with the living feeling breathing dynamical being of other living beings and things. Only if this is the case can our 'twin' be affected and altered, can it sense and anticipate, and only if this is the case can it reciprocate in a co-articulation of meaning and value. So let's examine in a little more detail what Merleau-Ponty might mean when he says that others are not fictions, that they are the flesh of my flesh; to do this we'll explore the notion of articulation through the lens of enkinaesthetic theory.

III   Enkinaesthetic Co-Articulation

There are three characteristics of enkinaesthetic theory which are crucial for this paper. The first is that there exists between us and other agents a prosody of resonance and fragmentation, in the form of interpersonal felt cadences that are "regulated by emotions of affection and enjoyment, expressed and given meaningful form by rhythms of modulated movement".[4] The second is that, through our actions and perceptions we inhabit the other's experience, which is to say that there exists an immanent intercorporeality in the prosody of our neuro-muscular entanglement. This notion is similar to Husserl's notion of *Paarung*, but we will also see that it is profoundly different. And, the third feature arises as an interplay of the first and second, for in that interplay we don't just enact and articulate our own meanings, we also bring forth, that is, articulate others' meaning. So, just as Varela claims about an object, that it is not 'out there' independently, but "arises because of your activity, so, in fact, you and the object are co-emerging, co-arising"[5], I claim that the co-articulation of meaning arises out of our enkinaesthetic co-emerging, co-arising. We'll proceed by examining these aspects in more detail.

---

4   Malloch & Trevarthen (2009), 2.
5   Varela (1999) 71-72.

Trevarthen observes that "There is an old and frequently rediscovered understanding that we come alive as subjects or persons *only in relation with others*, by [our] being innately sensitive to their actions toward us"[6]; I would add, and their being innately – and synchronously – sensitive to our actions towards them, for only with that co-responsiveness can we feel the other as immanent in our own being. Enkinaesthetic theory is a further rediscovery of this understanding, but it is also a development of it, bringing with it a means of drawing together how we articulate our concerns and the concerns of others within the sensitive community and reciprocity of living being, where each action, already characterised by its givenness and a saturated intentionality, engenders affect and that affect engenders action, not just within ourselves but within all life. Thus, it is through our enkinaesthetic entanglement that we experience the intercorporeal resonances, and the fractures and fragmentation of resonances, with those agents with whom, and those objects with which, we are in a perpetual community of reciprocal relations, within the experiential repertoire of the whole. As Sperry says "The experience of the organism is integrated, organised, and has its meaning in terms of coordinated movement"[7], but Sperry is too cautious and fails to mention that movement is affectively coordinated with the energic pitches, cadences, and tempos that characterise our polyphonic intertwining with other organisms and things [Stuart & Thibault 2015].

> Communication between similarly motivated and similarly formed subjects, with the same kind of brain and the same rhythms and forms of attending, evidently has evolved by brains taking up and engaging with – or resonating to – the timing, aim, and style of these intentional and sense-directed activities generated in other brains.[8]

Without embracing Trevarthen's emphasis on the brain, I am claiming that this communication is a natural, direct and unmediated apprehension of the other's experience in our own. This is not to say that our experience of the other's experience is from their perspective; that would be absurd. It is to say that when Merleau-Ponty speaks of others as real in our experience, as the flesh of my flesh, he is claiming

---

6 This is from an early draft of a paper sent in personal correspondence. It is worth noting that in that paper Trevarthen refers to Hutcheson (1755), Smith (1759), Buber (1937), Macmurray (1961), and Reddy (2008) who each make this 'rediscovery' from fascinating perspectives.

7 Sperry (1939), 295.

8 Trevarthen (2009), 12. Marvellous examples of this affective coordination are given in Validation Therapy, see: https://vfvalidation.org/web.php?request=index especially the validation breakthrough with Gladys Wilson https://www.youtube.com/watch?v=CrZX-z10FcVM

that the other is 'always "already there" [in my experience] before reflection begins' [Merleau-Ponty 1970, p.65], and that it is an always already there as the 'primordial being which is not yet the subject-being nor the object-being' (Merleau-Ponty 1970, p.65). In this way, in the day-to-dayness of the community and reciprocity of our affective *Mitseinwelt* we need neither to develop and implement some conjectural cognitive theory about the mind of the other, nor to perform some curious simulation of what the other might be thinking if we were them, and neither of these is required because we routinely transgress our own bodily boundaries spilling over into, that is, pervading the plenisentient bodily experience of the other, and they ours.

> [A]t the same time the other who is to be perceived is himself not a "psyche" closed in on himself, but rather a conduct, a system of behavior that aims at the world, he offers himself to my motor intentions and to that "intentional transgression" (Husserl) by which I animate and pervade him.[9]

Husserl argues that, through a process of corporeal analogising, by which he means recognising the other as having a body similar to our own, we co-present the other and understand it to be, not simply a body like an object, *Körper*, but as being an 'ensouled', psychic, or living body, *Leib*.[10] Analogising in this way couples ego and alter ego as *Paarung*, where the other, as psychically distinct from me, is, nevertheless, appresented as *Leib*. The emphasis for Husserl is on the auditory and visual perception of the other as having a moving, kinaesthetic body like my own, and it is in this observation that an 'intentional transgression' occurs, and I spontaneously appresent the other as another ego, an alter ego. Merleau-Ponty also emphasises the role of visual perception saying:

> Husserl said that the perception of others is like a "phenomenon of coupling" (*accouplement*). The term is anything but a metaphor. In perceiving the other, my body and his are coupled, resulting in a sort of action which pairs them (*action à deux*). This conduct which I am able only to see, I live somehow from a distance. I make it mine; I recover (*reprendre*) it or comprehend it. [ibid.]

---

9   Merleau-Ponty (1964b), 118.

10 The anomalous counter-factual cases of, for example, antisocial personality disorders, including psychopathy and sociopathy (DSM-5 301.7), of depersonalisation or derealisation disorders (DSM-5 300.6), of Cotard's delusion (DSM-5 297.1), and of Capgras syndrome, reveal our co-presentation of the other as *Leib* to be the non-anomalous one might say, natural, everyday response to other human beings. All references from DSM-5 2013.

Enkinaesthetic theory takes the intentional transgression further beyond Husserl's and Merleau-Ponty's conceptions, arguing that visual and corporeal analogising plays only a greatly diminished role in our grasp of the intentional arc of the other's current and future action, and that the 'alter' ego that we appresent whilst remaining visually other, is not affectively other; this appresented other is already there in its primordiality. We have no need for analogising the physical body with our own, recognising it as similar, and switching to seeing it as a living being. We are always already within the perpetual felt community and reciprocity of an enkinaesthetic field, where "field" refers to the domain within which a particular condition prevails – in this case a topologically complex, affectively-laden, intentionally-saturated *Mitseinwelt* of other beings and things. We dwell within our plenisentient intersubjective engagement with other agents, human and non-human, and this dwelling, this entangled enkinaesthetic experience, is a transcendental condition for the prenoetic affect, which makes alter ego identification, co-presentation, co-articulation, and co-action possible.[11] In these massively polyadic enkinaesthetic intertwinings, the simultaneous experience of our affect on others and their affect on us has an immanence in our being; we are, at one and the same time, both subject and object, and object not just in the gaze of the other, but also in our own sensed reflection in our memory, and in our anticipatory framing of ourselves within horizons of current and future possibilities.

In a phrase redolent of Merleau-Ponty's 'chiasm', Young (1980) uses "ambiguous transcendence" to describe the experiential inseparability of our being, at one and the same time, both subject and object; the dynamics of such a crisscrossing or 'intertwining' of the "touching [subject] and the tangible [object]", are fundamental for the success of living organisms within the enkinaesthetic field.[12]

> This can happen only if my hand, while it is felt from within, is also accessible from without, itself tangible, for my other hand, for example, if it takes its place among the things it touches, is in a sense one of them, opens finally upon a tangible being of

---

11 There is no opportunity in this current paper to do justice to the claim that there exist, and indeed that there *must* exist, intra- and inter-species enkinaesthetic resonances and fragmentations. In this regard I wholeheartedly commend to the reader Chapter 3, "Affect Attunement, Discourse Ethics Across Species", of Willett's *Interspecies Ethics*.

12 An alternative, not a counter-, example from nature might be an animal which can simultaneously be both predator and prey. Weasels prey on smaller mammals like mice and voles, but they are also prey for larger predators like foxes and owls; and fish are nearly all piscivorous which means that they eat fish but they are also likely to be eaten by larger fish. In both these cases weasel and fish are likely to be simultaneously the subject of their experience and aware that they might also be the object of the other's experience; they are both the touching and the tangible.

> which it is also a part. Through this crisscrossing within it of the touching and the tangible, its own movements incorporate themselves into the universe they interrogate, are recorded on the same map as it …[13]
>
> There is here no problem of the alter ego because it is not I who sees, not he who sees, because an anonymous visibility inhabits both of us, a vision in general, in virtue of that primordial property that belongs to the flesh, being here and now, of radiating everywhere and forever, being an individual, of being also a dimension and a universal.[14]

It is in this way, through the everywhere and forever radiation of resonances and fragmentations within our enkinaesthetic chiasm, that we bring forth, that is, co-articulate our, that is, collectively 'our', world. And, in this way, our 'own' world can never be brought forth or articulated, without the worlds of all others being brought forth or being articulated too; they are always already within our own articulation, immanent and never fully transcendent. Thus we are, at one and the same time, prenoetically a universal non-individuated being, and noetically (and visually) individuated.

All that I've said here is consistent with the direct perception theory[15] [Chemero 2006 & Gallagher 2008] proposed as an alternative to theory-theory [Carruthers & Smith 1996] and simulation-theory [Davies & Stone 1995] as the means by which we understand what is in another's mind. There is something so remarkably cumbersome about having to first establish a theory, in some third-person scientific manner, about what another may be thinking, or having to place myself rather awkwardly and time-consumingly in the other's 'shoes', so that I can have a first-person experience of the world from 'their' perspective, that makes one wonder why these theories have remained credible for so long. I am not here denying that there are occasions when the situation is massively complicated and I have to judge whether or not I am, for example, being deceived by a smile or humoured by an agreement, but in general this kind of judgement is made *post hoc* through some reflective analysis; however, crucially, the felt sense we have of doubting the sincerity of an interaction relies on our prenoetic openness to the radiating resonances and fragmentations within our enkinaesthetic chiasm. This too is consistent with Gibson's ecological approach to perceptual experience, the first tenet of which is that perception is

---

13 Merleau-Ponty, (1968), 133.

14 ibid., 142.

15 That enkinaesthetic entanglement is consistent with direct perception theory does not imply agreement with or acceptance of that theory. Direct perception is fine as far as it goes, but it doesn't go far enough. Enkinaesthesia offers immanence, experiential transgression, extended body theory, and co-articulation, all of which take direct perception into new territory.

direct and not for adding information to sensations; the second is that perception is for action guidance, not information gaining; and the third is that perception is of affordances, the action possibilities within a perceptual horizon. Within this horizon, which is, first and foremost, an enkinaesthetic, intentionally-saturated affective horizon "organism and environment enfold into each other and unfold from one another in the fundamental circularity [the enkinaesthetic community and reciprocity of co-articulation] that is life itself".[16]

Gendlin's process model [1966, 1992, 1997, & 2015] provides a means of elaborating this experiential enfolding and unfolding within what he refers to as the implicit intricacy of the organism's body-environment felt sense. I will provide a short summary of Gendlin's work[17] and then offer a practical extension of it through the technique of osteopathic manual listening [Sutherland; Gens & Roche 2014]. In this work we have an example of the naturally occurring co-articulation of meaning within the spontaneous enkinaesthetic appresentation of the other's experience in mine, in my experience in the other's.

IV The Felt Sense in Focusing and Osteopathic Manual Listening

The 'implicit' has to be felt in the body, but it is not only inside the body. Rather, it consists of body-environment interaction. "Interaction" comes first. Interaction has always already happened, even when we think about a separate environment and a separate body.[18]

One way in which we might begin to think about the implicit is as the givenness immanent in our experience. At first glance, this might seem sufficient, but it has echoes of individuation, as though it were somehow possible to separate the givenness of the individual's experience from its experiential situatedness and the

---

16 Varela, Rosch, & Thompson (1991), 217. Bracketed phrase my addition.

17 For a detailed account of the Process Model start here: http://www.focusing.org/process.html

18 Gendlin (2015).We might say that interaction is primary, but it would be no more true for Gendlin's model than it would be true to say it of relation in Leibniz's monadology (Leibniz [1991]); the existence of interaction or relation (respectively) presupposes a multiplicity of entities: bodies, monads, consciousnesses, depending on the ontological commitments of your system. What each model shares or, at least seems to share, is the implicit intricacy with which all things are interwoven. In each it is through consciousness, in each it is relational, in Gendlin's it is also interactional.

community and reciprocity of our being with other living bodies and things. And, it isn't just an 'implicit', it is an "implicit intricacy", and an "implicit interactional bodily intricacy" which characterises the proto-modal relationships of organisms in the practical everydayness of their lived being. In this way we might develop our understanding along enkinaesthetic lines, so that the "implicit intricacy" which is always already there too *is* our prenoetic enkinaesthetic experiential entanglement.

> There is an implicit interactional bodily intricacy that is first – and still with us now. It is not the body of perception that is elaborated by language, rather it is the body of interactional living in its environment. …We sense our bodies not as elaborated perceptions but as the body sense of our situations, the interactional whole-body by which we orient and know what we are doing.[19]

Gendlin's "Implicit intricacy" permeates the enkinaesthetic field, folding into, enfolding with, and unfolding from all things within a vast polyadic affective living landscape of articulating and co-articulating microcosmic intertwinings. Conceived thus, living being, *Leib*, transcends individual bodies and agents within this intricate enkinaesthetic web. As Gendlin says: "nature is an implicit intricacy"[20], implying, I will add, but I do not anticipate that Gendlin would disagree, the co-articulation of meaning within a values-realising non-individuated being. This dynamic Gendlin refers to as a "situational understanding", a kind of animal somatic grasp, and it has an ontological and experiential primacy to the "felt sense" that arises and pervades the co-activity or co-articulation of the organism-environment.

> If an animal hears a noise, many situations and behaviors will be **implicit** in its sense of the noise, places to run to, types of predators, careful steps, soundless moves, turning to fight, many whole sequences of behavior. Meanwhile the animal stands still, just listening. What it will do is not determined. Surely it won't do all the implicit sequences – perhaps not even one of just these but some subtler response …[21]

All living being in its implicit intricacy exists in this way within an enkinaesthetic field of affective enquiry and action. The implicit intricacy of "felt sensing" in its "situational understanding" is the articulation and co-articulation of non-propositional, pre-reflective, pre-conceptual, plenisentient interpretation, anticipation, and communication; all of which takes place within a horizon of action possibilities and comes already laden with the implicit non-propositional questions "how

---

19	Gendlin (1992), 352.
20	Gendlin (1997), 347.
21	Gendlin (1997) Chp. II, 7.

is my world now?", "how is it …becoming?". The plenisentient responses to these questions pervade the agent and perpetuate a continuously unfolding fresh horizon of action possibilities. It is this which constitutes the 'knowing' referred to by Gendlin, a 'knowing' which occurs in natural agents through an enkinaesthetic affective enfolding which enables the balance and counter-balance, the attunement and co-ordination of whole-body action through mutual reciprocal adaptation.

A felt sense of this kind cannot arise through our thinking ourselves into directing our attention towards an already existing object, as though somehow we are naturally cognitively and connatively separable from them, perceiving them in a successive order and even through distinct individuated senses; to do this is to immediately think ourselves out of an enkinaesthetic co-articulation of meaning, out into the alien world of propositional attitudes and body-environment-independent minds. For a felt sense we must proceed 'feelingly', [Gloucester in *King Lear*, Act 4, Scene VI] for only then can we articulate a somatic sense of our situation with its openness to action possibilities, and only then can we hope to experience a "felt shift" to a more distinctive semantic articulation.

"Thinking with the implicit" is a means of producing a felt sense or somatic articulation of something that possesses a semantics but which may not yet have words. "As it forms, the "feel" understands itself, so to speak."[22] The practice of focusing is characterised by enquiry and action, but not necessarily dynamic physical action in the form of movement of the body or limbs; rather it is an attempt to render articulable that which is felt, possibly only inchoately. Focusing is an act of values-realising in seeking though not striving, waiting without the intrusion of impatience or irritation, and an openness to letting a feeling come, and brought with it is a bodily sense of the fit or value of the feeling; in this way, in the community and reciprocity of the enquiry and anticipation, the inchoate begins to take shape. Gendlin says it has a feeling of "rightness or wrongness"[23], but in using any of these terms, 'fit', 'value', "rightness or wrongness" it must also be remembered that their aptness is not at the level of cognitive judgement, but at the level of how they sit within the body. We might think of finding the fit as akin, in some way, to the "a ha" moment that comes when we think "Now I understand.", "Now I know how

---

22   Gendlin (1997), 216 . "So to speak" is such an odd metaphor to use in this context where the central concern is with prenoetic somatic affection and its felt semantic articulation, and not with conceptualisation, judgement and verbalisation. But the phrase is in fairly common use to mean that the words, "the "feel" understands itself", are being used in a non-standard way.

23   ibid., 219.

to go on."[24], except that the accompaniment to the implicit fit is not a proposition but a prenoetic feeling or body-environment grasp. "A Felt Sense is a distinctly felt object which may now form and come as a bodily-felt "this"" [Gendlin 2015, fn 5], where 'this' has a growing clarity as a referent with an unfolding sense. We might refer to this unfolding sense as a process of interoception or intrasubjective enquiry, but that would be to fail to recognise and acknowledge the implicit intricacy of the enkinaesthetic body-environment as a condition for the articulation of meaning within an organism's horizons of current and future possibilities.

We will now examine the practice of osteopathic manual listening where we can find evidence of just such an unfolding sense, with its shift from a somatic to a semantic bodily-felt 'this'.[25] In addition, because the technique provides an express articulation of enkinaesthetic entanglement, we have a practical corroboration of the co-articulation of a bodily-felt 'this' which is immanent in our enkinaesthetic intentional transgression of and intertwining within the other's experience and in theirs within ours.[26]

Just as with traditional osteopathy, osteopathic listening is a practice which uses the hands, but unlike traditional osteopathy the hands are not used to manipulate the patient's body; instead the therapist's hands are the focal point of a synaesthetic listening-feeling process, the gentle touch – and even non-touch – of palpation, listening for rhythms and arhythms. Just as in focusing, there is, in the listening process, a seeking without striving, a waiting without impatience, and an openness to what presents itself, all of which can occur, in this non-traditional osteopathic method, without needing to be in constant tactile contact with the patient's body, and even without touching it directly at all. It is the non-necessity for touch which makes this form of osteopathic listening particularly intriguing, and especially because of its appreciation for the routine enkinaesthetic transgression of our bodily boundaries in which we pervade the plenisentient bodily experience of the other. In other words, osteopathic manual listening embraces the ambiguous transcendence of the everydayness of our folding into, enfolding with, and unfolding from the other within the processual co-articulation of our shared meanings.

> The defining characteristic of this listening process is that it essentially derives from the osteopath's ability to sense the inner space in which organic life develops – in other words, to sense this organic life itself, even though we are raised to believe that such a perception is impossible.

---

24 Wittgenstein (1958), 323.

25 See, for example, William G Sutherland.

26 For another example of enkinaesthetic intentional transgression of and intertwining within the other's experience see Stuart (2013).

> The common view is that, without using a scalpel, it is as impossible for us to have a knowledge of the interior life inside a living organism enveloped in skin as it is for us to see through a wall.[27]

The view that this perception is impossible is based on all manner of things including, but not limited to, the visual opacity of the appearance of the body,[28] but whilst vision may accompany manual listening, it is by no means necessary for the felt immanence of the other's experience in our own. The process is first and foremost an enkinaesthetic intertwining, a circle of the touched and the touching and what comes to light, that is, what is brought forth through the feeling shifting somatic sense.[29]

> There is a circle of the touched and the touching, the touched takes hold of the touching; there is a circle of the visible and the seeing, the seeing is not without visible existence; there is even an inscription of the touching in the visible, of the seeing in the tangible – and the converse; there is finally a propagation of these exchanges to all the bodies of the same type and of the same style which I see and touch – and this by virtue of the fundamental fission or segregation of the sentient and the sensible which, laterally, makes the organs of my body communicate and founds transitivity from one body to another.[30]

It is most unfortunate that Merleau-Ponty makes a 'virtue' "of the fundamental fission or segregation of the sentient and the sensible", for it is exactly that which

---

27 Gens & Roche (2014), 2.

28 The particular historical and cultural treatments of individuals as separable from communities in possession of distinctive souls has also played an important role. See, for example, Benedict's distinction between 'guilt' societies and 'shame' societies (Benedict [1989]).

29 Another example of the feeling shifting somatic sense and its enkinaesthetic co-articulation is given in Steinbeck's short story *The Chysanthemums* (Steinbeck [1952]) where the protagonist Elisa explains, to a travelling salesman, the sensitive palpating, enquiring and acting, touch of "planting hands".
"Did you ever hear of planting hands?"
"Can't say I have, ma am."
"Well, I can only tell you what it feels like. It's when you're picking off the buds you don't want. Everything goes right down into your fingertips. You watch your fingers work. They do it themselves. You can feel how it is. They pick and pick the buds. They never make a mistake. They're with the plant. Do you see? Your fingers and the plant. You can feel that, right up your arm. They know. They never make a mistake. You can feel it. When you're like that you can't do anything wrong. Do you see that? Can you understand that?"

30 Merleau-Ponty (1968), 143.

is being contested in manual listening, in focusing, and in the enkinaesthetic co-articulation of meaning which is possible precisely because of our ambiguous transcendence. In the former two cases, osteopathic manual listening and focusing, there must be a development of silence and openness in the psychic life of the practitioner, one in which the continuous chatter and play of words and images must be quieted, and once this quiet is established a felt sense of the other's bodily experience is brought forth. In this somatic awareness to receptivity a conscious, yet still somatic, co-articulation is forming, and "As it forms, the "feel" understands itself"[31], it shifts and, with a growing clarity, becomes a referent with a continuously unfolding sense, and so on until another felt shift occurs and the co-articulated meaning alters again. Throughout this process of openness and manual listening the experience of the participants is of a between, neither subject nor object. As an enkinaesthetically co-articulated ambiguous transcendence it is neither one nor other, but instead a prenoetically universal non-individuated being which brings with it the whole fabric of the perceptible world as its twin, the flesh of its flesh.

V   Conclusion

So, finally let's return to Mr Venus, who would have been able to surprise and charm Mr Wegg by re-articulating his skeleton, making sense of the individual unintelligible bones in the polyadic microcosms of their numerous intertwinings. We have begun to understand how Mr Venus could articulate this aspect of his world blindfold; his fingers and hands, in their enkinaesthetic enquiry and action, seeking, waiting, and letting come what arises somatically in the manual listening of his habitual trade. He feels for the extension, smoothness, concavity or convexity of the bones, and their somatic feel becomes a semantic felt sense, a felt 'this' which comes without conceptual interruption; as Gendlin would say: their feel understands itself; as I would say: their feel is an enkinaesthetically articulated grasp of a possibility of being. He works not just in the current moment but proceeds feelingly, anticipatingly within horizons of possibilities.

But this is to do with the articulation of bones, and our concern is with the co-articulation of felt meaning as outlined in the ambiguous transcendence of osteopathic manual listening, or in the practice of partnered focusing, or in the reciprocity and community of *Leib-Leib* enkinaesthetic entanglement. For that kind of affective co-articulation we needed to put feeling and sensing flesh on Mr

---

31 Gendlin (1997), 216.

Wegg's bones. With flesh on his bones he becomes Mr Venus's 'twin' once more and he too can proceed feelingly, trying to understand themselves and articulate their own meanings and values, and each bringing the other and the whole fabric of their perceptible world with them too.

Bibliography

Benedict, R. 1(989) *The chrysanthemum and the sword: patterns of Japanese culture*, Foreword by Ezra F. Vogel, Houghton Mifflin.

Buber, M. (1970) *I and thou*, translated with prologue and notes by Walter Kaufmann, Edinburgh : T. & T. Clark.

Carruthers, P. & Smith, P. K. (eds.) (1996) *Theories of theories of mind* , Cambridge: Cambridge University Press.

Chemero, A. (2006) "Information and direct perception: A new approach", in P. Farias & J. Queiroz (Eds.), *Advanced issues in cognitive science and semiotics*, pp. 59–72 Aachen: Shaker Verlag.

Davies, M. & Stone T. (eds.) (1995) *Folk Psychology: The Theory of Mind Debate*, Oxford: Blackwell Publishers.

Dickens, C. (1865) *Our Mutual Friend*, London: Chapman and Hall.

DSM-5 (2013) *Diagnostic and statistical manual of mental disorders: DSM-5*, American Psychiatric Association, Arlington, Va., London.

Gallagher, S. & Meltzoff, A. (1996) "The earliest sense of self and others: Merleau-Ponty and recent developmental studies", *Philosophical Psychology*, 9, pp.211–33.

Gallagher, S. & Marcel, A. J. (1999) "The Self in Contextualized Action", *Journal of Consciousness Studies*, 6 (**4**), pp.4–30.

Gallagher, S. (2008) "Direct perception in the intersubjective context",*Consciousness and Cognition*, 17 (**2**), pp.535–543.

Gendlin, E. T. (1966). "The discovery of felt meaning", in J.B. McDonald & R.R. Leeper (Eds.), *Language and meaning*, Papers from the ASCD Conference, The Curriculum Research Institute (Nov. 21-24, 1964 & March 20-23, 1965), pp. 45-62. Washington, DC: Association for Supervision and Curriculum Development. From http://www.focusing.org/gendlin/docs/gol_2039.html

Gendlin, E. T. (1992) "The primacy of the body, not the primacy of perception", *Man and World'*, 25 (3-4), pp.341–53.

Gendlin, E. T. (1997). A process model. New York: The Focusing Institute. (Also available at http://www.focusing.org/process.html).

Gendlin, E. T. (2015) "A New Way of Thinking–About Anything–and How to Write From It", *Synthesis Philosophica*, forthcoming.

Gens, J-C. & Roche, E. (2014) "The emergence of feeling in osteopathic manual listening", conference presentation, Consciousness and Experiential Psychology Section Conference, Sidney Sussex College, Cambridge, 4–6 September 2014.

Husserl, E. (1989) *Ideas Pertaining to a Pure Phenomenology and to a Phenomenological Philosophy, Second Book (Ideas II)*, trans. R. Rojcewicz & A. Schuwer, Dordrecht: Kluwer.

Hutcheson, F. (2005) *A system of moral philosophy, in two volumes* (1755), Intro. Daniel Carey, London ; New York : Continuum.

Ingold, T. (2000) *The perception of the environment: Essays on livelihood, dwelling and skill*, Abingdon, UK: Routledge.

Latour, B. (1999) "On Recalling ANT", The Sociological Review, 47, (**S1**), 15–25.

Latour, B. (2005) *Reassembling the Social: An Introduction to Actor-Network-Theory*, Oxford: Oxford University Press.

Leibniz, F. W. (1991) *G. W. Leibniz's Monadology, An Edition for Students*, trans. Nicholas Rescher, University of Pittsburgh Press.

Macmurray, J. (1961) *Persons in Relation*, London: Faber & Faber.

Malloch, S. & Trevarthen, C. (Eds.) (2009) *Communicative Musicality, Exploring the Basis of Human Companionship*, Oxford: Oxford University Press.

Merleau-Ponty, M. (1962) *Phenomenology of perception*, New York: The Humanities Press. Trans. Colin Smith, London: Routledge and Kegan Paul.

Merleau-Ponty, M. (1964a) *Signs*, Northwestern University Press, Evanston, Illinois.

Merleau-Ponty, M. (1964b) *The Primacy of Perception, And Other Essays on Phenomenological Psychology, the Philosophy of Art, History and Politics*, trans. James M. Edie, Northwestern University Press, Evanston, Illinois.

Merleau-Ponty, M. (1968) *The Visible and The Invisible*, ed. Claude Lefort, trans. Alphonso Lingis, Northwestern University, Studies in Phenomenology and Existential Philosophy, Evanston, Illinois.

Merleau-Ponty, M. (1970) *Themes from the Lectures at the Collège de France, 1952-1960*, trans. by John O'Neill, Northwestern University Press, Evanston, Illinois.

Pollatos, O., Gramann, K., & Schandry, R. (2007)"Neural systems connecting interoceptive awareness and feelings", *Human Brain Mapping* January, 28 (**1**), pp.9–18.

Reddy. V. (2008) *How Infants Know Minds*, Harvard University Press.

Shakespeare, W. (1969) *King Lear*, Penguin Books Ltd., Harmondsworth, Middlesex, London, pp.248–94.

Smith, A. (1976) *The Theory of Moral Sentiments* (1759), edited by D.D. Raphael & A.L. Macfie, Oxford : Clarendon Press.

Sperry, R. W. (1939) "Action current study in movement coordination" *Journal of General Psychology* 20, pp.295–313.

Steinbeck, J. (1952) "The Chrysanthemums", in *50 Great Short Stories*, edited by Milton Crane, Bantam Classics, Bantam Dell, New York, pp.337–48.

Stuart, S. A. J. (2010) "Enkinaesthesia, Biosemiotics and the Ethiosphere", in *Signifying Bodies: Biosemiosis, Interaction and Health*, The Faculty of Philosophy, University of Braga, Portugal, pp.305–30.

Stuart, S. A. J. (2012) "Enkinaesthesia: the essential sensuous background for co-agency", In: Radman, Z. (ed.) *The Background: Knowing Without Thinking*, Palgrave Macmillan, Chp. 8, pp.167–86.

Stuart, S. A. J. (2013) "The Union of Two Nervous Systems: Neurophenomenology, Enkinaesthesia, and the Alexander Technique", *Journal of Constructivist Foundations*, 8, (**3**), pp. 314–23.

Stuart, S. A. J. & Thibault, P. J. (2015) "Enkinaesthetic Polyphony as the Underpinning for First- order Languaging", Emotion in Language: Theory – Research – Application, edited volume, John Benjamins Publishing (forthcoming).
Sutherland, W. G. at http://www.cranialacademy.com/cranial.html
Trevarthen, C. (2009)"Embodied Human Intersubjectivity: Imaginative Agency, To Share Meaning", *Journal of Cognitive Semiotics*, IV (**1**), pp.6–56.
Varela, F., Thompson, E., & Rosch, E. (1991) *The Embodied Mind: Cognitive Science and Human Experience*, Cambridge, MA: MIT Press.
Willett, C. (2014) *Interspecies Ethics*, Columbia University Press.
Wittgenstein, L. (1958) *Philosophical Investgations*, Basil Blackwell Ltd., Oxford.
Young, I. M. (1980) "Throwing like a Girl: A Phenomenology of Feminine Body Comportment Motility and Spatiality", *Human Studies*, 3(**2**) pp.137–56.

Wenn Blicke sich kreuzen
Zur Bedeutung der Sichtbarkeit für zwischenmenschliche Begegnungen

Käte Meyer-Drawe

> *„In den besten Augenblicken, auch in den schlimmsten,*
> *wirkt man auf sich selbst nicht mehr wie man selber;*
> *sondern man verschwendet oder man erleidet irgendein –*
> *unwahrscheinliches Ich." (Paul Valéry)*

I Vorbemerkungen

Die folgenden Überlegungen konzentrieren sich auf die Bedeutung von Blicken in zwischenmenschlichen Begegnungen. Was zunächst einfach aussieht, weil wir alle einschlägige Erfahrungen auf diesem Gebiet haben, erweist sich bei genauerer Betrachtung als sehr schwierig. Dies soll in einleitenden Bemerkungen unter dem Titel „Blicke 1" zumindest angedeutet werden. Im nächsten Schritt wird die Rolle der Leiblichkeit für zwischenmenschliche Begegnungen hervorzuheben sein. Es folgt die Erörterung der damit zusammenhängenden Sichtbarkeit des Menschen, um abschließend kurz auf die Rolle der Blicke unter dem Titel „Blicke 2" zurückzukommen.

II Blicke 1

Jeder wird verstehen, was mit einem Blick gemeint ist, selbst wenn es sehr kompliziert ist, genau zu sagen, was ihn ausmacht. Auf jeden Fall reicht es nicht zu sehen, um zu blicken, und es ist etwas anderes, den Blick eines anderen zu sehen, als seine Beobachtung zu beobachten. Manche von uns haben Blicke brechen sehen. Im Sterben hört die Pupille auf, Spiegel des anderen zu sein. Wir ahnen, was tote Augen bedeuten, auch wenn die Worte fehlen wie überhaupt, wenn wir uns dem Blicken und seinem Verlöschen zuwenden. Blumenberg notiert in diesem Zusammenhang: „Eine akustische Fremdwahrnehmung ohne das Medium der Sprache wäre nicht möglich, wie sie optisch ohne das Medium der Sprache möglich ist."[1] Blicke sagen mitunter nicht nur mehr als tausend Worte, sie modulieren unser soziales Gewebe, ohne dass uns das explizit zu Bewusstsein kommen muss. Wir sind damit bei unseren Überlegungen vor eine verwickelte Aufgabe gestellt, nämlich jene, das Selbstverständliche verständlich zu machen, dem, was uns im alltäglichen Leben vertraut ist, eine gewisse Fremdheit zurückzuerstatten.

Stets blicken wir und werden angeblickt. Blicke sind flüchtig und gleichwohl prägend. Schon unmittelbar, nachdem wir zur Welt gekommen sind, werden sie auf uns gerichtet, ohne dass wir sie erwidern könnten. Das kommt ein wenig später. Unabhängig von der sprachlichen Artikulation, aber auch als deren Begleitung setzen Blicke Akzente. Wie vielfältig ihre Bedeutung ist, kann uns eine Auswahl an Redewendungen veranschaulichen: Es gibt indiskrete Blicke, verstohlene, laszive, verräterische, bösartige, verlorene, zudringliche, scheue und vielsagende.[2] Wir werden sie untereinander nicht verwechseln. Etwas kann in den Blick oder aus dem Blick geraten. Man kann durch Blicke würdigen, aber auch vernichten, weil sie höhnisch oder verächtlich sind. Man kann sie tauschen. Blicke können sich treffen, begegnen, verhaken oder kreuzen. Stets gibt es kleine Unterschiede. Der Blick kann hochmütig sein und jener, welcher ihn richtet, erscheint wie einer, „der die Welt von einem inneren Hochsitz aus betrachtet."[3] Zärtliche Blicke können tasten, wenn auch anders als unsere Hände. Falls sie bitter sind, dann schmerzen sie. Wenn nichts anderes zählt, dann kleben sie. Blicke können flackern, umherirren, sich konzentrieren oder auf etwas ruhen. Es gibt nachdenkliche und verzückte Blicke. Man kann sie abwenden oder zuwenden. Sie können gefangen genommen werden. Man kann durch Blicke zum Sprechen ermuntern oder zum Schweigen bringen. Man kann sich durch sie verschwören. *Kinder (auch viele Erwachsene) borgen sich den Bestätigungsblick des anderen wie die Erde das Licht der Sonne.*[4] Blicke können voll und leer sein.

Die Aufzählung könnte fortgesetzt werden. Bereits diese kleine Liste macht jedoch deutlich, wie Blicke unser Miteinander nicht nur begleiten, sondern ihm auch eine bestimmte Dramaturgie verleihen. Am Blick des anderen kann ich sehen,

---

1 Blumenberg (2006), 852.

2 Vgl. auch Waldenfels (2000), 384ff.

3 Mercier (1998), 19. Pascal Merciers (alias Peter Bieri) Buch kann auch als ein Buch über Blicke gelesen werden. Die vorliegenden Darlegungen verdanken ihm viele Anregungen.

4 Schenk (2009), 169.

was und wie er sieht. Vor allem aber spiegelt sein Blick meinen Anblick wider, den ich ihm biete. Blicke antworten damit auf etwas, das ich bin, aber nicht habe. Wenn ich angeblickt werde, erlebe ich eine subtile Entfremdung meiner Möglichkeiten, die durch den anderen gleichsam unter die Gegenstände der Wahrnehmungswelt gemischt werden.[5] Der Blick des anderen durchdringt mich auf eine Weise, die mir selbst versagt ist. Damit ist ein zentrales Problem angesprochen, das den roten Faden der nachstehenden Überlegungen bildet. Dem Blick des anderen ist etwas gegeben, was mir fehlt, dem Blick, den ich ihm schenke, öffnet sich etwas, das ihm versagt ist, nämlich sein lebendiges Gesicht im gemeinsamen Sprechen und Handeln.

III Leiblichkeit

Um blicken zu können, muss man sehen können. Allerdings verbirgt der Blick die Augen. Er scheint gleichsam vor sie zu treten.[6] Als leibliches Wesen bin ich eingetaucht in eine Welt, welche aus dem gleichen Stoff ist wie ich. Ich stehe ihr nicht gegenüber, sondern bin mit ihr verwoben. Das kann leicht in Vergessenheit geraten, wenn wir stets dem Denken den Zuschlag geben. Wir neigen dazu, uns selbst als Denkende, die Welt als gedachte und die anderen als Undenkbare zu verstehen und zu behandeln. Es gibt jedoch eine elementare Verbindung, „wo der Andere sich anders manifestiert als durch die Erkenntnis, die ich von ihm gewinne.“[7] Es stimmt wohl, dass Wahrnehmen nicht ohne Denken vollzogen wird, aber es reicht nicht zu denken, um wahrzunehmen.[8] Aufgrund unserer Erfahrungen sind wir stets über uns als Denkende hinaus. Das gilt auch und insbesondere für die zwischenmenschliche Begegnung.

„Im Körper *bedeutet* uns der Mitmensch er selbst; es ist nicht sein Körper, dem wir begegnen, sondern er selbst; so ist er in unserer Welt, *so* treffen wir ihn an, *so* wird ihm *begegnet* und so begegnet *er* von sich aus uns.“[9] Dabei schließen wir nicht vom Äußeren auf ein Inneres, sondern wir werden unmittelbar angesprochen und verfangen uns in ein Ausdrucksgeschehen, für das jede Erklärung zu spät kommt. „Ich muß mein Äußeres *sein*, und der Leib des Anderen muß er selbst sein. Dies Paradox, diese Dialektik von *ego* und *alter* ist möglich nur, wenn Ich wie der Andere

---

5 Vgl. Sartre (1995), 477. Zum Problem der Intersubjektivität bei Sartre vgl. Meyer (2008).
6 Vgl. Sartre (1995), 466.
7 Ebd., 457.
8 Vgl. Merleau-Ponty (1964), 51.
9 Langeveld (1968), 131.

aus der Situation, nicht unabhängig von jeglicher Bindung sich definieren, […]."[10] Maurice Merleau-Ponty entfaltet seine Philosophie der leiblichen Erfahrung, indem er sich grundsätzlich gegen den Primat des Bewusstseins richtet. Wir wenden uns niemals nur und nicht in erster Hinsicht unserer Welt, den anderen und uns selbst zu, um zu erkennen. Wir verhalten und benehmen uns vielmehr in Situationen, die ganz bestimmte Ansprüche an uns stellen. Dabei fungiert unser Leib auf der einen Seite als Nullpunkt aller Initiativen. Er selbst bleibt uns jedoch auf der anderen Seite nur sehr unvollkommen zugänglich. Er, der uns mit unserer Welt und den anderen verbindet, bleibt sich selbst merkwürdig fremd. Es handelt sich also um eine Täuschung, wenn wir meinen, wir seien uns selbst unmittelbar gegeben. Wir bedürfen des Umwegs über andere. Sie müssen uns durch ihr Verhalten zeigen, dass es uns gibt.[11] Von meiner Geburt und meinem Tod weiß ich nur durch andere, weil ich mich weder als zur Welt kommend noch als sie verlassend erlebe. Während andere um uns herumgehen und uns so prinzipiell von allen Seiten sehen können, sind wir immer nur in der Lage, von einer Seite aus zu sehen.[12] Es ist uns versagt, uns ohne Hilfsmittel von hinten zu sehen. Genau genommen sehen wir im Spiegel unsere Rückseite und nicht uns selbst von hinten. Auch können wir unseren eigenen Blick nicht erwidern, weil wir uns nicht gegenübertreten können. Der Leib ist verwickelt mit einer Welt, die ihm Geborgenheit schenkt und Fremdheit zumutet.

Die Einsicht in die Unmöglichkeit einer reinen Vernunft bedeutet zugleich die Ermöglichung sinnlicher Erfahrungen, die nicht länger im Schatten der Reflexion stehen. Mein Leib bedeutet meine gelebte Existenz, und doch gilt auch das Gegenteil, wie Blumenberg zu bedenken gibt: Mein „Leib habe sich gerade dadurch auszuzeichnen, nicht da zu sein."[13] Wir vergessen die Leiblichkeit in unserem Wahrnehmen und Handeln, wenn sie nicht aufgestört wird. Der Gesunde „*lebt seinen Leib, aber denkt nicht an ihn*, […]."[14] Wir konzentrieren uns auf die Ziele unserer Handlungen und nicht darauf, wie wir bei unseren Aktivitäten aussehen.

> Es gibt eine originäre Motorik, eine Bewegung, die nicht als Zweiheit von ‚es bewegt sich' und ‚ich weiß es' zu fassen ist, sondern als ‚ich bewege mich'. […] Von einer solchen Äußerung denkt keiner, sie sei sonderlich bemerkenswert, weil jeder so etwas irgendwann einmal sagt. Doch die Schwierigkeit liegt darin, zu *denken*, was dieses ‚ich

---

10   Merleau-Ponty (1966), 9.

11   Vgl. Blumenberg (2006), 757.

12   Vgl. ebd., 564.

13   Ebd., 669.

14   Ebd., 698.

bewege mich' heißt, ohne auf einen Dualismus hineinzufallen! Es gibt ein ,ich kann', das von vornherein mit dem ,ich denke' verbunden und ihm nicht unterstellt ist.[15]

Mit seiner Philosophie der Leiblichkeit widerstreitet Merleau-Ponty Auffassungen, die dem Leib eine Bedeutung nur dank der Reflexion zubilligen wollen. Unser Leib ist nach ihm „jener Bedeutungskern, der sich wie eine allgemeine Funktion verhält, jedoch existiert und der Krankheit zugänglich ist."[16] Philosophien, die dem Bewusstsein das letzte Wort geben, haben unsere Erfahrungswelt in ein Ideenkleid gehüllt.[17] Ein Indiz für die Vorherrschaft dieser Sichtweise ist unter vielem anderen die Vorrangstellung der Sprache.[18] Aber in der zwischenmenschlichen Interaktion geschieht weit mehr als der Austausch von Worten. Es gibt Blicke, die sich kreuzen, Stimmen, die aufeinanderprallen. Die Thematisierung der leiblichen Erfahrung kann nach Merleau-Ponty nur dann den Fallen des Cartesianismus entgehen, wenn sie eine dritte Dimension anerkennt, in der sich Natur und Geist durchdringen. Differenzierungen wie die von Körper und Geist entstehen allererst durch den Bruch mit der Lebendigkeit der ursprünglichen Verflochtenheit. In seiner Spätphilosophie ringt Merleau-Ponty unter dem Stichwort „Fleisch" um die Fassung dieses anonymen Bündnisses von Mensch, Mitmensch und Dingen, dieser vorreflexiven Verknüpfung, die durch das Denken zerstört wird, wenn es ihr zu nahe kommt.

Phänomenologische Analysen des Wahrnehmens hüten sich davor, sämtliche Sinnesempfindungen nach einem Muster zu deuten. Sie richten sich kritisch auf unsere Tradition, in welcher „der Gesichtssinn [nicht nur] vorzugsweise die Analogien für den intellektuellen Überbau geliefert [hat], er hat auch weithin als Modell der Wahrnehmung überhaupt und damit als der Maßstab für die anderen Sinne gedient."[19] Darüber gerieten die „vulgärere[n] Formen des Verkehrs mit der Zudringlichkeit der Dinge"[20] ins Abseits. Das gilt nicht nur für die sogenannten niederen Sinne wie Riechen und Schmecken, sondern auch für das Hören, das vom

---

15 Waldenfels (2000), 148.

16 Merleau-Ponty (1966), 177.

17 Vgl. Merleau-Ponty (1986), 61 und 67.

18 Ein weiteres Anzeichen ist die derzeitig herrschende Verzauberung durch die Hirnforschung. Selbst wenn die Neurobiologie keine Handlungs-, sondern eine Grundlagenwissenschaft ist, sparen populäre Autoren nicht mit praktischen Hinweisen, auch wenn die meisten eingestehen, dass sie guten Pädagogen nichts Neues sagen, dass sie jedoch die empirischen Beweise nachreichen könnten, was aufgrund der weitgehenden Unbekanntheit der Funktionsweise des Gehirns nicht stimmt. Vgl. Meyer-Drawe (2011).

19 Jonas (1973), 198.

20 Ebd., 198.

Standpunkt des Sehens überhaupt nicht zu verstehen ist.[21] Die Aufforderung „öffnet Eure Ohren" zeugt davon. Wie soll man etwas öffnen, das man gar nicht schließen kann? Man kann Ohren nicht ausweichen wie Augen. Es gibt kein Äquivalent für den Blick beim Hören. Etwas Vertrauliches unter „vier Augen" zu bereden, ist seltsam. Denn man kann nie die Ohren dritter oder mehrerer ausschließen. Selbst „Wände können Ohren haben". Ein Einverständnis kann im Austausch der Blicke hergestellt werden, ein Geheimnis durch sie jedoch nicht geteilt.[22] „Das Auge schweift umher, wählt aus, geht auf die Dinge zu, dringt ihnen nach, während das Ohr seinerseits von Schall und Wort betroffen und angegangen wird. Das Auge kann *suchen*, das Ohr nur *warten*. Das Sehen ‚stellt' die Dinge, das Hören wird gestellt; [...]."[23] Man spricht vom Licht der Wahrheit und von der Stimme des Gewissens. „Das Ohr in seiner Unbeweglichkeit und Ungerichtetheit verrät wenig davon, ob und mit welcher Intensität zugehört wird."[24] So ist das Nicht-Sehen weniger gewichtig als das Nicht-Hören; denn um etwas zu überhören, bedarf es der besonderen Anstrengung, weil sich Ohren von Natur aus nicht schließen. Simmel macht darauf aufmerksam, dass das Ohr an einem „Mangel an Reziprozität" leidet, „die der Blick zwischen Auge und Auge herstellt." Es sei „das schlechthin egoistische Organ [...], das nur nimmt, aber nicht gibt; [...]."[25]

Trotz aller Differenzen zwischen den Sinneswahrnehmungen ist es vor allem unsere unausweichliche Sichtbarkeit, das Gesehenwerdenkönnen, durch das wir als leibliche Wesen geschlagen sind. Das Hören des anderen ist mir nicht als Erfahrung gegeben, sein Sehen in seinem Blicken dagegen auf folgenreiche Weise.

IV   Sichtbarkeit

„Mein Körper ist eine gnadenlose Topie."[26] Das bedeutet, dass ich im Unterschied zur bloßen Utopie mit meinem Leib einen Ort in der Welt verwirkliche. Ich nehme einen Raum ein, an dem kein anderer Platz findet. Ich kann meinen Körper zwar ablehnen, aber nicht ablegen. Mein Körper mischt mich unter die Dinge und

---

21 Eine sehr differenzierte phänomenologische Analyse des Hörens bietet Annette Höhne (2005).

22 Vgl. Blumenberg (2006), 876f.

23 Blumenberg (2001), 163.

24 Blumenberg (2006), 876.

25 Simmel (1958), 487.

26 Foucault (2005a), 25.

Mitmenschen. Aber das ist nur die eine Seite. Auf der anderen Seite bedeutet er auch eine Utopie; denn er ist mir nur in Teilen sinnlich zugänglich. Insbesondere das Sehen gerät hier schnell an eine Grenze. Mir ist mein eigener Leib niemals vollständig präsent. Zwar ist mir alles durch den eigenen Leib allererst gegeben, aber er selbst ist durch mich nur teilweise wahrzunehmen. Wie ich aussehe, weiß ich nicht auf unmittelbare Weise. Beim Tasten verhält es sich umgekehrt: Die „Tastbarkeit des Eigenleibs reicht weiter als die Sichtbarkeit."[27] Unsere Sichtbarkeit hat stets das Nichtsichtbare bei sich.

> In gewissem Sinne ist er vollkommen sichtbar. Ich weiß, was es heißt, von jemand anderem angeschaut und von Kopf bis Fuß gemustert zu werden. Ich weiß, was es heißt, von hinten aufgespießt, mit einem Blick über die Schulter überwacht und überrascht zu werden, wenn ich es am wenigsten erwarte. Ich weiß, was es heißt, nackt zu sein. Und zugleich ist dieser doch so sichtbare Körper gleichsam in einer Unsichtbarkeit gefangen, von der ich ihn niemals zu befreien vermag. [...] Der Körper ist ein Fantom, das nur der Spiegelwelt mit ihren Trugbildern angehört, und das auch nur in Bruchstücken.[28]

So fasst Foucault diese Erfahrung zusammen. Spiegel scheinen mir zu zeigen, was ich nicht an mir selbst sehen kann. Aber: „[...] der Spiegel ist eine Utopie, weil er ein Ort ohne Ort ist. Im Spiegel sehe ich mich dort, wo ich nicht bin, in einem irrealen Raum, der virtuell hinter der Oberfläche des Spiegels liegt."[29] Ich sehe mich dort, wo ich nicht bin, und doch bin ich es, die ich sehe – oder? Foucault spitzt die Situation weiter zu und entzieht damit der Überzeugung von der unmittelbaren Selbstgegebenheit den Boden:

> Spiegel und Leiche weisen der zutiefst und ursprünglich utopischen Erfahrung des Körpers einen Raum zu. [...] Bedenkt man nun, dass Spiegelbilder sich in einem für uns unzugänglichen Raum befinden und dass wir niemals dort sein können, wo unsere Leiche sein wird, und bedenkt man, dass Spiegel und Leiche sich ihrerseits stets anderswo befinden, wird deutlich, dass nur Utopien die tiefgründige, beherrschende Utopie unseres Körpers in sich aufnehmen und einen Augenblick lang verbergen können.[30]

Das neuzeitliche Subjekt wird damit auf eine harte Zerreißprobe gestellt. Dass es ein unerschütterliches Fundament sicheren Erkennens sei, erweist sich als Täuschung. Es ist im Gegenteil äußerst fragil und vor allem sich selbst undurchsichtig, opak,

---

27 Blumenberg (2006), 826.
28 Foucault (2005a), 29f.
29 Foucault (2005b), 935.
30 Foucault (2005a), 35.

voller Schatten. Während das Erkenntnissubjekt mit Hilfe von Wissenschaft und Technik das Wissen von seiner Welt angehäuft und verlässlicher gemacht hat, herrscht im Hinblick auf die Erkenntnis seiner selbst Ratlosigkeit. „Wir bleiben uns nothwendig fremd", stellt Nietzsche fest, „wir verstehn uns nicht, wir *müssen* uns verwechseln, für uns heisst der Satz in alle Ewigkeit ‚Jeder ist sich selbst der Fernste', – für uns sind wir keine ‚Erkennenden'…"[31] Mit Kant ist er sich einig, dass Selbstbeobachtung unmöglich ist:

> Der Mensch ist gegen sich selbst, gegen Auskundschaftung und Belagerung durch sich selber, sehr gut vertheidigt, er vermag gewöhnlich nicht mehr von sich, als seine Aussenwerke wahrzunehmen. Die eigentliche Festung ist ihm unzugänglich, selbst unsichtbar, es sei denn, dass Freunde und Feinde die Verräther machen und ihn selber auf geheimem Wege hineinführen.[32]

Wie sich zeigen wird, machen sich Feinde wie Freunde durch ihre Blicke zu Verrätern. Blicke demaskieren, indem sie spiegeln.

In der eigenen Sichtbarkeit ist entscheidend, dass man aussieht, dass man gesehen wird und damit auch zu belangen ist. Mit dem Bewusstsein der eigenen Sichtbarkeit ist das Paradies auf immer verloren. Scham und Verlegenheit bekunden das Bewusstsein des Preisgegebenseins an den Blick von anderen. Nun wird auch der Abwesende berufbar. Gerät er unter Verdacht, braucht er ein Alibi, das bestätigt, dass er andernorts gesehen wurde. Weil ihm die eigene Sichtbarkeit jedoch selbstverständlich ist, wird der Unschuldige nicht an Augenzeugen denken. Der Schuldige dagegen inszeniert seine ununterbrochene Sichtbarkeit am anderen Ort als dem der Tat. Das Hiebfeste wird zum Verhängnis. Fortlaufend konstituiert das Gesehenwerden das Bewusstsein seiner selbst. Husserls knappe Bemerkung, dass der fremde Mensch „konstitutiv der an sich erste Mensch" sei,[33] trifft den Kern dieser Erfahrung. Das Selbst gewinnt seine Gestalt als eine Antwort auf dieses Gesehenwerden. Es richtet sich auf es ein, bis es sich schließlich auf das Gesehenwerdenwollen richtet.[34] Er macht sich zurecht. Dazu gehört, wie Ludwig Marcuse anmahnt, „nicht nur, daß man sich wäscht und kleidet – auch daß man, sobald man unter Menschen geht, sein einsames Gesicht gesellig verbirgt."[35] Unsere Sichtbarkeit verführt dazu, dass man unser Innerstes am Äußeren ablesen will. Sehen und Beobachten soll zum Durchschauen gesteigert werden. Mit der Sichtbarkeit wächst

---

31 Nietzsche (1988b), 247f.
32 Nietzsche (1988a), 318f. Vgl. auch Kant (1983), B X.
33 Vgl. Husserl (1977), 125.
34 Vgl. Blumenberg (2006), 786.
35 Marcuse (1969), 221.

jedoch auch das Unsichtbare. „Der Leib verhindert nicht nur, daß wir *unsichtbar* sind, sondern auch, daß wir *durchsichtig* werden müssen. Opazität ist das Korrelat der Visibilität."[36] Kein Licht ist ohne Schatten möglich. Im menschlichen Gesicht konzentriert sich diese Spannung von Preisgabe und Diskretion. Eine Wissenschaft vom Gesicht wurde schon oft ins Auge gefasst. Aber das Gesicht verwirklicht sich im Gesehenwerden und lässt sich nicht als Objekt isolieren. „Auf diesem Feld der untrüglichen Anzeichen stellt sich jeder Messung im Detail das halluzinierte Ganze eines lebendigen Wesens in den Weg."[37]

Mein eigenes Gesicht kenne ich nur als Bild, beispielsweise als Spiegelbild oder Selfie, als Porträt sowie Passbild oder vielleicht im Rahmen einer Videoaufnahme. Während ich mit einem anderen spreche, kann nur dieser wahrnehmen, welchen Ausdruck mein Gesicht annimmt, wenn mir etwas wichtig ist, ich um Zustimmung werbe oder mir etwas als fragwürdig erscheint. Meine Expressivität und meine Ausstrahlung gibt es nur für andere, nicht für mich selbst. In der Versagung, mein eigenes lebendiges Gesicht zu sehen, spitzt sich die „Zweiheit von Leib-Sein und Körper-Haben" zu, die besagt, „daß der *eigene Leib* Züge eines *Fremdkörpers* aufweist."[38] Diese Befremdlichkeit erlebt man etwa in dem Fall, dass man unerwartet seinem Spiegelbild begegnet. Das eigene Gesicht, das nicht mit der Begegnung seiner selbst rechnet, gleicht einer Maske, die mir nur von innen gegeben ist, ausgestattet mit einer Mimik, welche ich in den Gesichtern von anderen gelernt habe. Mit der Konzentration auf das eigene Bild beruhigt sich die Lage unmittelbar, indem mein Gesicht die Resonanz auf die anderen in seinem Standbild verliert. Kein Porträt befriedigt das narzisstische Begehren, und erst ein erheblicher Aufwand an Kunstfertigkeit eines Fotografen führt in seinen Augen zu der ersehnten „Natürlichkeit", in welcher man vielleicht eine Ähnlichkeit findet.

„*Woher wissen Sie das?*", fragt Roland Barthes.

Was ist dieses ,Sie', dem Sie ähnlich sehen oder nicht? [...] Wo ist Ihr Wahrheitskörper? Sie allein können sich immer nur als Bild sehen, niemals sehen Sie Ihre Augen, es sei denn verdummt durch den Blick, den Sie auf den Spiegel richten (mich würde nur interessieren, meine Augen zu sehen, wenn sie dich ansehen): [...].[39]

---

36 Blumenberg (2006), 789.

37 Strauß (2004), 51.

38 Waldenfels (2010), 163. Vgl. auch Blumenberg (2006), 683. In dieser Fremdheit zeigt sich die Wahrheit der cartesischen Differenz. Denn das *ego cogito* begegnet sich als *res extensa*.

39 Barthes (1978), 40.

Aber genau das ist uns entzogen. Man kann zwar seine Augen *sehen*, sich aber nicht in die Augen *schauen*. Der Blick in den Spiegel bleibt unerwidert. Er gewahrt einen Doppelgänger, der alles mitvollzieht, und zwar nicht etwa seitenverkehrt, wie immer wieder behauptet wird. So kommt uns die rechte Hand auf der rechten Seite entgegen und nicht auf der linken. Die Irritation rührt daher, dass wir von unserem Spiegelbild unbedacht erwarten, dass es wie ein anderer Mensch handele, der uns gegenübersteht und uns grüßt. Rechts bleibt gleichwohl rechts, und links bleibt links, auch oben und unten werden nicht vertauscht. „Wir gebrauchen das Spiegelbild richtig, aber wir sprechen darüber [...] falsch, als täte es selber, was effektiv wir es tun lassen [...].“[40]

Das Spiegelbild ist in räumlicher und zeitlicher Hinsicht unfasslich. Während das fotografische oder gemalte Abbild bleibt und sich unter die Dinge des Raumes mischt, wenn wir fortgehen, ist das Spiegelbild an unsere Gegenwart gebunden. Es erscheint und verschwindet mit uns, ein lästiger oder geliebter Doppelgänger. Tatsächlich müsste ich aber zu zweit sein, um mich selbst als Original zu sehen. Kann ich mich also im Spiegel beobachten? Nein, denn das „Spiegelbild meines Leibes folgt ständig meinen Intentionen, ihrem Schatten gleich, und wenn Beobachten heißt, den Gesichtspunkt abwandeln, den Gegenstand aber festhalten, so entzieht sich mein Leib auch im Spiegel meiner Beobachtung, [...].“[41] Das Spiegelbild versagt mir die Distanz, die ich suche, und verweigert das Original, statt es abzubilden. Es bleibt ein Fremdkörper. Es soll mich spiegeln. Ob es mich widerspiegelt, bleibt Spekulation.

So kommt es auch auf die Umgebung des Spiegels an. Eine warme Beleuchtung schmeichelt dem Bild. Die Konturen sind weich. Dunkle Augenringe werden zu einer matten Umrandung, welche die Augen betont. Ganz anders sieht man sich in einer weiß-grellen Ausleuchtung. Hier gibt es nur extreme Bilder: Entweder erscheint man als krebsrot oder kalkweiß. Die Konturen sind scharf. Die Spuren von Stimmungen werfen Schatten in den Linien, die sie hinterlassen. Das Spiegelbild im weißen Neonlicht ermuntert den klinischen Blick und ist geradezu eine Aufforderung, Maßnahmen zur Korrektur zu ergreifen. Auch die Tageszeit spielt eine Rolle. Der morgendliche Blick in den Spiegel zeigt ein Gesicht, das noch nicht strahlen kann oder will, ganz anders als das letzte Bild am Abend, weich gezeichnet durch ein Glas Wein und die Müdigkeit. Und da gibt es noch den wunderbaren Spiegel in dem Modeladen, der nicht plan ist und die eigene Silhouette schmal wiedergibt, an der einfach jedes Kleid wunderbar aussieht. Alles in allem sehen wir uns im Spiegel durchaus nicht, wie uns die anderen sehen, sondern wir werfen

---

40   Eco (1988), 31.
41   Merleau-Ponty (1966), 116.

einen eigenen Blick darauf, und dieser Blick ist alles andere als neutral. Er kann unerbittlich sein oder auch nachsichtig, gleichgültig wohl kaum. Jede Resonanz auf unser Spiegelbild ändert dieses. Niemals jedoch werden wir uns sehen, wie uns die anderen sehen. Normalerweise vergessen wir diesen Umstand, weil wir uns ja insbesondere für den Blick der anderen herrichten. Je länger wir allerdings darüber nachdenken, desto beunruhigender wird diese eigentümliche Entfremdung von uns selbst. Was heißt es, dass ich einen Anblick biete, der mir selbst verwehrt ist? Ich erscheine mir selbst immer nur in meinen Augen, selbst auf einer Fotografie oder in einem Film. Ich bin Anblick. Ich habe ein Gesicht, werde gesehen. Ich kann Blicke fangen. Blicke können sich jedoch auch auf mich heften oder durch mich hindurchgehen. Blicke kann man spüren. Im Kreuzfeuer der Blicke entsteht mein Anblick, über den ich selbst nicht verfüge.

Wir leben in einer Gesellschaft, die von Gesichtern beherrscht ist.[42] Wir sind umgeben von zahllosen Gesichtern. In Großformat blicken sie von den Werbeflächen zurück. Wahlplakate zeigen uns unsere Politikerinnen und Politiker mit makellosen Gesichtern. Kaum ein Buch können wir lesen, ohne dass wir uns das Gesicht des Autors oder der Autorin unter dem Klappentext anschauen müssen, oft in unverwüstlicher Jugend erstarrt. Die Allgegenwart der Gesichter beraubt sie ihrer Besonderheit. Wir gewöhnen uns nicht nur an sie, das makellose Standgesicht wird zur Normalität und lässt uns vergessen, dass es unser expressives Gesicht ist, auf das die anderen antworten, und dass es das andere Gesicht ist, das uns von Beginn unseres Lebens an spiegelt.

In der Entwicklung des Individuums ist das Gesicht der Mutter zumeist der erste Spiegel für den Säugling. Man spricht dabei vom Madonna-Phänomen, weil die Mutter nicht ihre eigenen Affekte mimisch wiedergibt, sondern jene des Säuglings nachahmt und ausprägt. Sie ist Spiegel und Echo in einem. Die lebenslange Bedeutung von Gesichtern in der zwischenmenschlichen Interaktion hat hier ihren Grund: Mutter und Kind spiegeln sich ineinander. Das wird insbesondere am ersten Lächeln deutlich, das ausdrücklich dem menschlichen Gesicht gilt. Dieses wird ungefähr im zweiten Lebensmonat zu einer privilegierten Erscheinung, und zwar zunächst nur in der Vorder- und nicht in der Seitenansicht.[43] Spiegelphänomene beherrschen die Interaktion. Nicht allein beim Füttern der eigenen Kinder, sondern auch bei fremden öffnen und schließen Erwachsene ihre Lippen in einem fiktiven Mitessen. Die mimetische Spiegelung betrifft auch die Höhe der Stimme und die Wahl der Worte, das Wuchern der Diminutive und die Rückkehr zur ansonsten längst vergessenen Echolalie.

---

42 Vgl. Macho (2011), 263ff.
43 Vgl. Spitz (1985), 69ff.

Es dauert, bis der Säugling sich selbst im Spiegel erkennt. Auf dem Arm der Mutter nimmt er zunächst die Mutter im Spiegel wahr, weil er den Vergleich hat. Aber woher sollte er wissen, dass er nicht auf einen Nebenbuhler neidisch zu sein braucht, der im Arm der Mutter liegt? Er hat doch nur sehr fragmentarische Erfahrungen von sich selbst, kann nicht einmal selbstständig auf eigenen Füßen stehen. Das Erkennen des eigenen Spiegelbilds wird mit Jubel begleitet.[44] Zugleich stellt sich aber auch eine Entfremdung ein, welche von nun an das Leben begleiten wird: Ich bin nicht allein das, als was ich mich unmittelbar erlebe. Ich bin auch noch ein Bild von mir, das sich andere machen können. Ich erscheine im eigenen sowie im fremden Blick, und beide sehen nicht dasselbe.

In diesem Fall behält der Spiegel magische Kräfte, indem er uns nämlich vorgaukelt, wir sähen uns in ihm, wie uns die anderen sehen. Tatsächlich jedoch sehen wir nicht das Bild, das die anderen von uns haben, sondern wir begegnen uns in unseren eigenen Augen. Sehen ist zudem noch sehr konventionell. Es ist auf ein Wiedersehen, auf Kontinuität aus. Deshalb wird unser Spiegelbild mitgezeichnet durch unsere Vertrautheit mit uns, durch Erinnerungsbilder, ähnlich wie das Gesicht des langjährigen Lebenspartners, das merkwürdigerweise immer weniger altert als das der Zeitgenossen. Das vertraute Gesicht ähnelt dem Spiegelbild. Es bleibt auf bestimmte Weise unzensiert und profitiert durch seine Geschichte. Es behält Züge des ehemaligen jungen Gesichts, und zwar nicht, weil die Falten oder Altersflecken fehlten. Die sind durchaus vorhanden, aber sie sind dem vertrauten Blick des anderen gleichgültig, weil dieser auf eine Ausstrahlung antwortet, welche die Jugend noch immer mit anwesend sein lässt.

Das Gesicht ist etwas Besonderes. So vertraut es sein kann, so fremd ist es zugleich. Das gilt insbesondere für das eigene Gesicht, das nur auf Umwegen zugänglich ist. So bedeutet etwa Schminken eine intime Befassung mit dem eigenen Gesicht, die keine fremden Blicke schätzt. „Man versenkt sich in die Zwiesprache mit einem überaus komplexen Adressaten: einem sich sehenden, sich durchsichtigen und sich doch auch undurchsichtigen Selbst."[45] Beim Schminken entsteht eines meiner Gesichter. Im Vordergrund steht dabei der Akt, das Zusammenspiel von Auge und Hand, aus der ein Gesichtsbild hervorgeht. Wir vergessen das geschminkte Gesicht meistens beim Verlassen des Badezimmers, weil das Schminken selbst das Wichtigste ist und wir danach dem anderen zugekehrt sind, eine Zuwendung, die möglich ist, weil wir von uns absehen. Im Spiegel entsteht wie beim Foto ein gespenstischer Eindruck dadurch, dass aus dem Subjekt des Betrachtens ein Objekt

---

44 Vgl. Meyer-Drawe (2000), 125.
45 Gehring (2006), 85.

der Betrachtung wird. Ich fühle selbst, wie ich zum Objekt werde, wie mir gleichsam das Bild die Lebendigkeit entzieht.[46]

Trotz aller Versagungen kommt uns das eigene Gesicht jedoch als so unverwechselbar vor wie ein Fingerabdruck. Ohne Umwege haben wir es niemals gesehen. Dennoch überkommt uns ein unangenehmes Gefühl, wenn wir für jemand anderen gehalten werden. Schrecklich ist es, wenn uns geliebte Menschen nicht wiedererkennen. Beide Male zweifeln wir eher am anderen als an uns selbst. Dabei haben wir streng genommen dazu keinen Grund. Nie haben wir gesehen, welchen Anblick wir bieten, während wir essen oder trinken. Niemals werden wir uns sehen, während wir mit einem anderen sprechen, wir verliebt blicken oder vor Wut zittern. Der andere ist uns dabei stets voraus. Wir sind wie unsere Mitmenschen und doch anders. Es gibt etwas Absonderliches, etwas Seltsames, das uns abweichen lässt. Doch ist diese Abweichung in ihrem vollen Ausmaß unzugänglich für uns selbst. Das blanke Entsetzen und die nackte Angst kennen wir nur aus den Gesichtern der anderen. Unser eigenes Gesicht ist uns selbst eigentümlich äußerlich, eine dem anderen zugewandte lebendige Larve. „Was mir fehlt, das ist dieses Ich, das du siehst. Und was dir fehlt, das bist du, den ich sehe."[47]

Das „Gesicht bewirkt, daß der Mensch schon aus seinem Anblick, nicht erst aus seinem Handeln verstanden wird. Das Gesicht, als Ausdrucksorgan betrachtet, ist sozusagen ganz theoretischen Wesens, es *handelt* nicht, wie die Hand, wie der Fuß, wie der ganze Körper; es trägt nicht das innerliche und praktische Verhalten des Menschen, sondern es *erzählt* nur von ihm."[48] Der Gesichtsausdruck ist nicht vom Gesicht abzulösen und etwa durch Sprache zu ersetzen wie manche Geste. „Ich freue mich" ist nicht dasselbe wie ein freudiger Gesichtsausdruck. „Du tickst nicht richtig" kann dagegen sehr wohl statt des Tippens mit dem Zeigefinger gegen die eigene Stirnseite gesagt werden. *Ein Gesichtsausdruck hat keine Stellvertreter.* Er moduliert die Interaktion. Man kann ihn nicht zurücknehmen. Ein entsetzter Blick kann mehr anrichten als Worte der Bestürzung. Im Umgang miteinander sind wir auf den Gesichtsausdruck angewiesen, was man insbesondere beim Telefonieren in Erfahrung bringen kann. Manche Äußerungen kann man dabei schlecht einschätzen, ironische etwa. Gleichzeitig gestikuliert das Gesicht des Sprechers am Hörer in die Leere und hört doch nicht auf damit. Vom Beginn seines Lebens an orientiert sich der sehende Mensch am anderen Gesicht. Daher verunsichert es, wenn das Gegenüber gar keine Resonanzen zeigt. Wir nehmen weder Zweifel noch Zustimmung wahr, wenn uns der andere im Gespräch zuhört. Ein Gefühl

---

46 Vgl. Barthes (1989), 22.

47 Valéry (1991), 308.

48 Simmel (1958), 485.

stellt sich ein, als würde man gegen eine Wand reden. Dem Reden fehlt bald der Schwung, die Emphase. Es trocknet aus. Gerade diesen für unser Miteinander wesentlichen Gesichtsausdruck enthält uns unser eigenes Spiegelbild vor. Weil er keine Stellvertreter hat, gibt es unser expressives Gesicht nur für den anderen.

Spiegelbilder sind Doppelgänger, die uns wie Schatten begleiten. Ohne unsere Anwesenheit gibt es sie nicht. Daher sind sie an unserer Sichtbarkeit für uns beteiligt wie wir an ihrer Existenz. Sie sind eher „Bildner"[49] als Abbilder. Sie lassen uns etwas sehen, das uns ohne sie nicht gegeben wäre. Doch verschaffen sie uns keinesfalls Gewissheit über uns selbst. Sie reflektieren auch nicht unsere Wirklichkeit. Im Zusammenspiel mit dem eigenen Blick entsteht ein Konterfei, dem nicht zu trauen ist. Dennoch kommt es uns näher als das gemalte Porträt, die Fotografie oder der Film.

> Er glaubte nicht an ein definitives Gesicht im Tode. Vom Augenblick an, wo er eingetreten ist, baut er die Substanz eines Gesichts ab, versetzt es mit dem unbestimmten Ausdruck der Verwesung. Der Mensch hat kein Gesicht für die Ewigkeit, es könnte nur eine Fratze sein. Dahinter verbirgt sich nichts mehr. Nur für die Vergänglichkeit haben wir ein Gesicht, so wie die Steine, aus dem Wasser gezogen, beim Trocknen ihren Glanz verlieren. Die Fixierung tötet das Lebendige noch einmal, der harmlose Schnappschuß nicht weniger als ein Schuß in die Brust.[50]

Spiegelbilder fangen unsere Leiblichkeit ein und lassen sich dennoch nicht berühren. Sie geben uns wieder als Sehende, die in ihnen sichtbar werden. Damit bewahren sie ihre Macht. Wir werden von ihnen in Bann gehalten. Selbst wenn wir ihnen aus dem Wege gehen, geschieht dies nicht aus Gleichgültigkeit. Sie widersetzen sich der strikten Ordnung nach Subjekt und Objekt. Sie spielen damit, dass das Sichtbare auf Unsichtbares verweist, Anwesenheit auf Abwesenheit. Spiegelbilder verbürgen keine Wahrheit. Eher nähren sie Zweifel, Unsicherheiten und Befremdlichkeiten.

V   Blicke

„Wie sich im Gesicht die Verflechtung von Visibilität und Opazität des Leibes konzentriert, so nochmals im Auge die des Gesichts. Der Unvermeidlichkeit seiner Funktion wegen bleibt das Auge noch bei den dichtesten Verhüllungen, Verschleierungen und Maskierungen ausgespart."[51] Der tiefe Blick in die Augen

---

49 Vgl. Lacan (1973).
50 Muschg (2003), 65.
51 Blumenberg (2006), 866.

eines anderen bringt die soziale Beziehung auf die kurze Distanz einer geraden Linie. Ein Ausweichen ist nur möglich, wenn das Visier fallen gelassen oder der Blick abgewendet wird. Wir spüren die Tiefe des Blickes. Wir fühlen Blicke, die in unserem Rücken auf uns geheftet werden. Blicke können uns erhöhen, aber auch erniedrigen. Blicke brauchen ein Gesicht. Augen allein blicken nicht. Augenlider sind wichtige Komplizen. Wie Schleier gestalten sie das Blickspiel mit.

Der erste Blick beherrscht das Kommende. Mit ihm kann alles falsch angefangen haben, aber auch richtig. Sympathie und Antipathie können hier ihren Ausgang nehmen.

> Im Blick *stellt* man den anderen. Wenn es heißt, daß der Blick eines Menschen es einem ‚verraten' kann, was er denkt und will, so heißt der Blick nicht daraufhin verräterisch, daß er als ein unwillkürlicher Ausdruck etwas kündet, sondern weil er der erste ungewollte Durchbruch eines Beginnens ist. Ähnlich wie die Wendung, daß jemand Miene macht … soviel hießt wie: er ist *schon dabei* zu … Denn was man will, ‚bestimmt' sich allererst im Verfolg eines Beginnens. Wobei aber nun im *Blick*, den man auffängt, sogleich ein freies Verhältnis von Mensch zu Mensch sich herstellt.[52]

Über Blicke kann man sich verständigen, wenn auch nicht erkennen. Durchschauen meint nicht Blicken. Es gibt im strengen Sinne keinen Zuschauer des Blickwechsels, der sich stets *zwischen* den Beteiligten ereignet. Das Auge selbst verschwindet im Blicken. Es bezieht sich nicht auf sich selbst. Der eigene Blick kann überdies nicht das Muster des fremden sein; denn es ist uns vollständig unzugänglich, weil wir nicht wissen, wie wir in der Blickkonfrontation aussehen.[53]

Die Angst gilt dem Blick, der alles sieht und mit dem sich kein anderer kreuzen kann. Unsere Sichtbarkeit gibt uns der Disziplinarmacht preis und damit der Auslieferung an das Gesehenwerdenkönnen, ohne selbst sehen zu können. Sartre schildert eine Kindheitserfahrung, die uns vielleicht nicht fremd ist:

> Ich hatte mit Streichhölzern gespielt und einen kleinen Teppich versengt; ich war im Begriff, meine Untat zu vertuschen, als plötzlich Gott mich sah. Ich fühlte Seinen Blick im Innern meines Kopfes und auf meinen Händen; ich drehte mich im Badezimmer bald hierhin, bald dorthin, grauenhaft sichtbar, eine lebende Zielscheibe. Mich rettete meine Wut: ich wurde furchtbar böse wegen dieser dreisten Taktlosigkeit, ich fluchte, ich gebrauchte alle Flüche meines Großvaters. Gott sah mich seitdem nie wieder an.[54]

---

52 Lipps (1941), 26.
53 Vgl. Blumenberg (2006), 867.
54 Sartre (1965), 59.

Der Blick, der alles sieht, ist nicht auf einen Wechsel aus. Er spioniert. Er bestimmt die Verhältnisse der Macht. Gesehenwerden ist jedoch nicht allein das Schicksal der gemeinen und üblen Subjekte, die sich dem Regime der Blicke nicht entziehen und sie doch auch nicht erwidern können. Zur Sichtbarkeit sind wir alle verdammt. Dieser Umstand lässt mitunter vergessen, dass Gesehenwerden auch ein Privileg sein kann. Man begrüßt in diesem Fall das Gesehenwerden, weil man eines Blickes gewürdigt wird.

Blicke, die sich kreuzen, sind keine Sache von Inspektoren und Introspektoren. Sie unterwerfen das Selbst nicht lediglich einer fremden Ordnung, sondern sie erneuern es. In der Erwiderung des Blicks lässt sich das Selbst in Anspruch nehmen als ein für beide Seiten unerreichbares Ich, das ohne Wort eminent beredt wird. „Blicke, die sich begegnen, erzeugen seltsame Beziehungen. Niemand vermöchte frei zu denken, wenn seine Augen nicht imstande wären, andere Augen, die ihnen folgten, zu verlassen. Sobald die Blicke einander festhalten, sind wir nicht mehr völlig *zwei*, und es wird schwer, *allein* zu bleiben."[55]

Literatur

Barthes (1978): Roland Barthes, *Über mich selbst*. Übers. von Jürgen Hoch. München [Paris 1975].

Barthes (1989): Roland Barthes, *Die helle Kammer. Bemerkung zur Photographie*. Übers. von Dietrich Leube. Frankfurt/M. [Paris 1980].

Blumenberg (2001): Hans Blumenberg, „Licht als Metaper der Wahrheit. Im Vorfeld der philosophischen Begriffsbildung". In: ders., *Ästhetische und metaphorologische Schriften*. Auswahl und Nachwort von Anselm Haverkamp. Frankfurt/M., 139-171.

Blumenberg (2006): Hans Blumenberg, *Beschreibung des Menschen*. Aus dem Nachlaß hg. von Manfred Sommer. Frankfurt/M.

Eco (1988): Umberto Eco, *Über Spiegel und andere Phänomene*. Übers. von Burkhart Kroeber. München [Mailand 1985/1987].

Foucault (2005a): Michel Foucault, „Der utopische Körper". In: ders., *Die Heterotopien. Der utopische Körper. Zwei Radiovorträge*. Übersetzt von Michael Bischoff. Mit einem Nachwort von Daniel Defert. Frankfurt/M. [Paris 2004], 23-36.

Foucault (2005b): Michel Foucault, „Von anderen Räumen [1967]". Übers. von Michael Bischoff. In: ders., *Dits et Ecrits. Schriften*. Bd. 4. Hg. von Daniel Defert und François Ewald unter Mitarbeit von Jacques Lagrange. Frankfurt/M., 931-942.

Gehring (2006): Petra Gehring, „Das Gesichtsbild als Akt". In: Christian Janecke (Hg.): *Gesichter auftragen. Argumente zum Schminken*. Marburg 2006, 79-93.

---

55 Valéry (1991), 308.

Höhne (2005): Annette Höhne, *Eine Welt der Stille. Untersuchungen zur Erfahrungswelt Gehörloser als Ausgangspunkt für eine phänomenologisch-orientierte Gehörlosenpädagogik*. München.

Husserl (1977): Edmund Husserl, *Cartesianische Meditationen. Eine Einleitung in die Phänomenologie*. Hg., eingeleitet und mit Registern versehen von Elisabeth Ströker. Hamburg.

Jonas (1973): Hans Jonas, „Der Adel des Sehens. Eine Untersuchung zur Phänomenologie der Sinne". In: ders., *Organismus und Freiheit. Ansätze zu einer philosophischen Biologie*. Übers. vom Verf. und von K. Dockhorn. Göttingen, 198-225.

Kant (1983): Immanuel Kant, „Anthropologie in pragmatischer Hinsicht". In: ders., *Werke*. Bd. 10. Hg. von Wilhelm Weischedel. 5. Auflage. Darmstadt, 395-690.

Lacan (1973): Jacques Lacan, „Das Spiegelstadium als Bildner der Ichfunktion, wie sie uns in der psychoanalytischen Erfahrung erscheint". (1949) In: ders., *Schriften I*. Ausgewählt und hg. von Norbert Haas. Übers. von Rodolphe Gasché et al. Olten/Freiburg im Breisgau [Paris 1966], 61-70.

Langeveld (1968): Martinus J. Langeveld, „Das Ding in der Welt des Kindes". In: ders., *Studien zur Anthropologie des Kindes*, 3., durchgesehene und ergänzte Auflage. Tübingen, 142-156.

Macho (2011): Thomas Macho, *Vorbilder*. München.

Marcuse (1962): Ludwig Marcuse, *Nachruf auf Ludwig Marcuse*. 2. Auflage. München.

Merleau-Ponty (1964): Maurice Merleau-Ponty, *L'Œil et l'Esprit. Préface de Claude Lefort*. Paris.

Merleau-Ponty (1966): Maurice Merleau-Ponty, *Phänomenologie der Wahrnehmung*. Übersetzt und durch eine Vorrede eingeführt von Rudolf Boehm. Berlin [Paris 1945].

Merleau-Ponty (1986): Maurice Merleau-Ponty, *Das Sichtbare und das Unsichtbare gefolgt von Arbeitsnotizen*. Hg. und mit einem Vor- und Nachwort versehen von Claude Lefort. Übers. von Regula Giuliani und Bernhard Waldenfels. Berlin [Paris 1964].

Mercier (1998): Pascal Mercier, *Der Klavierstimmer*. München.

Meyer (2008): Martin F. Meyer, „Das Problem der Intersubjektivität bei Sartre". In: Ulrike Barth (Hg.), *Jean-Paul Sartre – ein Philosoph des 21. Jahrhunderts?* Darmstadt, 137-152.

Meyer-Drawe (2000): Käte Meyer-Drawe, *Illusionen von Autonomie. Diesseits von Ohnmacht und Allmacht des Ich*. 2. Auflage. München.

Meyer-Drawe (2011): Käte Meyer-Drawe, „Der Mensch – eine neuronale Maschine?". In: Christian Sternad und Günther Pöltner (Hg.): *Phänomenologische und Philosophische Anthropologie. Orbis Phänomenologicus. Perspektiven. Neue Folge 28*, 77-94.

Muschg (2003): Adolf Muschg, *Sutters Glück*. Frankfurt/M.

Nietzsche (1988a): Friedrich Nietzsche, „Menschliches, Allzumenschliches I und II". In: ders, *Kritische Studienausgabe*. Hg. von Giorgio Colli und Mazzino Montinari. Bd. 2. 2. Auflage. München.

Nietzsche (1988b): Friedrich Nietzsche, „Zur Genealogie der Moral". In: ders, *Kritische Studienausgabe*. Hg. von Giorgio Colli und Mazzino Montinari. Bd. 5. 2. Auflage. München, 245-412.

Sartre (1983): Jean-Paul Sartre, „Die Wörter". In: ders., *Gesammelte Werke in Einzelausgaben*. In Zusammenarbeit mit dem Autor hg. von Traugott König. Autobiographische Schriften. Bd. 1. Übers. und mit einer Nachbemerkung von Hans Mayer. Reinbek bei Hamburg.

Sartre (1995): Jean-Paul Sartre, „Das Sein und das Nichts. Versuch einer phänomenologischen Ontologie." Hg. von Traugott König. Übersetzt von Hans Schöneberg und Traugott König. In: ders., *Philosophische Schriften*. Bd. 3. Reinbek bei Hamburg [Paris 1943].

Schenk (2009): Sylvie Schenk, *Parksünder*. Wien.

Simmel (1958): Georg Simmel, „Exkurs über die Soziologie der Sinne". In: ders, *Soziologie. Untersuchungen über die Formen der Vergesellschaftung.* 4. Auflage. Berlin, 483-493.

Spitz (1985): René A. Spitz unter Mitarbeit von W. Godfrey Cobliner, *Vom Säugling zum Kleinkind. Naturgeschichte der Mutter-Kind-Beziehungen im ersten Lebensjahr.* Übers. von Gudrun Theusner-Stampa. 8. Auflage. Stuttgart. [New York 1965].

Strauß (2004): Botho Strauß, *Paare, Passanten.* München.

Valéry (1991): Paul Valéry, „Zur Theorie der Dichtkunst und vermischte Gedanken". In: ders., *Werke. Frankfurter Ausgabe in 7 Bänden.* Bd. 5. Hg. von Jürgen Schmidt-Radefeldt. Frankfurt/M. [Paris 1957 und 1960].

Waldenfels (2000): Bernhard Waldenfels, *Das leibliche Selbst. Vorlesungen zur Phänomenologie des Leibes.* Hg. von Regula Giuliani. Frankfurt/M.

Waldenfels (2010): Bernhard Waldenfels, „Bewährungsproben der Phänomenologie". In: *Philosophische Rundschau.* Bd. 57. Heft 2, 154-178.

Druids[1] at Wayland's Smithy:
Tracing Transformations of the Sentient Body in Ritual

Thorsten Gieser

I   The Day after the Ritual

It was the day after our Samhain ritual at Wayland's Smithy.[2] It was a splendid day for November, warm, sunny and dry – in complete contrast to last night during the ritual. I went with two Druids – Chris and Richard – for a closer look at this place that I had not properly 'seen' before. I could not recognize it as the place of last night's ritual, it appeared so different. There was an earthen mound covered with

1   Contemporary Druidry is a spiritual movement – sometimes described as a 'nature religion' – which originated in England in the seventeenth century (see Hutton 2007). It sees itself as being in line with a tradition that goes back to the religion of the ancient Celts and their priesthood, the Druids. For that reason the British Isles are a sacred landscape to them. Archaeological sites, in particular, are highly regarded in Druidry as visible and tangible links to the past and the sacred. Moreover, Druids try to live in correspondence with a sacred Nature, in particular the cyclic pattern of the changing seasons, and perform rituals in order to achieve such a correspondence. Samhain (better known as Halloween) is one of the important days of the ritual calendar which marks the beginning of the dark days of autumn (the dying sun), of dying Nature and the dead/ the 'ancestors'.

2   Wayland's Smithy is a Neolithic chambered long barrow in Oxfordshire, England (see Darvill 2004). Used as an ancient burial site, it was constructed between 3590 and 3555 BC, with a second long barrow superimposed on the first sometime between 3460 and 3400 BC. It soon fell into disuse and became heavily damaged in the following millennia. Its current appearance is the result of a reconstruction following archaeological excavations in 1962-3. The name Wayland's Smithy was first mentioned in a Saxon chronicle dating from 955 AD and refers to the Saxon smith god who was said to reside there. This connection to a Saxon god in addition to its ancient use as burial ground makes this a very important site for Druidic practice. Such places are said to strengthen connections to the 'ancestors' and the spirits of the earth and/or the earth goddess. A ritual 'tuned in' into this spirit of place may thus open up a link between these powers and the participants.

grass, lined by a low dry stone wall, the entrance passage guarded by three huge standing 'façade stones' to the right and three to the left. Wayland's Smithy was surrounded by trees amidst a landscape of rolling fields, everything was dappled in a mosaic of patches of sunlight and shade, and birds were chirping. Now that we walked around it, touched the stones and laid in the grass while we talked about the place, it felt so solid and so part of the surrounding landscape. Although I had never experienced this place like I did now, I nevertheless felt some connection to it. Chris and Richard felt it, too. They had been here during many rituals; for them, it teemed with memories. The Druids moved around with an air of familiarity – unlike other visitors who came and went, just to have a look. They smiled at me, as they could see that I had now experienced the place as well. We shared a common experience of Wayland's Smithy – although I could not say that we shared exactly the same 'content' of experience.

Yet what I had experienced the night before was less important – what mattered was that I had opened myself, had allowed myself to be touched by this place and to resonate with it. That I had resonated with it was clear to the Druids from their own experience. I still felt a bit puzzled by last night, but I felt generally content – with me and the world. I felt somehow alive to the world, aware of the sunlight, the trees, the stones, the sky, I felt fresh despite the exertion of last night. Yet I also felt a bit disappointed because the ritual had failed to produce the kind of experience in me that would have provided me with answers to the questions the ritual leader had posed us: Who are you? Where do you come from? And what is your purpose in life? The spirit of place (Wayland's Smithy) and the spirit of time (Samhain) were to help us find answers. Yet all I got – so I thought – was a connection to this place and this season, a feeling of being alive to the world. But now I began to wonder as I lay on the grass next to the two Druids if I had been wrong in assuming that the ritual had failed. After all, a lot had changed since yesterday. I had been transformed…

In this essay I would like to advance a particular understanding of the body, the sentient body (*Leib*), in ritual theory. In this field, the body has so far primarily been conceived semiotically, i.e. as a passive recipient of meaning (constructed through/ by minds) or as a communicator of meaning, expressing in non-verbal, multi-modal means messages of propositional content (e.g. Leach 1976, Whitehouse 2006, Senft and Basso 2009). Ritual as a transformative experience has so far been comprehended as either transformative in terms of sociality (social identity, status, roles, see Strathern 1988, Houseman and Severi 1998) or as transformative in regard to states of consciousness (leading to so-called altered states of consciousness, ASC). In the latter case, the body is ritually manipulated via certain (archaic) techniques (of ecstasy, see Eliade 1975, Jilek 1982) in order to induce a change in the state of mind which in turn is supposed to lead to transformative experiences (*Erlebnisse*) such as visions, soul flights and so on. In neither case is the body central to the workings of the ritual. The real target for ritual action is the mind or soul which is regarded as a separate and superior domain (from the body) within the self.

Since the 1990s, a new phenomenologically inspired 'embodiment approach' to ritual has enriched our view of the body (e.g. Bell 1992, 1997).[3] It has been shown that and how the body, our bodily being-in-the-world, forms the basis of our experience and meaning-making. One of the most important insights of such an approach is that consciousness is always consciousness through the body, i.e. our consciousness can only gain access to the world, can only learn about the world, through our body and our senses. Consciousness and body are thus interlinked. However, what haunts phenomenology – at least the most influential exponents in ritual theory, Husserl and Merleau-Ponty – is a thinking that still employs a Cartesian dichotomy of mind and body, as evident in Thomas Csordas' work on ritual, for example. By invoking Merleau-Ponty's concept of the 'pre-reflective', he tries to find a way to study the embodied process of perception, i.e. he tries to "capture that moment of transcendence in which perception and objectification begin, constituting and being constituted by culture". His approach unites phenomenology (to capture the pre-reflective) and semiotics (to capture the 'reflective', the point when thinking and language sets in and re-works pre-reflective, embodied experience). The problem with this approach is that it locates the 'moment where culture begins' within the domain of semiotics, i.e. within the domain of thought and language. The Cartesian dichotomy remains, the body is still 'nature' (although now active and no longer passive), the mind still 'culture'.

A second problem pertains to a limitation in Merleau-Ponty's focus on perception which has been reproduced in phenomenologically-inspired ritual theory. Following the avenue laid out by Heidegger's work on *Dasein*, Merleau-Ponty was concerned with how we are-in-the-world via our body, i.e. how we perceive the world and engage with a world that to us appears meaningful from the very first moment, not as a meaningless, chaotic world which needs rational order and structure bestowed upon it by a reflective process of the mind. Merleau-Ponty thereby decentres the self; perception as an active process of meaning-making reaches out into the perceptual structures of the world, takes our consciousness through our body out there in the 'in-between' of perceiving subject and perceived objects. In this process, however,

---

3   It must be pointed out, however, that even Catherine Bell (2006, 538) sometimes construes embodiment in semiotic terms, as when she explains that "ritual is the expressed language of the body, a medium uniquely able to communicate messages, perform experiences, and create environments that are impossible with other media".

the body becomes 'transparent', the background to that which is foregrounded in our attention. In other words, our perceiving body disappears so that the world can appear to us. In Merleau-Ponty's work, the disappearance of the body seems inevitable. But it has been – rightly – questioned if that really needs to be the case. Is the living, perceiving body necessarily one that is oblivious to an awareness of itself being alive, feeling and perceiving itself?[4]

Following phenomenologists like Hermann Schmitz, Maxine Sheets-Johnstone and Richard Shusterman, I argue that we need a fuller appreciation of states of embodiment in order to trace the potential range of transformations inherent in human experience, in particular in ritual experience. As these hitherto under-represented dimensions of embodiment take us not outwards into the world but into ourselves, we will necessarily tread on ground that is commonly reserved for the mind or the soul. Hence, an exploration of these dimensions of embodiment questions the boundaries between body and mind just as the exploration of our being-in-the-world has questioned the boundaries between subject and object (see Gieser 2008a).

In the 'ethnographic miniature' with which I introduced this chapter I tried to convey a sense of the transformations I felt after having participated in a ritual at Wayland's Smithy. These were transformations of my whole self, a transformation of my way of being – not just of my soul or my body. Why should one need such a distinction? Hermann Schmitz (2011) followed the historical process of the con-stitution of the soul in the Ancient Greek world and sees a development towards an increasing distance between soul and body. According to Schmitz, the oldest known, that is, Homeric, anthropology does not conceive human beings with a central organising unit in a separate, (en)closed, private, inner world. Instead, the human being is described as a semi-autonomous concert of manifold motions, stirred by affects and divine inspirations and interventions. The body is not felt as a singularity but as a plurality of motions being felt in various locations within and around a body which is as open as the various souls inhabiting the person. Not before the philosophy of Sophocles does the soul (as psyche) become a bounded entity, locked up inside the body. Here, the rational capacity of the soul was supposed to be a means of controlling the body, preventing it from being controlled by things and beings outside itself (affects, divine influence). With Plato finally, the subject

---

4   We find the same problem in Andrew Strathern and Pamela Stewart's (2011, 389) recent definition of embodiment: "Embodiment is not the same as 'the body'. Embodiment refers to patterns of *behavior* inscribed on the body or enacted by people that find their *expression* in *bodily form*" (my emphasis). In other words, not the sentient body is in the foreground of this approach but the performing body.

becomes firmly located inside the soul, the body having become something (some *thing*?) outside of our self. This is the dualism of body and soul that Descartes later on developed into a full dichotomy. Schmitz draws the conclusion that the constitution of the modern person was only possible by a parallel withdrawal from our body. In accordance with Schmitz I argue that if we want to fully appreciate the whole range of embodiment we need to overcome this particular Western development and start thinking with a body that has retained its familiarity with the soul – a sentient body or *Leib*.

It is at this point that we need to enrich Merleau-Ponty's understanding of embodiment with Schmitz' conception of the *Leib*, i.e. what we can feel 'in the region' (within and around) of our material body without relying on our standard set of the Five Senses.[5] When our attention turns towards our own body, when we become aware that we are here, that we are now – that is, fully in the present – we can feel a range of states of bodily motions (*leibliche Regungen*). The term bodily motions is apt because our sentient body is dynamic (in a permanent state of becoming, we could say with Deleuze) and our attention is attracted to the various changes our body is undergoing in response to perceived changes in our surroundings. In Merleau-Ponty's work, the body is receptive and responsive to the world by adjusting its orientation to and grasp of the world. But Schmitz adds that the body's receptivity and responsivity also makes itself *felt* in some motion. At the most basic level, all motions register on a scale between a felt contraction (*Engung*) and expansion (*Weitung*).

Let me illustrate this with an example. How do we feel when a long held suspicion turns out to be untrue? To begin with, the suspicion might have held us in a grasp, tensed, our body alert, our chest feeling unpleasantly restricted, resulting in superficial, short breaths. When the suspicion turns out to be untrue, our tension relaxes, we take a deep breath that opens our chest and we feel the general contraction of our body releasing into an expansion. In this example, our body is grasped by an emotion/affect that results in a change of state between contraction and expansion. Another type of motion are holistic bodily motions (*Leibgefühle*), such as feeling tired, fresh, exhausted, hungry, passionate, etc. Furthermore, there are bodily movements as they are felt not seen, such as chewing, swallow-

---

5   Not unlike Schmitz' *Leibesphänomenologie*, Richard Shusterman (2008) has developed a *somaesthetics*, i.e. a philosophy that takes the awareness or perception (*aesthesis*) of the feeling and moving body (*soma*, otherwise what I call the *Leib* in this chapter) seriously. Interestingly, he proposes that this kind of somatic consciousness can become a 'discipline' (in Foucault's sense) for self-fashioning but also for acquiring embodied knowledge. Elsewhere (Gieser, unpublished), I argue that rituals can be seen as a discipline of this sort.

ing, breathing, grasping, walking (sensed kinaesthetically). For Schmitz, it is the dynamics of these motions which enable us to feel and understand ourselves as a self, i.e. I experience that it is 'I' who experiences and who *has* an experience. On this level, the 'I' is not separated out into body and mind, it is one whole sentient body/*Leib*. Only a secondary process of reflection then categorizes some of these motions into motions of the body (such as swallowing, breathing) and motions of the soul/mind (affect, passion).

Now that we have widened the scope of embodiment (as being-in-the-world) by including both our being-*towards-the-world* and being-*towards-our-body*, the body as background and as foreground, to a *Gestalt* experience, we can more easily trace and describe the transitions and transformations of the sentient body in ritual. Let me illustrate my approach by returning to my 'ethnographic miniature' and looking at the ritual that took place the day before.

II   Around the Fire

When Kieron announced this ritual at camp, ten people (including me) were interested in joining in, rather more than he expected. He had designed the ritual as a journey to the ancestors, to the centre of the earth and ourselves. As we got into the cars to drive to Wayland's Smithy it was already apparent that the ritual would take place in the rain. Nevertheless, no one wanted to cancel the ritual. We drove through the night (it was about seven o'clock in the evening) and stopped somewhere 'in the middle of nowhere'; at least, this is how it looked for those – like me – who had never been there before. Apparently we were in some kind of small woodland but nothing else could be seen as it was utterly dark by now. Kieron brought us to a place among the trees where he and his assistants had prepared a place to light a fire and put aside some firewood under a tarpaulin. He eventually managed to light the fire and then sat with us under the tarpaulin while the rain pattered loudly on this makeshift roof. Before he left us, he reminded us of the purpose of this ritual and told us to walk around the fire in meditation until we were called individually and guided to the Smithy.

And so we kept walking around and around the fire, our gaze fixed on the ground or on the feet of the person walking in front of us; to our left the yellow-red glaring blaze and to our right the black shapes of the trees against a pattern of shadows that merged with the darkness of the night in the distance; the left side of our bodies feeling the warmth of the flames, the right side feeling the cold of the rain. On each round we went through a cloud of smoke that stung the eyes but at least brought an intensive scent to our noses which otherwise noticed only the uniform, ambient smell of rain. All the while the rain poured down on our heads; the water dripped endlessly down our foreheads and slowly soaked through our 'water-proof' outdoor clothes until we shivered. My thoughts centred on the questions we were to ask ourselves in preparation of entering the sacred space of Wayland's Smithy: Who am I and what is my purpose in life? I repeated these questions again and again as I walked around the fire, hoping to open myself to find an answer later in the chambers of the long barrow.

When our body begins to move, it transforms and our experience of this sentient body transforms, too. At rest, our body has a certain materiality to it and a felt unity and singularity. But once it starts to move, we may experience the multiplicity of a fragmented body that is felt as a dynamic energy of movement.[6] Our legs do one thing, our arms another, our gaze reaches out in one direction, our ears in another.[7] This complexity of the moving sentient body needs a scaffold, i.e. a habit and/or a rhythm to stabilize itself and to re-establish its felt unity. Yet this felt unity of the moving body differs from the felt unity of a body at rest. The complex bundle of movements is a set of various motor intentions (Merleau-Ponty 2004) and relationships gathered around the body and incorporated into its bodily system of being-in-the-world. We could say with Merleau-Ponty (2003, 225) that the moving body is a 'libidinal body', a body which "asks for something other than the body-thing or than its relations with itself. It is in circuit with others".

This transformation of the body that comes with movement is illustrated in the first stage of the ritual described. We came to a 'non-place' in the middle of nowhere and a featureless darkness. While we crouched on the ground next to the firewood with a tarpaulin over our heads, a fire was lit for us that was to become the emergent 'pole of action' for the preparatory phase of the ritual. Our bodies, as yet without orientation, set themselves into motion and tried to make sense of

---

6   Maxine Sheets-Johnstone (2011, 131) describes it as follows: "From the start, what we find primordially there in self-movement is a felt unfolding dynamic and in virtue of that dynamic, a felt overall kinetic quality – a fleet swiftness, perhaps, or a sluggish heaviness, or a relaxed jauntiness, or an erratic intensity – or a constellation of qualities generated by a more intricate interplay of forces, an interplay that we might describe preeminently in terms of rhythmic complexity and abrupt directional changes, or in terms of constricted, jagged spatialities and alternately violent and fragile energies, for example".

7   The 'wholeness' of the body may be a material *a priori*, but not an experiential necessity. Think of Michel Serres' experience during a fire on a ship: "I am inside burnt to a crisp with only my frozen, shivering, blinded head outside. I am inside, ejected and excluded, and my head, arm and left shoulder are outside in the howling storm" (2008, 19).

and relate to the new surroundings. We walked around the fire, again and again, and our bodies found their rhythm and a balance of speed and distance (to the fire, to the other participants) so that we became like one body moving as a unit while protecting ourselves from the pouring rain and cold. Our bodies demanded some attention at the beginning; being in an unknown environment, mostly hidden in darkness, we had to concentrate on every move in order not to stumble and fall. Establishing a rhythm under these circumstances was not easy. Our attention turned first towards walking, feeling and responding to the uneven ground beneath our feet, then to our shoulder muscles tensed through permanently pulling up to offer at least an impression of protection against rain and cold, then to the flickering fire light in which it was hard to see and find our balance, then to the purpose of the ritual, all this repeatedly interrupted by stinging smoke.

In contrast to this ritual, our bodies move habitually in the world of everyday life (Merleau-Ponty 2004). They have established patterns of perception and movement which are guided and scaffolded by the space and objects around us. By habitually dealing with objects-in-space during our practices (which have certain intentions and goals to be accomplished) we gather a meaningful world around us, structured by our actions, bodily skills and bodily capacities. We are thus immersed in a texture of 'intentional threads' which pull or 'summon' us in ways that our bodies have learned to be receptive and responsive to. It is by these habits that our existence creates its own rhythm and thereby constitutes a certain 'style' (the 'pulse of existence', Merleau-Ponty 2004, 92) that we confer onto the world and the world in return on us. The ritual at Wayland's Smithy provides a kind of initiatory experience that pulls participants out of their known habits into the unknown. This method of 'disturbance' is quite common in contemporary Druidry, since Druids are critical of the habits of mainstream society, a society that – in their eyes – is governed by consumerism, environmental exploitation and the supremacy of technology. To break these habits and re-establish more 'natural' ways of living – and to live in Nature in particular – is one of the main concerns of Druids' 'nature religion'.

However, it must be said that by contrasting the 'naturalness' of the body with the 'artificiality' of (cultural) habits, Druids reproduce the very cultural conceptions of mainstream society that they mean to distance themselves from. Put differently, their ritual technique relates the 'inner nature' that is the body to the 'outer nature' of the environment, as if there could be an unmediated relationship – untainted by culture – from one nature to another. Nevertheless, this type of ritual is remarkable in its attempt to provide a framework for an experience, in this case an encounter with 'sacred Nature', with minimal (human/cultural) interference. It is not the ritual structure that should invoke the experience but the power of the encounter. Paralleling this sentiment, the phenomenologist Alphonso Lingis (1998, 119) reminds us that "reality weighs on us; we cannot be indifferent to it". In Druidry, Nature is conceived as alive with spirit, it is active and dynamic. The aim of rituals is to align oneself to this living spirit, to become resonant with its dynamics (e.g. the changing seasons) through interaction and involvement. As the Druid and writer Emma Restall Orr (2004:47) has put it:

> When we speak aloud, moving our body, our life energy flows from fingers to toes, through our brain and veins, affecting all that we are in our subconscious, in our bones. Only when the energy of our life-spirit is moving can we touch and be touched, better sensing and responding to the spirits around us.

Yet this spirit is not set within an inner domain of the self opposed to the body, it is in alignment with the 'natural body'. Thus a resonance can be achieved by spirit with and through the materiality of one's own body 'in touch' with the materiality of Nature. Regarding rituals this means that formalized ritual structure ideally helps Druids to achieve this resonance with Nature. But in the end, structures should not be needed as they do still interfere with a direct engagement of the self with Nature. In other words, Druidic rituals should not be conceived as a set of formalized, structured behaviour but as a stance, an attitude, a particular mode of experience and perception which gives rise to a feeling of being-in-the-world, of being part of Nature.

This in turn has consequences for our understanding of ritual space in Druidry. 'Space' – in ritual theory as in the work of Merleau-Ponty – tends to be a rather abstract phenomenon and is ill suited to describe what Druids try to relate to in ritual. Ritual space, for them, is not abstract geometrical space, nor is it geographical space in terms of the material topography of the landscape alone, nor is it exhausted in key features or key objects that give structure and character to a landscape (such as standing stones or a long barrow). Their ritual space is primarily an outdoor space, it is 'in the open' and what this entails is well summarized by Tim Ingold (2011, 121):

> To inhabit the open is not, then, to be stranded on a closed surface but to be immersed in the incessant movements of wind and weather in a zone wherein substances [materials] and medium [air, rain] are brought together in the constitution of beings that by way of their activity [such as rituals], participate in stitching the textures of the land.

The sentient body is thus immersed in this ever-transformative and ever-becoming 'weatherworld', a world of wind infused with flickering firelight in the midst of darkness, rain splashing down on our bodies and the ground where it mingles with the soil to become mud beneath and on our shoes, while more rain drips from our foreheads and a cold wind makes our bodies shiver. In this ritual space of the Druids there is no space for a nature-culture or mind-body dichotomy as the sentient body is busy engaging with the weatherworld which is all around, reaching down inside muscles, bones and the imagination.

III Betwixt and Between

Every now and then the 'Grove Mother' appeared as if from nowhere to guide one of us away from the fire and to Wayland's Smithy while the rest kept on walking. I think I was the sixth or seventh to be called by her. She took my wet arm and I stumbled clumsily into the darkness along a path she alone knew. After walking around a fire in otherwise complete darkness for such a long time, the whole world seemed to me to exist only visually as light and darkness, apart from the solid ground beneath my feet. I had problems moving between the trees as my vision seemed to lack depth and my eyes were still dazzled from the glare of the fire. So she had to guide me over or around the smallest of obstacles to bring me safely to the outer boundary of the Smithy and leave me there.

I could make out some blurry lights in the distance and walked towards them. As I came closer I saw that they were lanterns placed on the stones of what I presumed was the entrance area. Suddenly, out of nothing, a firm voice reverberated through the darkness and from out of the shadows appeared a black-robed figure with a cow's skull for a head. I froze on the spot. This spectre was now apparently trying to address me, but I could not make out a single word. Recovering from my initial shock, I recognised the voice of my friend Sean and, comforted by this knowledge, I could now understand what he was saying: "Who are you and why do you seek to enter this sacred place?" I cannot remember what I told him. I stammered something about trying to find out who I was and connecting to my ancestors. "Are you here to enter the womb or the tomb?" Being at an old burial site I thought a tomb would be appropriate and told the guardian so. "Then enter through the right entrance", he answered and, with a silent inviting gesture, stepped aside and let me pass.

However, I still saw everything in two dimensions only and I had not the slightest clue of the Smithy's interior. The lanterns did not help much as they bathed the scene in diffuse light and shadows. So I just stumbled forwards and promptly fell over a knee-high stone on the floor. With my eyes now close to the floor and walls I noticed that I was in a corridor with stone walls on both sides, open to the sky, and with stones protruding from the walls on each side at irregular intervals. I touched the stone walls and felt my way further along. The solidity of the stone against my hands brought a measure of stability back to my body, although I still perceived the stones more like small solid islands immersed in a sea of light and shadows.

Every ritual includes a phase that is neither here nor there but, to say it with Victor Turner (1977), 'betwixt and between'. The structures of everyday life dissolve or are even turned upside down, becoming 'anti-structure'. Social distinctions and hierarchies are smoothed out and replaced by a feeling of a common bond, of *communitas*, between the participants of the ritual. This is the phase of 'liminality', an experience of and beyond the margins of dissolving boundaries, including those between body and mind, subject and object, self and other. At first glance, this experience of liminality echoes Merleau-Ponty's (2000) conception of perception in/of the 'flesh', as he names his post-Cartesian way of describing the prereflective world of perceptual experience.[8] This world is made up of the 'flesh of the world' (the sensible) and the 'flesh of the body' (both sensible and self-sensing or sentient), "in a locus where they have not yet been distinguished, in experiences that have not yet been 'worked over', that offer us all at once, pell-mell, both 'subject' and 'object'" (Merleau-Ponty 2000, 130). Although we do not merge completely with the flesh, we do not experience ourselves as an 'I' who perceives, nor do we perceive distinguishable 'objects' opposed to ourselves. There is just the experience of an *Ineinander*, an 'intertwining', of me and the world immersed in sentience. This intertwining relies on a 'reversibility' between self and world. In prereflective perception,

> this bursting forth of the mass of the body toward the things... makes me *follow with my eyes* the movements and the contours of the things themselves, this magical relation, this pact between them and me according to which I lend them my body in order that they inscribe upon it and give me their resemblance ... (Merleau-Ponty 2000, 146)

In other words, reversibility refers to the adaptive process of a moving and perceiving body that is guided by and is responsive to the perceived. The body is thus transformed in a way that reflects the 'movements and the contours of the things', thereby inhabiting the 'style' of sentience into which both the sensible and the sentient are immersed.

We could say that in the experience of the flesh the world is truly incorporated (*einverleibt*). Walking around the fire for hours had given us time to establish a

---

8   It is in the notion of the flesh that Merleau-Ponty and Serres find their closest correspondence: "I only really live outside of myself; outside of myself I think, meditate, know; outside of myself I receive what is given, enduringly; I invent outside of myself. Outside of myself, I exist, as does the world. Outside of my verbose flesh, I am on the side of the world" (Serres 2008, 94).

rhythm and moving in a repetitive pattern was a form of incorporation. Our bodies made themselves 'at home' in this formerly unknown world around the fire and so we slowly became familiar with the look and feel of the fire, the ground beneath us and the surrounding impenetrable darkness. Yet 'incorporation' might be only one dimension of the intertwining and – as far as reversibility goes – needs to be supplemented by an opposing process of 'excorporation' or *Ausleibung* (Schmitz 2011), i.e. one's body reaches out and mingles with a dynamic, a rhythm, outside itself.

We were again – for a second time this night – confronted with a new, unknown world when the Grove Mother led each participant in turn from the world around the fire into the darkness of the night and into the woods. Here our bodily experience transformed from incorporation into excorporation. My senses were in disarray. Visually, I was dazzled by hours of staring into blazing flames. Haptically, I felt nothing of the material world apart from the ground beneath my feet and the guiding arm of the Grove Mother. When she left me alone, I lost my final hold on the world and mingled completely with the darkness, because my body had lost any feeling for its materiality, just as the world had lost its materiality. This experience of *Ausleibung* led to feelings of anxiety and uncertainty. I felt open and vulnerable; my body responded in an exaggerated way towards any changes in this weatherworld and I became agitated, tensed.

Initially, incorporation and excorporation may seem to denote a similar experience. However, the distinguishing element of excorporation is not only the direction of the dynamic (from the self out into the world). Whereas incorporation is a way of making yourself at-home-in-the-world, excorporation – as a marker of liminality – is an unsettling and disturbing experience which foregrounds the ambiguity and uncertainty of the perceptual process. The sentient body has difficulties perceiving and responding to the world unless given time to take on the external rhythm and correspond with it. In the ritual described this time was not given. Instead, I was suddenly grabbed by the presence of the Guardian. As I was still in a state of excorporation, though, I could not answer his questions fully from the centre of my 'rational mind'. I had to try to come back from the 'in-between world' to the centre of my self, i.e. to my body and to my consciousness. And this was the beginning of the next transformation that was to prepare me for the final stage of the ritual.

Only at this point did Wayland's Smithy begin to take form in my experience. Within the swirling dynamic of the weatherworld, the appearance of the Guardian introduced a material presence that foreshadowed the materiality of the sacred place that was the destination of the ritual. Coming slowly back into the materiality of my own body, my hands reached out, ready to touch. The light of the lantern helped me to feel my way along from stone to stone with my hands and my feet.

In other words, my body came back to me in fragments, as a complex bundle of activity of different body parts, still lacking a coherent feeling of itself in the sense of a *Leibgefühl*. Accordingly, this resulted in my experiencing Wayland's Smithy as neither 'object' nor 'place'; it existed only fleetingly, in fragments, as various parts of my body encountered various dimensions of it, slowly building up to a 'sphere of its presence' (*Sphäre ihrer Anwesenheit*) as Gernot Böhme (2013) would phrase it.

IV   Buried in Stone

I arrived at a cave-like entrance that opened up into a small stone-walled chamber with a niche opposite me and curtains covering passageways or niches to my left and right. I drew the curtain to the right and behind it was a dark, narrow and low chamber with an old mattress on the floor. A voice from behind told me to lie down there and close the curtain. The floor was covered in mud, both stone walls and floor were cold and rough as I went down on my knees to enter the chamber. I felt the presence of the stone around me and touched the wall with my feet as the chamber was too small to lie down straight. With the curtains closed it really felt like being buried, enclosed in tons of stone and earth.

Then I heard the voice again, right beyond the curtain. It was Kieron's and he told me to relax and go deep into the earth to find my roots and thereby also myself. Then I heard the monotonous boom of a drum and the eerie wail of a didgeridoo as I tried to get into a trance state. I felt myself sinking into the earth, deeper and deeper, blackness all around me, silently repeating the question I wanted to have answered.

But nothing came. Perhaps I lacked concentration or needed more time, but as my time was eventually up I had to leave. Outside, the Grove Mother was already waiting for me to take me back to the fire. I re-entered that world of night, rain and fire and watched the remaining participants being called up. Those who had been in the Smithy now stood around the fire, a bit further away, or leaned against a tree, pensive and silent. Only when the last one was led away did we dare to speak and the hours of waiting and tense anticipation disappeared as we talked and laughed with one another. We could not believe that we had just spent the last four hours walking around a fire in pouring rain. It was good though, as everyone nevertheless agreed.

The ritual had led me from the open weatherworld alive with wind, rain, light and darkness into an enclosed material world of solid, hard stone. In this world of hardness and solidity I became a material, touching body again, aware of my skin, of the boundaries of my body as they came into contact with the stones around me which 'ob-jected' me, i.e. opposed me, stood over against me. In this process of objectification my consciousness retreated into my body and I became my self again. The world of stone had enclosed me and thereby had enclosed my soul inside my body again. I was no longer 'in the open' and no longer felt 'open'. In contrast, I felt rooted again, firmly located and oriented within this place that I now perceived as the sacred mound of Wayland's Smithy.

Hermann Schmitz (2011) might have recognized my experience as an experience of *primitive Gegenwart* (primitive presence/present) and a transformation of *personale Regression* (personal regression) into *personale Emanzipation* (personal emancipation). In its purest, ideal form, the experience of *primitive Gegenwart* is an experience of my self and nothing beside. In such an experience, nothing seems to exist that is not me, for instance when a shock of pain fills my awareness. Seen as a process (rather than as a defined, instantaneous event) it is characterised by a gathering of the self as the sentient body contracts (like in a hard encounter with stone, for example). This gathering encompasses everything that belongs to me, is my own (*das Eigene*), as opposed to what does not belong to me, differs from me, is Other (*das Fremde*). This experience thus counters the experience of a decentred self that is in-between, that 'regresses' into its being-in-the-world. Through bodily contraction I gain a 'feeling of myself' (*Sichspüren*) as self which bears the potential for me to become 'self-conscious'. This presence of my self (an 'emancipated' self) in turn serves as the basis for subsequent processes of identification with entities other than myself (both in the sense of 'identifying with' and 'to differ from').

What Schmitz describes here in terms of common human experiences is of particular relevance in a ritual context. In Victor Turner's ritual theory, for example, the phase of liminality is followed by the phase of 'reaggregation' or 'reincorporation'. The first term is similar to the gathering of the self while the second term refers to the body becoming body again. Both terms thus emphasize a need for rituals to first lead us away from our centre and then to return to it. However, it is impossible to return to the self we were before because we have transformed in myriad ways along the way: "Men are released from structure into *communitas* only to return to structure *revitalized by their experience* of *communitas*" (Turner 1977, 129, my emphasis). This ritual rebirth of the self is thus bound up with a revitalization of the sentient body: It is not just an experience in the sense of 'life as told'[9], it is first and foremost an experience that is *felt*, perhaps as invigorating, as exciting, as enjoyable, as meaningful because one is literally touched by the experience, as one can

---

9   Edward Bruner (1986, 6) distinguishes 'life as lived' from 'life as experienced' and 'life as told', thereby calling attention to the transformations that necessarily take place when moving from one level to the other.

feel the various ways the sentient body responds and is moved. These feelings are, I believe, one of the main reasons why many Druids long for participating in rituals.

These same feelings are also the markers of an 'effective' ritual (see Sax *et al.* 2010). Ritual efficiency, in Druidry, relies on a congruence and correspondence between the participants and the (natural) world around them which has to be made visible and tangible. Thus a ritual which failed (see Hüsken 2007) is one where all these factors are missing. By now, we have seen that the Samhain ritual at Wayland's Smithy might have been effective, although it had not provided me with any form of propositional answers to the questions underlying the ritual. I presume that my problem in the last stage of the ritual lay in an inherent paradox of the ritual design which in turn reflects a paradox in contemporary Druidry's conception of the sacred. On the one hand, Druidry as a 'nature religion' encourages ritual participants to reach out and engage with sacred Nature which is essentially a bodily and sensual activity. To paraphrase Michel Serres once again: "The senses are nothing but the mixing of the body, the principal means whereby the body mingles with the world and with itself, overflows its borders" (Connor 2008, 3). In this sensual encounter the sacred in Nature is revealed and Druids have the opportunity to correspond and resonate with it. On the other hand, Druidry has a Neoplatonic legacy which was introduced into Druidic organizations in the nineteenth century via the practice and study of Hermetic magic. In the Hermetic interpretation, the sacred was to be found not 'out there', via the body and the senses, but 'deep inside', in a 'world of ideas', accessed through the soul. In Druidry – as in other contemporary Neopaganisms – this world of ideas is the Otherworld, a sacred world in touch with the world as we know it but peopled by gods, nature spirits and ancestral spirits. In a psychoanalytically inspired interpretation favoured by Druids, this Otherworld can be accessed through the unconscious and the imagination. The purpose of certain rituals therefore is to allow participants to cross the boundaries to the Otherworld and engage in imaginative journeys which account for 'real' and sacred experiences while the body stays still and the senses are withdrawn from the world (see Gieser 2008b).

The ritual at Wayland's Smithy was problematic because it tried to unite two ultimately contrary developments. At the heart of the ritual was a withdrawal from the outer world into the Otherworld through the spirit of the place in the stone chamber. However, in the hours before, participants were trying their best to get in touch with the sacred in Nature through their sentient bodies. These are two incompatible modes of engagement with the sacred and I at least was not able to switch from one mode to the other as there was not enough time for me to readjust. My sentient body was still too active and engaged to allow my 'soul' – as a particular structure of inner experience – to come forward for an imaginary experience of the Otherworld.

We can now see how tracing the transformations of the sentient body in ritual enabled us to evaluate the transformative capacity of ritual, to discuss questions of ritual efficacy and to address various structures of experience without reducing them to 'altered states of consciousness'. We have thus described the manifold relationships the sentient body (including its potential to become a complex, multiple body) is engaged in without presupposing an *a priori* coherence or essence of a body (as a thing or *Körper*). We have criticised the disappearance of the body as common in performative approaches to ritual and contrasted this to acknowledging not only the living, feeling body but also the body being felt. This has enabled us to open up the range of potential bodily experiences of *being-in-the-world* and to emphasize how self and world affect and resonate with each other. Conceiving the world of ritual not as 'ritual space' but as a transforming (weather)world has further implications for our understanding of ritual dynamics, particularly in the acknowledgement of how the dynamics of the natural environment introduce their own 'agency' into the ritual. We could go as far as saying that the concept of the sentient body seriously undermines the dichotomization of ritual agency and structure, especially when sensual intermingling, affectivity and correspondence precludes a simplified conception of an agent as an autonomous individual. Finally, tracing the transformations of the sentient body in ritual shows how 'experience' may turn into *an* experience (*Erlebnis*), that "like a rock in a Zen sand garden, stands out from the evenness of passing hours and years" (Turner 1986, 35). And although these *Erlebnisse* are 'good to think with', remembered and talked about (Bruner 1986) we have grounded them in a prereflective understanding of felt motions of a sentient body that is alive to its environment.

Bibliography

Bell (1992): Catherine Bell, *Ritual Theory, Ritual Practice*. Oxford.

Bell (1997): Catherine Bell, *Ritual: Perspectives and Dimensions*. Oxford.

Bell (2006): Catherine Bell, "Embodiment". In: Jens Kreinath, Jan A.M. Snoek & Michael Stausberg (eds), *Theorizing Rituals. Vol. I: Issues, Topics, Approaches*. Leiden, 533-543.

Böhme (2013): Gernot Böhme, *Atmosphäre: Essays zur neuen Ästhetik*. Berlin.

Bruner (1986): Edward M. Bruner, „Experience and its expressions". In: Victor W. Turner & Edward M. Bruner (eds), *The Anthropology of Experience*. Urbana, 3-32.

Connor (2008): Steven Connor, "Introduction". In: Michel Serres, *The Five Senses: A Philosophy of Mingled Bodies (I)*. London, 1-16.

Csordas (1990): Thomas Csordas, *Embodiment as a Paradigm for Anthropology*. Ethos, Vol. 18 (1): 5-47.

Darvill (2004): Tim Darvill, *Long Barrows of the Cotswolds and Surrounding Areas*. Stroud.

Eliade (1975): Mircea Eliade, *Schamanismus und archaische Ekstasetechnik*. Berlin.

Gieser (2008a): Thorsten Gieser, "Embodiment, emotion and empathy: A phenomenological approach to apprenticeship learning". In: *Anthropological Theory*, Vol. 8 (3): 299-318.

Gieser (2008b): Thorsten Gieser, *Experiencing the Lifeworld of Druids: A Cultural Phenomenology of Perception*. PhD Thesis, University of Aberdeen, Aberdeen.

Gieser (unpublished): Thorsten Gieser, *With Druids in Stonehenge: A Cultural Phenomenology of Perceiving and Knowing in Ritual Atmospheres*.

Houseman and Severi (1998): Michael Houseman and Carlo Severi, *Naven or the other self*. Leiden.

Hüsken (2007): Ute Hüsken (ed.), *When Rituals Go Wrong: Mistakes, Failures, and the Dynamics of Ritual*. Leiden.

Ingold (2000): Tim Ingold, *The Perception of the Environment: Essays in Livelihood, Dwelling and Skill*. London.

Ingold (2011): Tim Ingold, *Being Alive: Essays on Movement, Knowledge and Description*. London.

Jackson (1989): Michael Jackson, *Paths Towards a Clearing: Radical Empiricism and Ethnographic Inquiry*. Bloomington.

Jilek (1982): Wolfgang Jilek, *Indian Healing: Shamanic Ceremonialism in The Pacific Northwest Today*. Surrey.

Leach (1976): Edmund Leach, *Culture and Communication, the logic by which symbols are connected: An introduction to the use of structuralist analysis in social anthropology*. Cambridge.

Locke (2007): Patricia M. Locke, "The Liminal World of the Northwest Coast". In: Suzanne L. Cataldi and William S. Hamrick (eds.), *Merleau-Ponty and Environmental Philosophy: Dwelling on the Landscapes of Thought*. Albany, 51-66.

Merleau-Ponty (2000): Maurice Merleau-Ponty, *The Visible and the Invisible*. Evanston.

Merleau-Ponty (2003): Maurice Merleau-Ponty, *Nature: Course Notes from the College de France*. Evanston.

Merleau-Ponty (2004): Maurice Merleau-Ponty, *Phenomenology of Perception*. London.

Restall Orr (2004): Emma Restall Orr, *Living Druidry: Magic Spirituality for the Wild Soul*. London.

Sax, Quack and Weinhold (2010): William S. Sax, Johannes Quack & Jan Weinhold (eds), *The Problem of Ritual Efficacy*. Oxford.

Schmitz (2011): Hermann Schmitz, *Der Leib*. Berlin.

Senft and Basso (2009): Gunter Senft and Ellen B. Basso (eds.), *Ritual Communication*. Oxford.

Serres (2008): Michel Serres, *The Five Senses: A Philosophy of Mingled Bodies (I)*. London.

Sheets-Johnstone (1966): Maxine Sheets-Johnstone, *The Phenomenology of Dance*. Madison.

Sheets-Johnstone (2011): Maxine Sheets-Johnstone, *The Primacy of Movement*. Amsterdam.

Shusterman (2008): Richard Shusterman, *Body Consciousness: A Philosophy of Mindfulness and Somaesthetics*. Cambridge.

Strathern (1988): Marilyn Strathern, *The Gender of the Gift: Problems with Women and Problems with Society in Melanesia*. Berkeley.

Strathern and Stewart (2011): Andrew J. Strathern and Pamela J. Stewart, "Embodiment and Personhood". In: Frances E. Mascia-Lees (ed.), *A Companion to the Anthropology of the Body and Embodiment*, Oxford, 388-402.
Turner (1977): Victor Turner, *The Ritual Process: Structure and Anti-Structure*, Ithaca.
Turner (1986): Victor Turner, "Dewey, Dilthey, and Drama: An Essay in the Anthropology of Experience". In: Victor W. Turner & Edward M. Bruner (eds), *The Anthropology of Experience*. Urbana, 33-44.
Whitehouse (2006): Harvey Whitehouse, "Transmission". In: Jens Kreinath, Jan A.M. Snoek & Michael Stausberg (eds), *Theorizing Rituals. Vol. I: Issues, Topics, Approaches.* Leiden, 657-669.

2
Zwischen Text und Leib: Körpermanipulationen als Einschreibungen?

Körper als Text?
Körper, Rituale und die Grenzen einer Metapher

Andreas Ackermann

Für Sula & Wolfgang – Gefährten in schwierigen Zeiten

I   Einleitung: Über (nicht-)metaphorisches Einschreiben

Der Titel dieses Bandes verwendet die Metapher des Einschreibens, die den Körper
als eine Art Leerstelle, Text oder Speicher, mithin als Objekt denkt. Damit schließt er
an eine Vorstellungstradition an, die vielleicht am radikalsten bei Kafka umgesetzt
worden ist. Seine Erzählung *In der Strafkolonie* (1914) berichtet über die Vollstreckung
eines Urteils, bei der dem Delinquenten das Gebot, das er übertreten hat, auf den
Leib geschrieben wird. Im konkreten Fall lautet es: „Ehre Deinen Vorgesetzten!"
Der Vorgang des Einschreibens wird mit einer gigantischen Tätowier-Maschine
vollzogen, deren Nadeln sich während zwölf Stunden immer tiefer in den Körper
des Verurteilten eingraben, bis er schließlich stirbt. Ist dieser Vorgang an sich schon
grauenvoll, so wird das Grauen noch dadurch gesteigert, dass der solchermaßen
Gepeinigte sein Urteil nicht kennt: „Es wäre nutzlos, es ihm zu verkünden. Er erfährt es ja auf seinem Leib."[1] Der Verurteilte weiß also weder, dass bzw. wessen er
angeklagt ist, noch dass er verurteilt wurde und hatte somit auch keine Gelegenheit,
sich zu verteidigen. Das ist nach Ansicht des Richters (der gleichzeitig auch sein
Henker ist) unnötig, denn: „Die Schuld ist immer zweifellos."[2] Erst im Moment des
Todes erfährt der Delinquent im wahrsten Sinne des Wortes sein Urteil, das er mit
seinen Wunden entziffert: „Es ist allerdings viel Arbeit; es braucht sechs Stunden
zu ihrer Vollendung."[3] Allerdings, so der gleichermaßen unheimliche Schluss der

---

1   Kafka (2007), 170.
2   Kafka (2007), 171.
3   Kafka (2007), 176.

Erzählung, scheint die Zukunft dieser Praxis in der Strafkolonie ungewiss, der Richter – der gleichzeitig auch der Vollstrecker des Urteils ist – kämpft gegen ihre sich abzeichnende Abschaffung und legt sich schließlich, als er die Vergeblichkeit seiner Bemühungen erkennen muss, selbst unter die Maschine.

In Kafkas Text klingt bereits eine Vielzahl von Themen an, die für den Kontext der Einschreibungs- bzw. Textmetapher von Bedeutung sind: die Machtasymmetrie von Einzelnem und Gesellschaft, die körperliche Disziplinierung bzw. Bestrafung im Rahmen eines Rituals, der – in diesem Falle besonders grausame – gesellschaftliche Zugriff auf den Körper des Einzelnen, der zudem für eine lange Zeit im Unklaren über das, was mit ihm geschieht, gelassen wird. Von Nietzsche (Genealogie der Moral, § II-3) stammt das Diktum, „nur was nicht aufhört, weh zu thun, bleibt im Gedächtniss" und viele der blutigen Einschnitte in den Körper, wie sie gerade bei Initiationsritualen vorkommen, dürften in diesem Zusammenhang von Schmerz und Erinnerung ihren Grund haben: Körperveränderungen sind mnemotechnische Hilfsmittel, mit denen zentrale Wahrheiten in der kulturellen Überlieferung unterstrichen werden.[4] Allerdings klingt mit dem Suizid des Richters bzw. Henkers bei Kafka auch die Möglichkeit einer selbstbestimmten Auslieferung an die Maschine an, was insofern interessant ist, als die meisten der zeitgenössischen Formen der Körperveränderung wie Tätowierung, *Piercing* oder *Branding* das Resultat einer individuellen Entscheidung sind, die gerade als Ausdruck eigener Persönlichkeit gedeutet werden will.[5] Dass der Grund des Suizids des Richters letztlich im drohenden Bedeutungsverlust des (Bestrafungs-)Rituals liegt, berührt schließlich ein weiteres Thema dieses Sammelbandes.

Kafkas *Strafkolonie* ist aber vor allem deshalb wichtig für die folgende Argumentation, weil darin der Prozess der Einschreibung nicht in einem metaphorischen, sondern einem buchstäblichen Sinne vollzogen wird. Er eignet sich deshalb vorzüglich als Ausgangspunkt für die Diskussion einer kulturwissenschaftlichen Perspektive auf den Körper, die durch die Vorstellung des Einschreibens geprägt ist und den Körper – bzw. weitgehender noch: Kultur – als Text analysiert, was in zweierlei Hinsicht problematisch erscheint. Zum einen tendiert die Verwendung des Einschreibens im metaphorischen Sinne dazu, sich weniger für die tatsächlichen Manipulationen des Körpers durch unterschiedliche Techniken der Körperveränderung zu interessieren. Der Vorgang beispielsweise der Abtrennung der männlichen Vorhaut vollzieht sich aber alles andere als metaphorisch. Dies scheint gerade dann wichtig, wenn normative Ansprüche begründet oder widerlegt werden sollen, was letztlich nur in anthropologischer Hinsicht geschehen kann. Dabei ist

---

4   Streck (2000), 132f.
5   Kasten (2006) 234-236; Stirn (2006).

die Beschneidung nur ein Beispiel für eine Vielzahl von sowohl historisch als auch kulturvergleichend universal verbreiteten Techniken der Körperveränderung, die vor allem im Kontext von Ritualen vorgenommen werden. Diese Techniken involvieren Subjekte, die Schmerz erfahren, unter Umständen Stufen eines rituellen Todes und der Wiedergeburt durchlaufen, und definieren über den Körper die Beziehung zwischen dem Selbst und der Gesellschaft. Der Körper ist dann nicht nur der Ort, an dem Kultur sowohl im metaphorischen wie buchstäblichen Sinne eingeschrieben wird, sondern auch der Ort, an dem das Individuum definiert und in die kulturelle Landschaft eingepasst wird. Gleichzeitig erlauben z. B. *Tattoos*, Narben, Brandzeichen und *Piercings* Individuen, die jeweils eigene Biographie auf die Körperoberfläche zu schreiben.[6] Diese Praktiken verbinden Vergangenheit und Gegenwart, sie sind Ausdruck von Grenzziehungen bzw. -überschreitungen, zwischen Individuum und Gesellschaft, zwischen unterschiedlichen Gruppen und zwischen Menschen und dem Göttlichen. Zum anderen wird die ausschließliche Sicht auf den Körper als *Oberfläche* bzw. Leerstelle der tatsächlichen Komplexität des Verhältnisses von Körper bzw. Selbst und Kultur nicht gerecht. Wie dieser Beitrag zu zeigen versucht, ist der Körper nicht *Ausdruck*, sondern *Voraussetzung* der Kultur. Daraus resultiert letztlich auch die Bedeutung des Körpers für das Ritual.

Im Folgenden soll daher ein ideengeschichtlicher Abriss die kulturwissenschaftliche Perspektive auf den Körper skizzieren, wobei das Augenmerk auf Körper und Ritual generell sowie die Text- bzw. Einschreibungsmetapher konkret gerichtet wird. In zuspitzender Absicht wird dabei eine Art „kopernikanischer Wende" postuliert, die von einer Periode des Körperverständnisses als Objekt bzw. Ausdruck von Kultur überleitet zu einer Auffassung des Körpers als „Leib" oder Subjekt bzw. als Bedingung von Kultur. Diese Wende lässt sich vor allem durch den Versuch charakterisieren, den cartesianischen Dualismus von Körper und Geist zu überwinden. Dieser Dualismus geht auf Descartes' These zurück, dass sich der immaterielle Geist und der greifbare Körper (bzw. die Materie allgemein) grundlegend unterscheiden, auch wenn sie gemeinsam die menschliche Existenz bilden. Darüber hinaus räumt Descartes dem Geist auch einen höheren Stellenwert ein, der Körper bzw. die Materie sind lediglich sekundär („cogito ergo sum"). Sein Ideal ist das körperunabhängige individuelle, geistige Selbst, während der Körper lediglich das abgegrenzte Behältnis dieses Selbst darstellt. Descartes zufolge ist der Körper Teil der Natur und unterliegt daher mathematischen Gesetzen, d. h. der cartesianische Dualismus etabliert damit auch ein materialistisches bzw. mechanistisches Weltbild. Letztlich resultieren aus dem cartesianischen Dualismus noch weitere konzeptuelle Gegensätze, die nicht nur tief im modernen Welt- und Selbstbild

---

6   Schildkrout (2004), 338; vgl. Kasten (2006) 240f.

verankert sind und daher oft als natürliche, universale und damit unvermeidliche und unlösbare Widersprüche gelten, sondern auch in den Kulturwissenschaften eine große Rolle spielen. Teresa Platz nennt in diesem Zusammenhang die Unterscheidung von Natur und Kultur, Individuum und Gesellschaft sowie aktivem, beobachtendem Subjekt bzw. passivem, beobachtetem Objekt. Auch die Trennung und entsprechende Bewertung von Vernunft und Leidenschaft bzw. Gedanken und Emotionen sind mit dem cartesianischen Dualismus verbunden.[7]

In diesem Zusammenhang ist es auch wichtig, darauf hinzuweisen, dass das im Folgenden vorgestellte Konzept des *Embodiment* nicht einfach mit Verkörperung gleichzusetzen ist. *Embodiment* bezeichnet in der englischen Philosophie die körperlichen Aspekte der menschlichen Subjektivität und wird vor allem mit der Phänomenologie von Merleau-Ponty in Zusammenhang gebracht.[8] Eine umstandslose Übersetzung dieses Begriffs als „Verkörperung" ist insofern problematisch, als er impliziert, dass der Körper nicht nur sowohl als eigenschaftsloser Träger von Kultur erscheint, sondern auch als materieller Körper einer geistigen, immateriellen Kultur gegenübersteht, so als existiere Kultur völlig unabhängig von leibhaftigen Akteuren und werde von diesen lediglich verkörpert.[9] Das in diesem Beitrag vorgestellte Konzept des *Embodiment* von Thomas Csordas will jedoch genau das Gegenteil, nämlich die konzeptuellen Dualitäten von Geist und Körper sowie von Subjekt und Objekt aufheben. Daher ist im Folgenden mit Platz jeweils von *Embodiment* die Rede, wenn diese philosophische Bedeutung gemeint ist. Daneben ist auch die begriffliche Unterscheidung von Körper und Leib wichtig. Der Begriff „Körper" hat seine Wurzeln im lateinischen Wort „corpus" und bezeichnet gewöhnlich den objektivierten Körper und wird auch für Tiere und Lebloses bzw. den Leichnam verwendet. Im modernen westlichen Sprachgebrauch *hat* der Mensch einen materiellen, biologischen Körper, über den der immaterielle Geist bzw. das Bewusstsein verfügen kann. Der Begriff „Leib" hat dieselbe Sprachwurzel wie „Leben", d. h. der Leib wird dadurch definiert, dass er die Welt erlebt („erleibt"), er wird als Wahrnehmungs- und Handlungspotential erfahren. Der Mensch *ist* ein Leib, in dem Körper und Geist untrennbar verbunden sind. *Embodiment* könnte im Deutschen daher als „kulturell wahrnehmender und handelnder Leib bezeichnet werden"[10], denn es bezieht sich auf den Leib und nicht auf den Körper. Das englische Adjektiv „embodied" wird dementsprechend als „leiblich" übersetzt. Wird

---

7 Platz (2006), 15f.

8 Audi (1999), 258.

9 Platz (2006), 9f.

10 Platz (2006), 10.

das Wort „body" im Sinne von Leib verwendet wird, wird es auch entsprechend wiedergegeben, andernfalls das Wort Körper verwendet.[11]

## II	Der Körper als Ausdruck der Kultur

Da der Körper sich aus cartesianischer Perspektive auf der Seite der Natur und des Individuums befindet, hatte er lange einen schweren Stand in den Sozial- und Kulturwissenschaften, die dazu tendieren, Kultur bzw. Gesellschaft als sein Gegenteil zu konzeptualisieren.[12] In *Die elementaren Formen des religiösen Lebens* (1912) konstruiert Émile Durkheim (1858-1917), der Begründer der französischen Soziologie und Vordenker der Année-Schule (so genannt nach der ebenfalls von ihm gegründeten Zeitschrift *Année Sociologique*) den Menschen in doppelter Hinsicht: zum einen als biologisches, individuelles Wesen mit eingeschränktem Wirkungsbereich, gleichzeitig aber auch als soziales Wesen, das die moralischen Regeln und intellektuellen Vorstellungen der Gesellschaft darstellt und damit sich selbst transzendiert.[13] Dies geschieht laut Durkheim dadurch, dass Sozialität die Individuen in intellektueller Hinsicht mit kollektiven Repräsentationen versieht, die ihre Wahrnehmung ordnen und ihren Gedanken Ausdruck verleihen, und auf moralischer Ebene Regeln vermittelt, die ihr Verhalten organisieren und ihren persönlichen Interessen Grenzen setzen.[14] Diese sozialen Phänomene sind jedoch zu komplex, als dass der Einzelne begreifen könnte, wie sie ihn von außen beeinflussen, d. h. die Gesellschaft verschleiert die Prinzipien und Prozesse des sozialen Lebens, und zwar in der Form des Sakralen, des radikalen Anderen. Die Trennung von sakral und profan hält Durkheim dementsprechend für die universal und absolut geltende Grundunterscheidung des sozialen Lebens.[15] Wenn das Sakrale die kollektive Repräsentation der Gesellschaft darstellt, dann sind soziale Phänomene Tatbestände, die außerhalb des individuellen menschlichen Willens existieren und auch nicht aus ihm entstanden sind, d. h. das Individuum erlebt sich als in sie hineingeboren und durch sie geprägt.[16] Durkheims Annahme, „dass die sozialen Phänomene eine Ideenordnung bilden, die nicht auf die Fähigkeiten des Einzelnen

---

11 Platz (2006), 10f.

12 Platz (2006), 16. Vgl. Turner (1997).

13 Durkheim (1981), 27.

14 Platz (2006), 17.

15 Platz (2006), 18.

16 Platz (2006), 19.

reduzierbar ist und die es dem Menschen ermöglicht, seine Welt zu ordnen und zu verstehen",[17] stellt seitdem eine der wesentlichen Grundlagen der Sozial- und Kulturwissenschaften dar. Vor allem aber bildet sie die Grundlage der Vorstellung, dass der menschliche Körper lediglich die physische Voraussetzung von Gesellschaft bildet, gleichsam ein Rohmaterial, das von der Gesellschaft geformt wird.

Mit Robert Hertz (1881-1915) widmet sich ein weiterer Vertreter der Année-Schule als erster explizit dem menschlichen Körper. In seinem Aufsatz zum Vorrang der rechten Hand (1909) argumentiert er, dass die Tendenz zur Privilegierung der rechten Hand durch kulturelle Konzepte verstärkt wird. Dies wird z. B. deutlich an Ausdrücken wie „das Recht", „das ist rechtens", „er ist seine rechte Hand" oder „er wurde gelinkt". Mit der rechten Hand werden ‚heilige' Dinge getan, während die linke, ‚unreine' Hand für profane Aktivitäten benutzt wird.[18] Unklar bleibt allerdings noch, wie das Soziale auf den Körper einwirkt, denn von der Physis her könnten seiner Meinung nach beide Hände die gleichen Fähigkeiten erwerben.[19] Auch Marcel Mauss (1872-1950), Neffe und Schüler Durkheims, bewegt sich innerhalb des cartesianischen Dualismus und reproduziert ihn letztlich sogar, indem er zwei getrennte Publikationen zu den Körpertechniken (1935) einerseits und zur Person (1938) andererseits vorlegt. Dennoch deutet sich bei ihm schon die Notwendigkeit an, den Gegensatz von Körper und Geist, Subjekt und Objekt, Individuum und Gesellschaft zu überdenken. Mauss' Begriff der Körpertechniken bezieht sich auf „die Weisen, in der sich die Menschen in der einen wie der anderen Gesellschaft traditionsgemäß ihres Körpers bedienen".[20] Er verweist darauf, dass es nicht nur zwischen Kulturen, sondern – historisch gesehen – auch innerhalb einer Kultur unterschiedliche Formen z. B. des Schwimmens, militärischen Marschierens, Laufens, Sitzens etc. gibt. Sein diesbezügliches „Aha-Erlebnis" hat Mauss anlässlich eines Aufenthaltes in einem New Yorker Krankenhaus, wo ihm der Gang der Krankenschwestern merkwürdig vertraut vorkommt, und zwar, wie er herausfindet, aufgrund von amerikanischen Filmen, die er gesehen hat. Zurück in Paris fällt ihm die Häufigkeit dieser Gangart nun auch bei jungen Französinnen auf und er schließt daraus, dass die amerikanische Gangart in Frankreich durch das Kino verbreitet wird.[21] Die Gesamtheit der Körpertechniken einer Kultur nennt er „Habitus" – ein Begriff, den später Pierre Bourdieu weiter ausarbeiten wird. Wichtiger noch ist jedoch, dass Mauss den Körper als „das erste und natürlichste

---

17 Platz (2006), 20.

18 Platz (2006), 21.

19 Hertz (1973), 22.

20 Mauss (1989a), 199. Eine aktuelle Perspektive bietet Hirschauer (2004).

21 Mauss (1989a), 202.

technische Objekt und gleichzeitig technische Mittel des Menschen" bezeichnet.[22] Damit ist bereits formuliert, dass der Körper sowohl von der Kultur geprägt wird, als auch das Mittel, mit dem die Kultur diese Arbeit verrichtet, darstellt. Wenn aber die soziale Welt aus von ihr geformten Körpern besteht und diese gleichzeitig die Welt formen, dann kann der Körper nicht ausschließlich biologisches Objekt sein. Der Einsicht, dass Gesellschaft aus der Sicht des Körpers erklärt werden kann, lässt Mauss allerdings keine Konsequenzen folgen.[23]

In den folgenden Jahrzehnten richtet sich die sozial- bzw. kulturwissenschaftliche Aufmerksamkeit vor allem auf die Analyse von Gesten, nonverbaler Kommunikation und Proxemik (z. B. Hall 1968). Das Interesse am Körper bleibt dabei der Analyse von Kommunikation als kulturellem Prozess untergeordnet, der Körper wird primär als Mittel der Kommunikation begriffen. Man interessiert sich nicht für Körperlichkeit *per se*, sondern untersucht mit einem linguistischen Modell unterschiedliche Formen von Körpersprache.[24] Die ersten Ansätze einer Ethnologie des Körpers finden sich in den Publikationen von Mary Douglas (1966, 1970) und einem von John Blacking herausgegebenen Sammelband (1977). Dabei gilt das Hauptinteresse dem Körper als einem Symbolsystem bzw. Ausdrucksmedium der Gesellschaft, d. h. der Körper wird zunehmend unter semiotischen Gesichtspunkten analysiert. Vor allem zwei theoretische Positionen bestimmen in dieser Phase die ethnologische Perspektive auf den Körper: der Strukturalismus nach Claude Lévi-Strauss und die Vorstellung von Kultur als Text, die von Clifford Geertz in die Ethnologie eingeführt wird und letztlich zu einer „linguistischen Wende" führt.

Der französische Ethnologe Claude Lévi-Strauss (1908-2009) ist einer der bekanntesten und einflussreichsten Vertreter des Strukturalismus. Lévi-Strauss ist vor allem von den Erkenntnissen der Linguistik beeinflusst: er übernimmt die Bedeutung der Sprachfiguren der Metapher und der Metonymie von Roman Jakobson sowie die Idee der Willkürlichkeit von Ferdinand de Saussure. Dementsprechend untersucht er Gesellschaft als ein sprachähnliches System von Zeichen, bei dem die sozialen Strukturen den Individuen ebenso unbewusst bleiben wie die Grammatik ihrer Muttersprache. Es ist der Wissenschaftler, der aus seiner Beobachter-Position Modelle konstruiert, die die einzelnen Elemente und ihre Beziehungen zueinander systematisch darstellen. Saussure geht davon aus, dass Sprache ein abstraktes Prinzip ist („langue"), das unabhängig vom Kontext analysiert werden kann; dem Akt des eigentlichen Sprechens („parole") kommt daher nur eine nachgeordnete Bedeutung zu. Überträgt man ein solches Modell auf Kultur,

---

22 Mauss (1989a), 206.
23 Platz (2006), 23.
24 Csordas (1999) 176.

gerät man zwangsläufig in eine intellektualistische Position, d. h. man privilegiert ein intellektuelles Prinzip gegenüber einer äußeren Welt und argumentiert somit entlang des cartesianischen Dualismus.[25] Bei Lévi-Strauss findet sich dann auch das Bild des beschriebenen Körpers: „Die Maori-Tatauierung soll nicht nur eine Zeichnung in das Fleisch, sondern auch alle Traditionen und die Philosophie der Rasse in den Geist ritzen".[26]

Ganz ähnlich formuliert die sowohl von Durkheim als auch strukturalistisch beeinflusste britische Ethnologin Mary Douglas (1921-2007): „Was in das menschliche Fleisch geschnitten wird, ist ein Bild der Gesellschaft."[27] In *Reinheit und Gefährdung* (1966) beschreibt Douglas im Zusammenhang mit ihrer These, dass die im dritten Buch Mose aufgeführten Speisegebote die Kategorien der israelitischen Kultur zusammenfassen, den Körper als Modell, mit dem soziale Klassifikationssysteme veranschaulicht werden können: „Die Funktionen seiner verschiedenen Teile und ihre Beziehungen zueinander bieten anderen komplexen Systemen eine Fülle von Symbolmöglichkeiten."[28] Aufgrund seiner festen Grenzen einerseits, die eine Person von der anderen materiell abgrenzen, und seiner Körperöffnungen andererseits, die wiederum die Ambivalenz von Innen und Außen symbolisieren, eignet er sich gut für die Darstellung gesellschaftlicher Strukturen.[29] Diese These entwickelt Douglas in *Ritual, Tabu und Körpersymbolik* (1970) weiter, indem sie versucht, unterschiedliche Formen der Besessenheit mit korrespondierenden Formen sozialer Organisation zu erklären. Wie Durkheim unterscheidet sie dabei zwischen sozialem und physischem Körper, wobei der soziale Körper bestimmt, wie der physische Körper wahrgenommen wird. Der Körper ist damit „das mikrokosmische Abbild der Gesellschaft, ihrem Machtzentrum zugewandt und in direkter Proportion zum zu- bzw. abnehmenden gesellschaftlichen Druck ‚sich zusammennehmend‘ bzw. ‚gehenlassend‘."[30] Er bringt universale Bedeutungsgehalte insofern zum Ausdruck, „als er als System auf das Sozialsystem reagiert und dieses systematisch zum Ausdruck bringt".[31] Dabei wird das Verhältnis zwischen dem individuellen physischen Körper und dem sozialen Körper durch das in einer bestimmten Kultur verwendete „Klassifikationsgitter" sowie den jeweils auf den Einzelnen ausgeübten „Druck der

---

25 Platz (2006), 28.

26 Lévi-Strauss (1967), 281.

27 Douglas (1988), 153.

28 Douglas (1988), 152.

29 Douglas (1988), 152, 160.

30 Douglas (1986), 109.

31 Douglas (1986), 123.

Gruppe" („grid and group") bestimmt.[32] D. h. für Douglas ist der Körper Abbild, Träger oder Ausdrucksmedium sozialer Bedeutungen. Wie ihre Unterscheidung zwischen physischem und sozialem Körper deutlich macht, folgt sie der Trennung zwischen Kultur und Körper bzw. Geist und Körper. Csordas verweist in diesem Zusammenhang auf eine Ambivalenz: Einerseits lassen sich die beiden Körper als physiologischen (natürlichen) bzw. sozialen (oder kulturellen) Aspekt des menschlichen Körpers begreifen. Andererseits kann sich die Unterscheidung auf das Verhältnis zwischen dem individuellen und dem gesellschaftlichen Körper (im Durkheimschen oder metaphorischen Sinne) beziehen. Douglas' Feststellung: „Die ‚zwei Körper' sind einmal das Selbst und zum anderen die Gesellschaft"[33], scheint jedenfalls die letztere Interpretation nahezulegen.[34]

Von nicht zu überschätzender Bedeutung für die kulturwissenschaftliche Auffassung vom Körper ist das Werk Michel Foucaults (1926-1984), der eine diachrone Perspektive auf den Körper richtet und zeigt, wie sich das Verständnis des Körpers in Europa im Laufe der Epochen verändert hat. In *Nietzsche, die Genealogie, die Historie* beispielsweise beschreibt Foucault den Körper u. a. als „eine Fläche auf dem [sic] die Ereignisse sich einprägen" und sieht die Aufgabe der Genealogie darin, zu zeigen, dass der Körper von der Geschichte geprägt wird.[35] Aus seiner Sicht ist eine Geschichte der Gefühle, des Verhaltens und des Körpers auf die Analyse der Mechanismen von Macht und Herrschaft angewiesen. Dabei kann es sich um die Herrschaft von außen – Thema seiner Studien zu Klinik (1973) und Gefängnis (1977) – oder die Herrschaft von innen, aus sich selbst heraus – Thema seiner Untersuchung zur Sexualität als Ort der Selbstvergegenwärtigung des modernen Individuums (1978, 1985 und 1986) – handeln. Damit verschiebt Foucault das Verständnis vom Körper als einem Symbolsystem, das Metaphern produziert, durch die Macht konzeptualisiert wird bzw. als Konsequenz langfristiger Transformationen menschlicher Gesellschaften, hin zu einem Ausdruck von tieferliegenden strukturellen Arrangements von Wissen und Macht. Seine Annahme, dass der physische Körper bereits ein Produkt der Gesellschaft ist, geht noch einen Schritt über Douglas' These hinaus, dass die Wahrnehmung des Körpers sozial festgelegt sei. Obwohl Foucault damit entscheidend dazu beiträgt, den Körper nicht nur als ein biologisches, sondern auch als ein kulturelles und historisches Phänomen zu begreifen, bewegt er sich doch letztlich innerhalb des cartesianischen Dualismus. Catherine Bell hat darauf hingewiesen, dass der Foucaultsche Körper als Produkt

---

32 Douglas (1986), 111.
33 Douglas (1986), 123.
34 Csordas (1999), 176f.
35 Foucault (1971), 174.

bzw. Objekt der diskursiven Macht letztlich abstrakt und ahistorisch bleibt und somit paradoxerweise ein körperloses Subjekt in der Wissenschaftstheorie darstellt, nämlich eine Widerspiegelung der Idee, dass der menschliche Körper lediglich die Hülle für den Geist sei.[36]

Der französische Ethnologe und Soziologe Pierre Bourdieu (1930-2002) versucht dagegen, die Positionen des „Objektivismus" (wie er den Strukturalismus bezeichnet), der ihm zufolge absolutes Wissen suggeriere, und des Subjektivismus (zu dem er Existentialismus, Interaktionismus und Ethnomethodologie zählt), der die Illusion eines unmittelbaren Verständnisses verbreite, zu vermitteln. Mit seinem Versuch, die Kluft zwischen Theorie und Praxis, zwischen objektiven sozialen Strukturen und subjektivem Erleben der sozialen Praxis zu überwinden, stellt sich Bourdieu außerhalb des cartesianischen Dualitäts-Postulates. In diesem Zusammenhang spielt der Körper eine essentielle Rolle, denn in ihm fällt für Bourdieu der vermeintliche Gegensatz von objektiven Strukturen und subjektivem Erleben zusammen (d. h. der Körper wird von Bourdieu als Leib aufgefasst). Bourdieu legt seinen *Entwurf einer Theorie der Praxis* erstmals 1972 vor und überarbeitet seine Theorie 1980 in *Sozialer Sinn*.[37] Für ihn liegt die Logik der Praxis vor allem darin begründet, dass sie sich den Anschein von Unmittelbarkeit und Selbstverständlichkeit gibt. Da dies ebenso für die wissenschaftliche Praxis gilt, bedarf es eines „doppelten epistemologischen Bruchs" (Platz), um dieser Logik auf die Spur zu kommen. D. h. der Wissenschaftler muss die zu untersuchende Praxis mit den objektiven Bedingungen konfrontieren, die soziales Handeln bestimmen, jedoch gleichzeitig auch die theoretischen Grundlagen seiner eigenen objektivierenden Analyse hinterfragen.[38]

Das zentrale Prinzip, das soziale Praktiken produziert und reproduziert, bezeichnet Bourdieu als „Habitus". Der Habitus ist das „Körper gewordene Soziale",[39] in den die Denk- und Sichtweisen einer Gesellschaft sowie ihre Wahrnehmungsschemata und Prinzipien des Urteilens bzw. Bewertens eingegangen sind.[40] In diesem Zusammenhang spricht Bourdieu aber nicht von Prozessen der Einschreibung, sondern der Inkorporierung oder Einverleibung, d. h. er denkt nicht an eine Aufnahme bzw. Verinnerlichung sozialer Strukturen, sondern fasst diesen Prozess ganz leiblich, sodass die sozialen Strukturen in Fleisch und Blut übergehen und zu einem handelnden Leib werden.[41] Diese Einverleibung der objektiven sozialen

---

36 Bell (2006), 536f.

37 Der Originaltitel lautet allerdings zutreffender *Der praktische Sinn* (Platz 2006, 53).

38 Bourdieu (1987), 52.

39 Bourdieu/Wacquant (1996), 161.

40 Krais/Gebauer (2002), 5.

41 Platz (2006), 62.

Strukturen findet mittels einer „stillen Pädagogik" statt, bei der das soziale Umfeld von Anfang an mit Ermahnungen („Halt dich gerade!", „Nimm das Messer nicht in die linke Hand!") und mit rituellen Praktiken und Diskursen zwischen Kind und Welt tritt.[42] Gleichzeitig „zeigen die Kinder für die Gesten und Posituren, die in ihren Augen den richtigen Erwachsenen ausmachen, außerordentliche Aufmerksamkeit" und eignen sich diese mimetisch an.[43] Die einverleibten Strukturen des Habitus stellen jedoch nur die eine Seite des komplexen Verhältnisses dar, das die soziale Welt ausmacht. Die andere Seite bilden die externen, objektivierten Strukturen der sozialen Felder. Dabei handelt es sich um ein dialektisches Verhältnis, denn „ohne leibliche Akteure gibt es keine soziale Praxis und ohne soziale Praxis existieren auch keine objektiven Strukturen".[44] Bourdieus Analyse zufolge stiftet der sozial geprägte Leib vor allem Sinn. Da die sozialen Agenten[45] sich in der Gesellschaft orientieren, ihr soziales Praxisfeld für sinnvoll und bedeutsam halten, reproduzieren sie dieses auch als etwas Selbstverständliches bzw. Notwendiges. Mit Platz lässt sich dieser Sinn, d. h. die vom sozial geprägten Leib gelebten Bedeutungen und Werte, auch als „Kultur" bezeichnen.[46]

Die Auffassung vom Körper als Oberfläche bzw. Ausdruck der Kultur verstärkt sich noch im Zuge der sogenannten „Linguistischen Wende" („linguistic turn"). Grundlage dieser Entwicklung ist die Konzeptualisierung von Kultur mittels der von Paul Ricoeur (1913-2005) entlehnten und von Clifford Geertz (1926-2006) in die Ethnologie eingeführten Text-Metapher. In seinem wohl berühmtesten Text ‚*Deep Play': Bemerkungen zum balinesischen Hahnenkampf* (1972) behauptet Geertz, dass die Kultur eines Volkes aus einem Ensemble von Texten bestehe, die der Ethnologe bemüht ist, „über die Schultern derjenigen, für die sie eigentlich gedacht sind, zu lesen".[47] Diese Vorstellung von Kultur als Symbolsystem, das als Text gelesen werden kann, führt mit der sogenannten „Writing Culture"-Debatte innerhalb der Disziplin zu einer noch wesentlich radikaleren Schlussfolgerung. Diese lautet, dass all das, was wir als Kultur erkennen, letztlich ein Artefakt ethno-

---

42 Bourdieu (1987), 128.

43 Bourdieu (1979), 190.

44 Platz (2006), 74.

45 Die Beibehaltung des französischen „*agent*" (statt der üblichen Übersetzung mit „Akteur") resultiert aus der von Bourdieu intendierten Konnotation des „Agenten" als für eine Organisation oder (fremde) Macht Handelnden: „Die handelnden Individuen handeln nicht nur für sich, sondern – als soziale Wesen – immer auch für die Gesellschaft, das heißt über sie vermittelt macht sich Gesellschaft wirksam" (Krais/Gebauer 2002, 84).

46 Platz (2006), 77.

47 Geertz (1987), 259.

graphischer Praxis ist, mit anderen Worten: ein Resultat von Genre-Konventionen, die letztlich durch als „Ethnographien" bezeichnete Texte produziert werden.[48] Im Zusammenhang mit der Anziehungskraft von zuerst Strukturalismus und später dann Post-Strukturalismus haben diese Entwicklungen nicht nur zur Konsequenz, dass Methoden der Literaturkritik Einzug in die Ethnologie halten, sondern auch, dass die Ethnologie innerhalb der Kulturwissenschaften relevant wird: kulturvergleichende Ethnologie und vergleichende Literaturwissenschaft begreifen sich als verwandte Fächer und es kommt zu einer intensivierten Kommunikation zwischen Historikern, die mit Texten arbeiten und Ethnologen, die mit der Text-Metapher arbeiten (z. B. Darnton 1984).

Diese an sich begrüßenswerte, und in ihren Anfängen durchaus fruchtbare Entwicklung führt allerdings dazu, dass im weitesten Sinne semiotische Ansätze den kulturwissenschaftlichen Diskurs zu großen Teilen bestimmen. Textualität wird mit einer Formulierung von Thomas Csordas zu einer „hungrigen Metapher", die Kultur bzw. den Körper praktisch verschluckt, was z. B. in Redewendungen deutlich wird wie „Körper als Text", „die Einschreibung von Kultur in den Körper" oder „den Körper lesen". Letztlich entsteht der Eindruck, so Csordas, dass viele zeitgenössische Forscher die Text-Metapher nicht länger als Metapher, sondern im wörtlichen Sinne verwenden.[49] Auf epistemologischer Ebene vollzieht sich damit ein Wandel, in dessen Verlauf der Begriff der „Erfahrung" zugunsten von „Repräsentation" völlig aus den Kulturtheorien bzw. dem theoretischen Diskurs überhaupt verschwindet: Repräsentation bezeichnet Erfahrung nicht länger, sondern konstituiert diese. Damit wird zwar die Lücke zwischen Sprache und Erfahrung geschlossen und ein Dualismus eliminiert, aber eben nicht, indem er aufgehoben wird, sondern, wie Csordas kritisiert, indem Erfahrung auf Sprache, Diskurs oder Repräsentation *reduziert* wird. Dies ermöglicht zwar die Kritik bestimmter Repräsentationen, aber nur um den Preis einer Abschirmung der Wissensform Repräsentation von jeglicher Kritik. D. h. es wird schwer, nach den Grenzen der Repräsentation zu fragen bzw. ob es irgendetwas jenseits der Repräsentation gibt, ohne sich gleichzeitig dem Vorwurf des Essentialismus auszusetzen.[50]

---

48 Vgl. Clifford/Marcus (1986). Für einen Debatten-Überblick siehe Berg/Fuchs (1993).
49 Csordas (1999), 182.
50 Csordas (1999), 183.

III   Der Leib als Voraussetzung von Kultur: Embodiment[51]

Im Zuge der Postmoderne vollzieht sich ein weiterer Paradigmenwechsel, der häufig als „Performative Wende" („performative turn") bezeichnet wird und in dessen Zusammenhang eine Fülle von Publikationen zum Thema Körper in den Kulturwissenschaften allgemein und der Ethnologie speziell entsteht.[52] Der Körper wird dabei zum Ort, an dem Konzepte von Person, Zugehörigkeit und Identität verankert, produziert und verhandelt werden.[53] Die eingangs apostrophierte „kopernikanische Wende" hin zu einer Auffassung des Körpers als Subjekt bzw. existentieller Bedingung von Kultur ist Ausdruck dieser Entwicklung, die sich vor allem durch drei Aspekte charakterisieren lässt. *Erstens* wird die Tendenz, den Körper als Objekt, als Rohmasse bzw. „tabula rasa" zu betrachten, der die Gesellschaft ihre Codes aufdrückt, aufgehoben zugunsten eines Verständnisses vom Körper als Quelle von Handlungsmacht („agency") und Intentionalität, der die Welt vermittels intersubjektiven Engagements in sich aufnimmt und sie bewohnt. *Zweitens* sieht sich die Spaltung von Körper und Geist und daraus resultierenden konzeptionellen Dualismen wie etwa Subjekt und Objekt, „sex" und „gender"[54] oder Körper und Verkörperung einer zunehmenden Kritik ausgesetzt. Dies führt u. a. dazu, dass die biologische Natur des Körpers als stabiles Substrat menschlicher Existenz ganz grundsätzlich in Frage gestellt wird. *Drittens* schließlich wird der ausschließlich semiotische Blick auf die Kultur bzw. den Körper und die damit einhergehende Ausblendung körperlicher Erfahrung in Frage gestellt.

In diesem Kontext hat der amerikanische Ethnologe Thomas Csordas sein Konzept des *Embodiment* entwickelt. Wie er betont, geht es ihm dabei keineswegs darum, Repräsentation als methodologische Figur abzulösen, sondern ihr einen „dialogischen Partner" an die Seite zu stellen.[55] Seine grundsätzliche These lautet, dass der Körper – der von Csordas im Sinne von „Leib" aufgefasst wird – kein *Forschungsobjekt* im Verhältnis zur Kultur darstellt, sondern das *Subjekt* der Kultur bzw. ihre „existentielle Voraussetzung" („existential ground") bildet.[56] Bei seinem Versuch einer Überwindung des cartesianischen Dualismus stützt sich Csordas im Wesentlichen auf zwei Impulsgeber, zum einen auf Bourdieu und sein bereits

---

51 Ich danke Thorsten Gieser für seine Anregungen zu diesem Thema.

52 Ausführliche Literaturangaben finden sich u. a. bei Csordas (1999), Farnell (1999), Lock (1993), Platz (2006).

53 Polit (2013), 216f.

54 Z.B. Butler (1997).

55 Csordas (1999), 183.

56 Csordas (1990), 5.

vorgestelltes Konzept des Habitus und zum anderen auf das Konzept des Vorobjektiven, das von dem französischen Philosophen Maurice Merleau-Ponty (1908-1961) entwickelt worden ist. Obwohl die beiden Positionen – hier der Phänomenologe, dort der „dialektische Strukturalist" (Csordas) – methodologisch inkompatibel sind, treffen sie sich doch im Paradigma des *Embodiment*. Beide wenden sich gegen dualistische Auffassungen: Während Bourdieu sich gegen die Trennung von Struktur und Praxis richtet, möchte Merleau-Ponty die Dualität von Subjekt und Objekt auflösen. Und beide versuchen, diese Dualitäten nicht zu vermitteln, sondern mithilfe des gelebten Leibes in eins fallen zu lassen.[57] Mit Bourdieus Habitus kann Csordas zeigen, dass der Leib von Anfang an sozial geprägt ist; Merleau-Pontys Theorie der Wahrnehmung wirft Licht auf den Prozess, in dem Selbst und Kultur simultan entstehen. Da diese für das Verständnis des *Embodiment*-Konzeptes essentiell ist, werden im Folgenden einige ihrer Grundgedanken erläutert.

Der Leib spielt die zentrale Rolle in Merleau-Pontys Theorie, ist er doch die Bedingung der Möglichkeit aller Wahrnehmung und damit die notwendige Grundlage der Existenz des Menschen als „In-Welt-sein".[58] Merleau-Ponty beginnt seine *Phänomenologie der Wahrnehmung* (1945) denn auch mit einer Kritik der herkömmlichen Auffassung von Wahrnehmung als einer intellektuellen Handlung, „bei der die Sinne passiv Eindrücke einer gegebenen Objektwelt als exakte Bilder im Geist aufnehmen".[59] Diese sogenannte „Konstanzhypothese" geht cartesianisch von einer gegebenen Welt einerseits und einem reinen Bewusstsein, das die Welt wahrnimmt, andererseits aus und suggeriert zudem, dass Wahrnehmung und Stimulus übereinstimmen. Dass dies gerade nicht zutrifft, lässt sich beispielsweise mit dem Phänomen der Zeugenaussage illustrieren, bei der verschiedene Personen dasselbe Ereignis ganz unterschiedlich beschreiben, sodass ungewiss scheint, ob alle dasselbe gesehen haben. Wenn der Wissenschaftler daher von einem vorgegebenen Objekt ausgeht, so hat er bereits abstrahiert und bemerkt nicht, dass tatsächlich ganz Unterschiedliches wahrgenommen wird. Für Merleau-Ponty ist die Welt also nicht die „Summe determinierter Gegenstände", die nur noch wahrgenommen werden müssen, sondern ein „Horizont von möglichen Objektwerdungen", in dem jedes Ding als unendlich viele mögliche Perspektiven gegeben ist.[60] Jede Perspektive gibt

---

57 Csordas (1990), 7, 8.

58 Merleau-Pontys eigentliche Formulierung wird als „Zur-Welt-sein" übersetzt. Sie bezieht sich auf Heideggers „In-der-Welt-sein", aber im Sinne eines besonderen Modus, nämlich einer „Hingebung" an die Welt (vgl. Merleau-Ponty 1966, 7, Fn. d). Im Sinne einer einfacheren Lesbarkeit wird auf diese Differenzierung aber verzichtet.

59 Platz (2006), 44.

60 Merleau-Ponty (1966), 381.

einen Teil des Objekts preis, aber in keiner ist es erschöpfend gegeben. Die Welt ist vor dem Prozess der Wahrnehmung unbestimmt und unvollendet, sie legt nicht von vorneherein fest, was sinnlich wahrgenommen werden soll, wie dies geschieht und welche Perspektive dabei eingenommen wird.[61]

Der Umstand, dass die Objektwelt nicht gegeben, sondern das Ergebnis einer Reflexion ist, die mit der Wahrnehmung beginnt, hat zur Folge, dass Welt, Leib und Bewusstsein nicht getrennt voneinander betrachtet werden können, da sie einander bedingen.[62] Welt und Selbsterkenntnis entstehen erst in der Hinwendung des Leibes zur Welt.[63] Mit anderen Worten: Wahrnehmung beginnt im Leib und endet durch den Prozess der Reflexion mit Objekten.[64] Für eine Analyse ist das Verständnis des jeder Objektwerdung zugrunde liegenden Unreflektierten oder Vorobjektiven unabdingbar. Dieses Vorobjektive besteht in der ursprünglichen Erfahrung des Leibes mit der Welt. Da der erste Kulturgegenstand, den der Leib erkennt, ein anderer Leib ist, ist diese Erfahrung jedoch keinesfalls vorkulturell, sondern vielmehr von Anfang an sozial eingebettet und somit intentional. D. h. der „Leib erkennt sich also über das Andere, das vor allem andere Menschen, aber auch kulturelle Objekte und damit die ganze kulturelle Welt einschließt".[65] Für Merleau-Ponty ist Subjektivität also Voraussetzung von Erkenntnis und sie ist zugleich auch schon immer eine Intersubjektivität, denn der Prozess der Entstehung von Selbst und Welt ist ursprünglich und vorobjektiv sozial: „Das Soziale ist je schon da, ehe wir es erkennen oder darüber urteilen".[66] Wahrnehmung findet demnach vor einem gemeinsamen kulturellen Hintergrund statt. Merleau-Ponty zufolge ist der Prozess der Objektivierung jedoch niemals abgeschlossen, denn „das Erkennen und damit Erzeugen eines kulturellen Objekts durch den Leib in der Wahrnehmung beeinflusst das Selbst".[67] In dem Maße, in dem dem Leib die Welt bewusst wird, ändert sich seine Perspektive auf die Welt. Dies wiederum führt zu einer veränderten Wahrnehmung der Welt, sodass die Objektwerdung als andauernder Reifungsprozess beschrieben werden kann, bei dem Selbst und Kultur sich ständig neu konstituieren.[68]

---

61 Platz (2006), 46.

62 Vgl. Böhme (2003).

63 Merleau-Ponty (1966), 6-9.

64 Platz (2006), 45.

65 Platz (2006), 50.

66 Merleau-Ponty (1966), 414.

67 Platz (2006), 50.

68 Platz (2006), 50.

Mit Rückgriff auf Merleau-Ponty begreift Csordas Wahrnehmung als eine grundlegend körperliche Erfahrung, bei der der Körper nicht Objekt, sondern Subjekt ist, sodass *Embodiment* die Voraussetzung dafür ist, dass wir überhaupt Objekte haben (also die Realität objektivieren). D. h. Kultur wohnt nicht nur Objekten und Repräsentationen inne, sondern auch den körperlichen Wahrnehmungsprozessen, durch die diese Repräsentationen entstehen.[69] Dies bedeutet umgekehrt, dass der Körper gleichzeitig als eine Quelle von Repräsentationen à la Douglas bzw. als Produkt von Repräsentation à la Foucault *und* als Voraussetzung des In-der-Welt-seins konstruiert werden kann. Csordas verändert Merleau-Pontys Ansatz jedoch insofern, als er den Prozess der Objektivierung vor allem als *Selbst*objektivierung fasst, bei dem ein abgelehnter oder verkannter Teil des Selbst als kulturelles Objekt veräußerlicht wird. Ihn interessiert also vor allem die Ebene der vorobjektiven Erfahrung, auf der Kultur (von Wahrnehmung bestimmt) und Selbst (von Kultur bestimmte Wahrnehmung) entstehen.[70] Die Beschäftigung mit *Embodiment* dient damit dem Zugang zu Kultur und Selbst, geradeso wie Textualität als Zugang zu Kultur und Selbst dienen kann.[71]

IV   Performanz & Transformation: Aspekte einer Ritualdefinition

Folgt man den Autoren einer aktuellen Einführung zum Thema Ritualforschung, dann ist die Frage „Was ist ein Ritual?" letztlich sinnlos, denn es gibt weder „*das* Ritual, losgelöst von allen historischen, regionalen oder sprachlichen Kontexten"

---

69 Csordas (1999), 183.

70 Platz (2006), 83.

71 Csordas (1999), 184. Ein Beispiel für einen solchen Zugang gibt Turner (2011) im Hinblick auf die südamerikanischen Kayapo. Abschließend sei darauf hingewiesen, dass das Embodiment-Konzept auch methodologische Konsequenzen insofern hat, als es der von Fabian (2001, 11) konstatierten „Entkörperlichung des Wissens" in der Ethnologie entgegensteuert und den „Kampf zwischen Beobachtung und Teilnahme" (Fabian 2001, 163) obsolet werden lässt. Csordas zufolge reicht es eben nicht aus, über Körper bzw. Phänomene der Verkörperung zu forschen, es muss vielmehr darum gehen, mit dem Körper zu forschen und diesen konsequent als Mittel der Wissensproduktion zu nutzen (Csordas 1999, 184). Zu den wenigen Ethnologen, die diese Fragestellung aufgegriffen haben, gehören Paul Stoller (1989), Michael Jackson (1983, 1989), Judith Schlehe (1996), Mario Bührmann (2010) und Georg Spittler (2001).

noch „ein Wesen des Rituals".[72] Die Verwendung des Begriffs „Ritual" (wie übrigens auch „Religion"[73]) geschieht keineswegs wertfrei und geschichtsübergreifend, sondern ist Ausdruck kulturell und historisch geprägter Diskurse. Die universale Geltung beanspruchenden Konzepte von „Religion" und „Ritual" sollten eine vergleichende Einordnung fremder kultureller Praktiken ermöglichen, wobei ihr partikularer Hintergrund weitgehend unreflektiert blieb. Die Prozesse von Kolonialisierung und Globalisierung verstärkten die Reichweite dieser Diskurse und führten dazu, „dass ein spezifisch konnotierter Ritualbegriff [...] nicht zuletzt von Vertretern der verschiedene religiösen Traditionen selbst übernommen wurde, was wiederum auch Transformationen kultureller Praktiken und im religiösen Selbstverständnis anstieß".[74] Die Entstehung der akademischen Fächer Soziologie, Ethnologie und Religionswissenschaft ist auch Ausdruck dieser Entwicklungen, und berühmte Begründer dieser Disziplinen sind zugleich auch die ersten europäischen Ritualtheoretiker: William Robertson Smith (1889), James George Frazer (1890), Arnold van Gennep (1909) oder Émile Durkheim (1912).[75]

Seitdem hat es eine stetig wachsende Zahl von konkurrierenden und einander ablösenden Versuchen gegeben, *Ritual* wissenschaftlich zu definieren, und auch außerhalb der Wissenschaft hat die Verwendung des Begriffs signifikant zugenommen. Es erscheint daher wenig sinnvoll, auf eine endgültige Definition des Begriffs „Ritual" hinzuarbeiten, wohl aber geraten, eine Grundlage interdisziplinärer Verständigung anzubieten. In diesem Sinne schlägt die Ritualforscherin Catherine Bell folgende – in der Praxis sich durchaus überlagernde – Kategorien rituellen Handelns vor: (1) Übergangsrituale, (2) Kalender- oder Gedenkrituale, (3) Austausch- und Kommunionsrituale, (4) Heilungsrituale, (5) Rituale des Miteinander-Essens, Fastens und des Feierns sowie (6) politische Rituale.[76] Darüber hinaus geht sie von drei generellen Ritual-Konzeptionen aus: (a) als Ausdruck des paradigmatischen Wertes von Tod und Wiedergeburt; (b) als Mechanismus, der das Individuum in die Gruppe integriert und eine soziale Entität konstituiert sowie (c) als Prozess sozialer Transformation.[77] Auf der Basis von Brosius / Michaels / Schrode lässt sich zudem eine Liste charakteristischer Aspekte von Ritualen erstellen.[78] Dazu gehören:

---

72 Brosius/Michaels/Schrode (2013), 10.

73 Vgl. z.B. Bowie (2006), 18-25.

74 Brosius/Michaels/Schrode (2013), 11.

75 Brosius/Michaels/Schrode (2013), 11.

76 Bell (1997), 4.

77 Bell (1997), 89.

78 Brosius/Michaels/Schrode (2013), 13-17.

1. Performanz, Repetition und Formalität: Rituale sind durch eine förmliche, stilisierte, teilweise stereotype Performanz gekennzeichnet, d.h. sie bestehen überwiegend aus wiederholten und wiederholbaren bzw. (mimetisch) nachahmbaren Handlungen, die u.U. in Ritualkomplexen zusammengefügt und Skripten festgehalten sind. Diese Formalität gibt bestimmte Reaktionen vor und standardisiert und kanalisiert damit die Emotionen der Beteiligten. In diesem Zusammenhang ist häufig von „verkörpertem" Wissen die Rede, das sich von sprachlich vermittelbarem Wissen unterscheide und gerade auch im Ritual zum Tragen komme.[79] Dabei wird z. B. auf Catherine Bells (1992) Konzept der „rituellen Beherrschung" („ritual sense") verwiesen: Menschen, die mit bestimmten rituellen Praktiken aufgewachsen sind oder sie schon lange praktizieren, üben ihre Handlungen ähnlich unreflektiert aus wie ein Autofahrer, der ‚schlafwandlerisch‘ von einem Gang in den anderen schaltet. Handlungen können in Fleisch und Blut übergehen, sodass sich das Knie gleichsam ‚automatisch‘ beugt, wenn eine katholisch sozialisierte Person – gläubig oder nicht – eine Kirche betritt. Mitglieder einer Ritualgemeinschaft können sich so gegenseitig erkennen, auch wenn die entsprechenden Regeln keinesfalls immer explizit und sprachlich kommunikabel sind, sondern durch Mimesis und Imitation implizit zum Teil der eigenen Persönlichkeit werden. Naturalisierung schließlich führt dazu, dass es so scheint, als *müsse* man in bestimmten Momenten den Körper in eine bestimmte Position bringen (z. B. niederknien) oder eben eine bestimmte Handlung (z. B. sich bekreuzigen) ausführen. Diese Art des „praktischen Glaubens" ist laut Bourdieu kein „Gemütszustand", sondern ein „Zustand des Leibes".[80] Ganz ähnlich lassen sich Erinnerung und kinästhetische Erfahrung als verkörperte Form der kulturellen Wissensübertragung einordnen, etwa in der Form von „gesellschaftlicher Erinnerung" (Connerton 1989) oder als „Archiv" bzw. „Repertoire" (Taylor 2003). Bei der Übertragung, Reproduktion und Weiterentwicklung dieses Wissens spielen Rituale eine zentrale Rolle.

2. Rahmung und Beschluss: Rituale sind ‚gerahmt‘, d. h. sie finden zu besonderen Zeiten und an bestimmten Orten statt, mitunter aufgrund eines förmlichen Beschlusses, um die Ritualhandlungen auf einen bestimmten Zweck ausrichten zu können; Rahmung und Beschluss markieren ‚normale‘ Handlungen als besondere Handlungen, die sich vom Alltag abheben.

3. Transformation: „Rituale intendieren Veränderungen" und zeigen Wirkung, d. h. in ihrem Vollzug geschieht etwas, das weder trivial noch ohne weitrei-

---

79 Polit (2013), 218.
80 Bourdieu (1987), 126.

chende Konsequenzen ist.[81] Lebenszyklische Rituale können zum Beispiel den Wechsel des Status oder der Kompetenz einer beteiligten Person bewirken. Nach dem Ritual ist man ein Anderer, z. B. ein Erwachsener, eine Ehefrau oder Promovierter. In diesem Zusammenhang spielen kulturelle Ordnungszeichen, die eine Überhöhung der Handlungen und ihre Normativität ausmachen, eine zentrale Rolle – beispielsweise Herrschaftszeichen, Metaphern und Medien der Überlieferung, mit denen auf ideale oder überpersönliche Wertvorstellungen Bezug genommen wird. Rituale aktualisieren Sinn- und Bedeutungsfragen sozialen Handelns, um sie dann weitgehend auszuschalten und in Habitus, Gewohnheit oder Struktur zu transformieren. Sie „formen damit ein kulturelles Gedächtnis, bei dem das richtige und angemessene Verhalten nicht unbedingt jedes Mal neu ausgehandelt oder legitimiert werden muss".[82] Sie tragen damit zu kognitiver Entlastung bei und verschaffen sozialen Systemen eine effektive Ordnungsstruktur, wozu auch die Ausblendung der Tatsache gehört, dass soziales Handeln einem kontinuierlichen, u.U. gewaltsamen Wandel unterworfen ist. „Als ‚Wesen' des Rituals gilt stattdessen das Kontinuierliche, oft gleichgesetzt mit dem ‚Ursprünglichen' und daher ‚Authentischen'".[83] Aufgrund ihrer Öffentlichkeitswirkung sind Rituale zudem „probate Mittel, Macht und Herrschaft nach außen hin sichtbar zu machen".[84]

4. Ritual/Ritualisierung: Das Kriterium der Normativität ermöglicht auch die fallweise Unterscheidung zwischen Ritualen im engeren Sinne und routinisierten Alltagshandlungen oder Ritualisierungen, wie sie etwa von Erving Goffman (1967) untersucht worden sind. Auch bei Ritualisierungen finden sich Merkmale von Ritualen, wie etwa förmliche, repetitive oder performative Handlungsmuster; Normativitäts-Bezüge fehlen aber. Ein weiterer Unterschied besteht darin, dass Rituale gestaltete und inszenierte, d.h. bewusst durchgeführte Handlungen darstellen, auch wenn unbewusste Anteile in ihnen wirksam sind. Bei Ritualisierungen hingegen ist der bewusste Anteil nicht oder nur wenig vorhanden.

5. Körper: Da Rituale in erster Linie Handlungen sind, d.h. eine Form des bewussten und zielgerichtete Einwirkens des Menschen auf seine Umwelt, ist die Frage nach der Rolle des Körpers essentiell. Ihr soll deshalb im folgenden Abschnitt aus der Perspektive des *Embodiment*-Konzeptes nachgegangen werden.

---

81 Müller (2010), 384.
82 Brosius/Michaels/Schrode (2013), 16.
83 Brosius/Michaels/Schrode (2013), 17.
84 Brosius/Michaels/Schrode (2013), 17.

V   Körper im Ritual

Wenn Rituale als Handlungen verstanden werden, die nicht-triviale soziale bzw. kulturelle Veränderungen intendieren und der kulturell wahrnehmende und handelnde Leib als Voraussetzung der Kultur, dann wird auch die unabdingbare Rolle des Körpers im Ritual deutlich. Unmittelbar einsichtig ist dies bei Ritualen, die den Körper tatsächlich in integrativer bzw. abgrenzender Absicht manipulieren, wie dies klassischerweise bei *Übergangs-* bzw. *Initiationsritualen* der Fall ist. Im Sinne Bells sind diese nicht nur Ausdruck des paradigmatischen Wertes von Tod und Wiedergeburt, sondern auch ein Mechanismus der Integration bzw. sozialer Transformation und haben schon früh das Interesse der Ritualforschung auf sich gezogen. So hat der französische Ethnologe und Ritualforscher Arnold van Gennep bereits zu Anfang des 20. Jahrhunderts seine wegweisende Arbeit über „Übergangsrituale" (1909) vorgelegt, in der er sich u. a. mit all jenen kulturellen Praktiken der Körperveränderung beschäftigt, die „auf eine für alle sichtbare Weise die Persönlichkeit eines Individuums verändern".[85] Diese sind äußerst vielfältig, letztlich wird der menschliche Körper, wie van Gennep es ausdrückt, „wie ein einfaches Stück Holz behandelt", Hervorstehendes abgeschnitten, Wände durchbohrt und Glattes eingeritzt.[86] Dies betrifft vor allem solche Organe, „die, wie die Nase oder die Ohren, durch ihr Hervortreten den Blick auf sich lenken und aufgrund ihrer histologischen Beschaffenheit allen möglichen Behandlungen unterzogen werden können, ohne das Leben oder die Handlungsfähigkeit des Individuums zu beeinträchtigen."[87]

Dabei lassen sich drei grundsätzliche Vorgehensweisen unterscheiden: (1) die Manipulation des Körpers durch u. a. Durchbohren, Spalten, Ausweiten, Verengen und Verschließen; (2) die Addition von Narben auf die Haut sowie die Injektion von Farbe oder Insertion von materialen Gegenständen in bzw. unter die Haut; schließlich (3) die Amputation von Teilen des Körpers. Für van Gennep inszenieren Übergangsrituale den simultanen Prozess von Integration und Abgrenzung, von Grenzziehung und Grenzüberschreitung, indem sie das Individuum mit Hilfe eines Trennungsritus aus der undifferenzierten Menge der Menschen herauslösen und es gleichzeitig so an eine bestimmte Gruppe angliedern. Da die Operation unauslöschliche Spuren hinterlässt, ist die Integration endgültig.[88] Bei den Iatmul in Neuguinea beispielsweise werden die Körper der männlichen Jugendlichen

---

85 Gennep (1986), 76.

86 Gennep (1986), 76.

87 Gennep (1986), 77.

88 Gennep (1986), 76.

durch wiederholte Schnitte in einem langwierigen und äußerst schmerzhaften Prozess skarifiziert, d. h. mit Narben versehen. Diese Narben verweisen auf das Krokodil des lokalen Schöpfungsmythos, das auf den Grund des Urmeeres taucht, Schlamm emporholt und diesen zu einer Insel formt. Die Skarifizierung auf Brust und Rücken der Initianden ahmt die Struktur einer Krokodilshaut nach und deutet in ihrem Verlauf Gliedmaßen und Schwanz dieses Tieres an. Die auch als „Biss des Krokodils" bezeichneten Narben stehen als Beweis dafür, dass die Jungen vom Krokodil-Ahnen verschlungen wurden, um später als erwachsene, fertige Männer wiedergeboren zu werden.[89]

Neben der Unterscheidung zwischen Heranwachsenden und Erwachsenen ist diejenige zwischen Männern und Frauen entscheidend. Vor diesem Hintergrund verwundert es nicht, dass das Interesse einer klaren Trennung – die im Sinne Mary Douglas' für ‚Ordnung' sorgt – sich auch in Initiationsritualen niederschlägt. Um die Anteile des bei der Geburt vorhandenen jeweils anderen Geschlechts abzutrennen, sind Initiationen häufig mit Praktiken wie Beschneidung der Geschlechtsteile, Ohrlochstechen oder dem Durchstechen der Nasenscheidewand, einer Tatauierung oder eben der Skarifizierung verbunden. Die dabei verursachten Blutungen werden (vor allem aus männlicher Perspektive) oft als ein Ablassen des mütterlichen Blutes interpretiert. Der Junge wird so von allen weiblichen Anteilen ‚gereinigt' und damit zum erwachsenen Mann.[90] Ein samoanisches Sprichwort lautet: „Wenn ein Mädchen geboren wird, so wird es die Schmerzen des Gebärens ertragen müssen; wenn ein Junge geboren wird, so wird er die Schmerzen der Tatauierung ertragen müssen."[91] Beider Schmerzen wiegen also einander auf, wobei als verbindendes Moment der Verlust des Blutes gilt. Das Interesse an einer klaren Trennung gilt aber auch hinsichtlich der Unterscheidung zwischen Mensch und Tier bzw. Kultur und Natur. Diese Markierung kann etwa mit Hilfe der Zähne vorgenommen werden. So wollen die Bewohner der Nil-Gebiete, die die unteren Schneidezähne herausschlagen, nicht „wie Esel essen", während in anderen Teilen Afrikas Zahnextraktionen damit begründet werden, dass man sich vom Raubtier unterscheiden möchte.[92] Auf Bali wird die nicht natürliche Zahnstellung als Zeichen der Unterscheidung zwischen menschlichen und nicht menschlichen Wesen interpretiert. In Zähnen werden besondere Lebenskräfte vermutet, bei ‚wilden' Zähnen eine frei schweifende

---

89 Raabe (2006).
90 Raabe (2006).
91 Meesenhöller (2006).
92 Streck (2000), 131f.

Sexualität. Das Ziel des Rituals der Zahnfeilung besteht also in einer Markierung der Beherrschung dieser Leidenschaften.[93]

In den sogenannten *Besessenheitsritualen*, die zu den Heilungsritualen gehören und ein Individuum in eine bestimmte religiöse Gruppe integrieren, bei denen es aber auch zu sozialen Transformationsprozessen kommen kann, spielt die Verkörperung von Gottheiten, Dämonen, Ahnen oder anderen spirituellen Kräften durch ausgewählte Teilnehmer eine zentrale Rolle. Der Ritual-Diskurs geht davon aus, dass die Gottheiten Besitz von den Gastgeberkörpern ergreifen, um zu den Menschen zu sprechen oder in menschlicher Gesellschaft zu tanzen. In diesem Sinne „Besessene" haben u.U. die Kraft, zu heilen oder Aussagen über Dinge zu machen, die gewöhnlichen Menschen verborgen bleiben. Obwohl häufig stark reguliert, sind solche Rituale von Europäern häufig als „wild" und „chaotisch" beschrieben worden, denn die Bewegungen, Bewegungsradius, Gesten und gesprochene Worte folgen unausgesprochenen, eben leiblichen, Regeln, die nicht ohne Weiteres vom Beobachter gesehen bzw. erfragt werden können. Entgegen der klassischen Interpretations-Tradition der Ritualforscher handelt es sich bei dieser Art göttlicher Verkörperung auch nicht um eine vordergründig symbolische bzw. metaphorische Handlung. Die Akteure des Rituals begreifen sich nicht als Schauspieler, sondern sie und die anderen Teilnehmer gehen von einer realen Anwesenheit des Göttlichen aus. Häufig ist es gerade diese Präsenz einer Entität im Körper des Mediums, die für die Gültigkeit und Wirksamkeit des Rituals bürgt. Das Konzept des *Embodiment* ermöglicht es dagegen, solche Zustände „als Teil des rituellen Repertoires einer Gruppe und alltägliche religiöse Handlungen zu verstehen".[94] Dies lässt sich mit zwei Beispielen aus Csordas' Forschung zu Heilungsritualen bei den zur Pfingstbewegung gehörenden charismatischen Christen in den USA illustrieren, sowie dem in diesen Gemeinschaften ebenfalls verbreiteten „Zungenreden".

Das erste Beispiel entstammt einem Geistheilungs-Ritual, bei dem durch einen charismatischen Prediger Dämonen ausgetrieben werden. Die geschieht in der Regel dadurch, dass verschiedene Typen böser Geister im Gebet benannt und aufgefordert werden, ihre jeweiligen ‚Gastgeber' unter den Teilnehmern zu verlassen. Im Moment ihres Ausfahrens produzieren diese Geister eine physische Manifestation, bei der es sich – je nach Dämon – um hyänenartiges Geschrei, Gebrüll, Fauchen, Rülpsen, Husten, Erbrechen, aber auch auf dem Boden wälzen oder Augenrollen handeln kann. Csordas berichtet von einem Teilnehmer, der – als der Ritualleiter fragt, wer vom Dämon der Masturbation erlöst werden möchte – sich gemeinsam mit 15 bis 20 anderen Anwesenden plötzlich erhebt. Als der Prediger während des Gebets

---

93 Rein (2006).
94 Polit (2013), 220.

dem Dämon zu verschwinden befiehlt, reißen die Stehenden plötzlich ihre Arme heftig nach oben, wobei sich ihre Hände unwahrscheinlich weit nach hinten biegen; gleichzeitig empfinden sie ein leicht elektrisierendes Gefühl. Nach einer gewissen Zeit sinken ihre Hände dann wieder nach unten. Csordas' Gewährsmann gibt an, von dem Vorgang überrascht worden zu sein und nicht gewusst zu haben, was jeweils als Nächstes passieren würde. Schmerzen habe er dabei keine empfunden.[95]

Zur Interpretation dieses Vorgangs unterscheidet Csordas zwischen Dämonen als kulturellen Objekten einerseits und ihren empirischen Manifestationen als konkreten Selbstobjektivierungen der Teilnehmer andererseits. Als kulturelle Objektivationen sind die bösen Geister intelligente, nichtmaterielle Wesen der Hölle, die den Menschen bedrohen, unterdrücken und von ihm Besitz ergreifen können (eine entsprechende „Dämonologie" klassifiziert die über 150 verschiedenen Geister nach Arten und Symptomen). Die Dämonen repräsentieren mit ihren negativen Eigenschaften letztlich das Gegenbild zum kulturell definierten idealen Selbst. Dabei handelt es sich Csordas zufolge jedoch bereits um den zweiten Schritt im Prozess der kulturellen Objektwerdung. Denn zuerst einmal nehmen die Teilnehmer keine Dämonen in sich wahr, sondern den Eindruck, die Kontrolle über einen Gedanken, ein Verhalten oder ein Gefühl verloren zu haben. Erst der Heiler macht diesen Kontrollverlust zu einem kulturellen Objekt, indem er ihn entweder als Dämon oder als emotionales Problem identifiziert. Vor diesem kulturellen Hintergrund der objektiven Strukturen können die beschriebenen Manifestationen als Beispiele für den leiblichen Prozess der Selbstobjektivierung analysiert werden. Die Teilnehmer erleben die dämonischen Manifestationen spontan und ohne vorgegebenen Inhalt, was auf das vorobjektive Element des Prozesses verweist. Den Möglichkeiten dieser Manifestationen werden durch den gemeinsamen Habitus jedoch Grenzen gesetzt. Das von Merleau-Ponty beschriebene vorobjektive leibliche Wissen besteht in diesem Fall im Überschreiten einer Toleranzschwelle: ein Gefühl, ein Gedanke oder ein Verhalten wird als unkontrollierbar empfunden. Zur Heilung wird diese Selbstobjektivierung durch die Manifestation des abgelehnten Zustands in bestimmten „leiblichen Bildern" („embodied images") wie eben schreien, erbrechen, auf dem Boden wälzen oder Hände in die Luft recken. Die mit diesen Bildern verbundenen Bedeutungen thematisieren Kontrollverlust und Erlösung von dem, was die Kontrolle übernommen hat. Dass bereits das Vorobjektive kulturell ist, zeigt sich Csordas zufolge daran, dass Gesundheit in den USA mit Metaphern für Selbstkontrolle und Befreiung von Druck belegt wird.[96]

---

95 Csordas (1990), 14f.
96 Csordas (1990), 14f.

Hinzuweisen ist aber auch auf die erlebte Unbestimmtheit, wie mit den Dämonen umzugehen ist. Dies zeigt sich zum einen daran, dass unbestimmt ist, bis zu welchem Grad jemand von einem Dämon beeinflusst wird. Es gibt diesbezüglich zwar Abstufungen von Bedrohung über Unterdrückung bis Besessenheit, aber keine objektiven Kriterien, dies in der Praxis festzustellen. Die Praxis vollzieht sich auf der Ebene der vorobjektiven Intersubjektivität zwischen dem Teilnehmer und dem Heiler, der den Dämon empathisch oder intuitiv erahnt. Zum anderen ist unbestimmt, welche Dämonen auftreten und wie viele. Diese sind zwar in der Regel hierarchisch um einen Hauptgeist organisiert, doch in der Praxis ist die Identifikation der Dämonen ein offener Prozess, da sie in beliebiger Zahl erscheinen können, wobei die Bandbreite der Möglichkeiten wiederum durch den Habitus bestimmt wird.[97] Mit dem Habitus lässt sich auch erklären, warum bestimmte Dämonen mit bestimmten Zeichen in Verbindung gebracht werden, die sich entsprechend bei der Austreibung manifestieren. So wird Masturbation von Csordas' Informanten als stark verpönte aber gleichzeitig zwanghafte (und damit dämonische) Handlung empfunden. Die spontane kollektive Geste der in die Luft geworfenen Arme kann als kraftvolles „Hände weg!" interpretiert werden, was durch die nach hinten gebogenen Handflächen noch betont wird. Dass diese Beugung nicht als schmerzhaft empfunden wird, obwohl sie stärker ausfällt als ‚natürlich' wäre, stimmt mit dem Konzept der Erlösung von den Fesseln böser Geister überein, das anstelle des Konzepts einer Bestrafung für begangene Sünden von den Beteiligten verwendet wird. Gleichermaßen ist das elektrisierende Gefühl kein strafender Schock, sondern eine Verkörperung spiritueller Kraft.[98]

Csordas zufolge zeigt sich hier, was Bourdieu mit willkürlicher Notwendigkeit der Kultur meint, die durch den Habitus ‚natürlich' wird. Im religiösen Diskurs der Pfingstkirchen erscheint es als natürlich, dass sich das Böse in bestimmten Verhaltensweisen, Gedanken etc. zeigt und sich auch in einer ganz bestimmten Weise bei der Austreibung manifestiert. Ebenso selbstverständlich erscheint es, solch ein Verhalten in Form einer Heilung auszutreiben und nicht zu bestrafen.[99] Die physischen Manifestationen der Dämonen sind Ausdruck einer „multisensorischen Bildsprache" („multisensory imagery"), eine leibliche Manifestation der Erfahrungen der Teilnehmer in verschiedenen kulturell bestimmten Bildern, die nicht ausschließlich visueller Natur sein müssen.[100] Diese Bilder stellen keine inneren

---

97   Csordas (1990), 17.
98   Csordas (1990), 18.
99   Platz (2006), 85f.
100 Csordas (1990), 13.

Erfahrungen dar, sondern objektivieren und erzeugen ein leibliches Heilen.[101] D. h. der Habitus bietet bestimmte Dispositionen, wie der Leib sich selbst erkennen und dies objektivieren und ausdrücken kann. Laut Csordas offenbaren die spontanen religiösen Bilder den Habitus; sie zeigen, was die Handelnden nicht erkennen und was ihre Handlungen objektiv an andere und an die Strukturen anpasst, aus denen ihr Habitus als Erzeugungsprinzip der Praktiken hervorgegangen ist.[102] Durch diese leiblichen Bilder werden Dispositionen des Habitus in rituellen Handlungen manifestiert. Da den Handelnden verborgen bleibt, dass sie diese Bilder in Form von Dispositionen teilen, werden diese unvermeidlich als von Gott erzeugt verkannt („inevitably misrecognized"). Csordas schließt daraus, dass der Leib ein Prinzip ist, das sich nicht weiter reduzieren lässt, insofern der Leib die existentielle Voraussetzung von Kultur darstellt.[103]

Das zweite Beispiel behandelt nicht die Austreibung von Dämonen, sondern im Gegenteil das Einfahren des Heiligen Geistes in einzelne Teilnehmer. Diese Manifestation göttlicher Macht soll die Ungläubigen zur Konversion bewegen und den Glauben der bereits Gewonnenen verstärken. Zu den in diesem Zusammenhang verbreiteten leiblichen Bildern gehören das schnelle Flattern oder Vibrieren von Armen und Händen sowie somatische Empfindungen wie Leichtigkeit oder Schwere, Kraft oder Liebe, die durch den Körper fließen, Hitze und Kribbeln. Es kann zu ansteckendem Lachen oder Weinen in der Versammlung kommen. Viele Teilnehmer „ruhen im Geist" („rest in the spirit"), d. h. erfahren eine motorische Dissoziation, bei der eine Person vom Heiligen Geist übermannt wird und in einen ohnmachtsähnlichen Zustand gerät, was als entspannender und verjüngender Moment der Gegenwart Gottes erlebt wird. Wichtiger noch ist das „Wort des Wissens" („word of knowledge"), eine Form der Offenbarung, die als göttliches Geschenk des Wissens über Personen oder Situationen verstanden wird, das nicht durch menschliche Kommunikationskanäle erworben wurde, sondern als spontaner Gedanke oder Bild erfahren wird.[104]

Das von Csordas beobachtete Heilungsritual besteht aus abwechselnden Phasen von kollektiven Gebeten, religiösen Gesängen, Heilungs-Gebeten und Vorträgen. Während der ersten Phase des Gebetes erhalten die Prediger durch das Wort des Wissens die Eingebung, dass Gott Menschen mit Rücken-, Atemwegsproblemen und Arthritis heilen wolle. Personen mit diesen Leiden werden nach vorne gebeten, damit die erfahrenen Prediger ihnen die Hände auflegen und für sie beten können.

---

101 Csordas (1990), 22.
102 Csordas (1990), 23.
103 Csordas (1990), 23.
104 Csordas (1990), 18f.

In der nächsten Runde sind alle Teilnehmer eingeladen sich abwechselnd als Heiler bzw. Heilungsbedürftige zu beteiligen. Die Versammlungsleiter erklären, dass einige Personen im Publikum eine Schwere in Oberkörper und Kopf spüren werden, Hitze in Gesicht oder Lippen oder ein Kribbeln in den Händen. Diese Personen werden gebeten, ihre Hände mit den Handflächen nach oben in Gebetshaltung auszustrecken und sich so erkennen zu geben; die Umstehenden sollen ihnen im Gebet die Hände auflegen, um die Manifestation göttlicher Kraft zu stärken und unter sich weiterzugeben. Die Teilnehmer werden eingeladen, das Wort des Wissens selbst zu erfahren und zusammen mit denjenigen zu beten, die auf das von ihnen identifizierte Problem reagieren.[105] Während eine semiotische Analyse in diesen Phänomenen vor allem Zeichen oder Repräsentationen ausmachen würde, die im Geist des Heilers entstehen, konstatiert die phänomenologische Sicht, dass der Heiler mit allen seinen Sinnen, d. h. mit seinem Leib, einem anderen Leib Aufmerksamkeit schenkt. Es handelt sich also um eine intersubjektive leibliche Erfahrung, die leibliches Wissen erzeugt.[106] Dies geschieht, ohne dass abstraktes Wissen involviert ist, weil Heiler und Heilungsbedürftige denselben Habitus teilen, dieselben Dispositionen, den Leib zu objektivieren.[107] Csordas spricht in diesem Zusammenhang von „somatischen Aufmerksamkeitsmodi", die er als „kulturell verfeinerte Weisen der Achtsamkeit auf den und mit dem Leib" bezeichnet, in „Umgebungen, die die leibliche Präsenz anderer einschließen".[108] Der Mensch achtet also mit seinem Leib auf andere Leiber, die wiederum auf ihn achten.[109]

Aus der Perspektive des *Embodiment* lässt sich aber auch *Sprache*, die typischerweise in den Bereich von Linguistik, Semiotik und Textanalyse fällt, analysieren. Dies zeigt Csordas am Beispiel des „Zungenreden" (*Glossolalie*). Aus semiotischer Perspektive gilt *Glossolalie* als Störung der Welt der Bedeutungen, wodurch sie einerseits kulturellen Wandel ermöglicht und andererseits bedrohlich auf Außenstehende wirkt. Diese Interpretation will Csordas dadurch erweitern, dass er *Glossolalie* mit Hilfe von Merleau-Ponty als leibliches Phänomen betrachtet.[110]

---

105 Csordas (1990), 19.

106 Csordas (1999), 152.

107 Platz (2006), 89f.

108 „somatic modes of attention [...] culturally elaborated ways of attending to and with one's body in surroundings that include the embodied presence of others." Csordas (1993), 139.

109 Csordas (1993), 244. Mannschaftssportarten, Tanz, Magersucht und Hypochondrie sind weitere Beispiele Csordas' für somatische Aufmerksamkeitsmodi außerhalb des Ritualkontextes.

110 Csordas (1990), 24f.

In der Lesart Csordas' setzt Merleau-Ponty eine „verbale Geste mit immanenter Bedeutung" als den Ursprung der Sprache.[111] Merleau-Ponty widerspricht damit der Annahme, dass Sprache Denken repräsentiert und argumentiert vielmehr, dass sie eine Form der Hinwendung des Leibes zur Welt sei. Vermittels Sprache nimmt der Leib bereits vorobjektiv eine bestimmte Perspektive gegenüber der Welt ein und objektiviert sie aus dieser Perspektive. Durch *Glossolalie* objektiviert der Leib eine religiöse bzw. sakrale Welt, denn die Ritual-Sprache wird als Gabe Gottes angesehen.[112] Gerade anhand von *Glossolalie* lässt sich Csordas zufolge Sprache generell als leibliche Handlung erkennen, bei der das Selbst unausgereifte Gefühle und Gedanken objektiviert. Da eine semantische Bedeutung kaum zu erkennen ist, überwiegt beim Zungenreden die Bedeutung der Gesten. Demzufolge interpretiert Csordas *Glossolalie* als Teil des kulturellen Prozesses der Selbstobjektivierung.[113] Gestik, emotionaler Ausdruck und Sprache gehören zusammen, da der Mensch mit ihnen die biologische Welt transzendiert und eine soziale Welt über sie legt.[114] Die Bedeutung der Sprache transzendiert die biologisch gegebenen Sprechwerkzeuge auf zweierlei Weise: einerseits dadurch, dass es viele unterschiedliche Sprachen gibt und andererseits dadurch, dass Sprache Menschen miteinander verbindet. Sprache verweist darauf, dass der Mensch als Person und nicht lediglich als Objekt existiert. Auch dabei lässt Glossolalie die leibliche Bedeutung von Sprache deutlich werden: Obwohl ihre Zeichen ohne Bedeutung sind, gibt es verschiedene Ausformungen und ist sie für andere verständlich. Csordas zufolge wird das Sakrale in der leiblichen Erfahrung konkret, d. h. religiöse Erfahrung manifestiert sich in bestimmter, vom Habitus gegebener Form, die für andere verständlich ist. Die Unbestimmtheit von Glossolalie liegt Csordas zufolge nicht nur in ihrer semantischen Unverständlichkeit. Der Zungenredner kann sich auch passiv oder aktiv fühlen, seine Eingebungen intim oder autoritär von Gott erhalten, das Reden privat oder kollektiv praktizieren etc.[115] Dieser Unbestimmtheit werden durch den Habitus Grenzen gesetzt, die sich in diesem Fall nach Generation und gesellschaftlicher Klasse bestimmen.[116]

Wie diese Beispiele zeigen, lassen Rituale den kulturell wahrnehmenden und handelnden Leib ,anschaulich' werden. Dies gilt vor allem für Rituale, bei denen der biologische Körper offensichtlich manipuliert wird, aber auch für solche, in denen der biologische Körper in einen ,anderen Zustand', z. B. der „Besessenheit" gerät.

---

111 Csordas (1990), 25.

112 Csordas (1990), 25.

113 Csordas (1990), 26.

114 Csordas (1990), 26.

115 Csordas (1990), 26.

116 Csordas (1990), 29.

Gerade im letzteren Fall wird deutlich, dass eine kulturwissenschaftliche Analyse gewinnt, wenn sie über einen ausschließlich semiotischen Ansatz hinausgeht und neben dem Konzept der Repräsentation auch das der Erfahrung miteinbezieht. Die Rolle des Körpers erschöpft sich eben nicht darin, dass er gesellschaftliche Normen und Werte ‚verkörpert‘ – seine Leiblichkeit, wie sie z. B. in den somatischen Aufmerksamkeitsmodi deutlich wird, ist vielmehr Voraussetzung der Kultur.

VI   Resümee: Rituale ohne Körper?

Wie im Vorangegangenen deutlich geworden sein sollte, greift die Metapher des Einschreibens bzw. des Textes im Blick auf den Zusammenhang von Körper bzw. Selbst und Kultur zu kurz, weil Körper nicht lediglich *Repräsentationen* sind, sondern vor allem die existentielle *Voraussetzung* von Kultur. Dies zeigt sich gerade bei Ritualhandlungen, die Transformationsprozesse intendieren und sich dazu des Körpers bedienen. Dem Konzept der Repräsentation bzw. der Textualität sollte daher das Konzept des *Embodiment* bzw. des „In-der-Welt-seins" an die Seite gestellt werden. Darüber hinaus ist der kulturell wahrnehmende und handelnde Leib unabdingbar für das Ritual, weil durch ihn Performanz und Transformation (bzw. die damit verbundenen Normen und Werte) sowohl subjektiv als intersubjektiv erlebbar werden. Diese These mag vor dem Hintergrund der scheinbar schwindenden Rolle ‚traditioneller‘ (will sagen: „überlebter") Rituale im Zusammenhang mit der Ausprägung ‚moderner‘ Identitäten gewagt erscheinen. Strukturalistische wie poststrukturalistische Theoretiker haben sich letztlich wenig für Rituale interessiert. Catherine Bell verweist in diesem Zusammenhang auf Foucault, dessen Analyse des modernen Diskurses über Sexualität sich zwar auf mittelalterliche Beicht- und Bußrituale stützt, die Rituale selbst aber zugunsten von stärker diskursiven „Wahrheitsspielen" in den Hintergrund treten lässt.[117] Für Judith Butler wiederum stellt das Ritual lediglich einen Fall wiederholbarer sozialer Konventionen dar, die es zu dekonstruieren gilt. Bell rät jedoch zur Vorsicht bei der Verwendung analytischer Werkzeuge, die aus einer säkularen Kultur mit eher wenigen Ritualen und gering geschätzten Konventionen hervorgegangen sind.[118] Sie verweist auf noch nicht lange zurückliegende Ereignisse, die die Frage nach dem Zusammenhang von Körper und Ritual auf eindrückliche Weise aktuell scheinen lassen. Nach der Zerstörung des World Trade Centers in New York am 11. September 2001 wurde

---

117 Bell (2006), 536f.
118 Bell (2006), 540.

beispielsweise mit äußerster Sorgfalt versucht, Trauerrituale zu entwickeln, die sowohl eine unmittelbare Trauer als auch das langfristige Gedenken ermöglichen sollten. Trotzdem brachten die Angehörigen immer wieder ihren Schmerz darüber zum Ausdruck, dass sie keine Körper beerdigen konnten. Die Bergung des Leichnams ist offensichtlich der vom Herzen benötigte Beweis dafür, so Bell, dass ein geliebter Mensch *tatsächlich* tot ist. Ohne den Körper lässt sich der Prozess der emotionalen Bewältigung nur schwer abschließen, ungeachtet der Durchführung von abschließenden Übergangsriten.[119] Vor diesem Hintergrund verwenden ‚moderne' Gesellschaften viel Mühe darauf, Leichen zu bergen, seien es die von z. B. in Afghanistan oder im Irak getöteten Soldaten oder bei Flugzeugabstürzen ums Leben gekommenen Passagiere. Die betroffene Luftfahrtgesellschaft versucht dann, Angehörige so nahe wie möglich an die Absturzstelle zu bringen; über das Fernsehen verbreitete Bilder von Seekarten und auf das Wasser gestreute Blumen markieren den unsere Vorstellung offensichtlich tröstenden Ort der Toten: Dort ruht ein Körper.[120]

Literatur

Audi (1999): Robert Audi (Hg.), *The Cambridge Dictionary of Philosophy*. 2. Auflage, New York: Cambridge University Press.

Barthes (1986): Roland Barthes, *The Rustle of Language*. Berkeley: University of California Press.

Bell (1992): Catherine Bell, *Ritual Theory, Ritual Practice*. New York: Oxford University Press.

Bell (1997): Catherine Bell, *Ritual Perspectives and Dimensions*. New York: Oxford University Press.

Bell (2006): Catherine Bell, "Embodiment". In: Kreinath, Jens / Snoek, Jan / Stausberg, Michael (Hg.), *Theorizing Rituals: Topics, Approaches, Concepts*, Leiden/Boston: Brill, 533-544.

Berg / Fuchs (1993): Eberhard Berg / Martin Fuchs (Hg.), *Kultur, soziale Praxis, Text. Die Krise der ethnographischen Repräsentation*. Frankfurt am Main: Suhrkamp.

Blacking (1977): John Blacking (Hg.), *The Anthropology of the Body*. London: AP.

Böhme (2003): Gernot Böhme, „Leibliches Bewusstsein". In: Hauskeller, Michael (Hg.), Die Kunst der Wahrnehmung. Beiträge zu einer Philosophie der sinnlichen Erkenntnis. Baden-Baden: Die Graue Edition, 35-50.

Bourdieu (1979): Pierre Bourdieu, *Entwurf einer Theorie der Praxis: auf der ethnologischen Grundlage der kabylischen Gesellschaft*. Frankfurt am Main: Suhrkamp.

---

119 Bell (2006), 541f.
120 Bell (2006), 542.

Bourdieu (1987): Pierre Bourdieu, *Sozialer Sinn: Kritik der theoretischen Vernunft*. Frankfurt am Main: Suhrkamp.

Bourdieu / Wacquant (1996): Pierre Bourdieu / Loic Wacquant, *Reflexive Anthropologie*. Frankfurt am Main: Suhrkamp.

Bowie (2006): Fiona Bowie, *The Anthropology of Religion. An Introduction*. 2. Auflage, Oxford: Blackwell Publishing.

Brosius / Michaels / Schrode (2013): Christiane Brosius / Axel Michaels / Paula Schrode, „Ritualforschung heute. Ein Überblick". In: Dies. (Hg.), *Ritual und Ritualdynamik. Schlüsselbegriffe, Theorien, Diskussionen*. UTB, 9-24.

Böhme (2003): Gernot Böhme, „Leibliches Bewusstsein". In: Hauskeller, Michael (Hg.), Die Kunst der Wahrnehmung. Beiträge zu einer Philosophie der sinnlichen Erkenntnis. Baden-Baden: Die Graue Edition, 35-50.

Butler (1997): Judith Butler, Körper von Gewicht. Die diskursiven Grenzen des Geschlechts. Frankfurt am Main: Suhrkamp.

Clifford / Marcus (1986): James Clifford / George E. Marcus (Hg.), *Writing culture: The Poetics and Politics of Ethnography*. Berkeley: University of California Press.

Connerton (1989): Paul Connerton, "Bodily Practices". In: Ders., *How Societies Remember*. Cambridge: Cambridge University Press, 72-104.

Csordas (1990): Thomas J. Csordas, "Embodiment as a Paradigm for Anthropology". In: *Ethos* 18/1, 5-47.

Csordas (1993): Thomas J. Csordas, *Somatic Modes of Attention. In Cultural Anthropology* 8/2, 135-156.

Csordas (1994a), Thomas J. Csordas, "The Body as Representation and Being-in-the-World". In: Ders. (Hg.), *Embodiment and Experience. The Existential Ground of Culture and Self*. Cambridge: Cambridge University Press, 1-24.

Csordas (1994b), Thomas J. Csordas, "Words from the Holy People: A Case Study in in Cultural Phenomenology". In: Ders. (Hg.), *Embodiment and Experience. The Existential Ground of Culture and Self*. Cambridge: Cambridge University Press, 269-290.

Csordas (1999), Thomas J. Csordas, "The Body's Career in Anthropology". In: Moore, Henrietta L. (Hg.), *Anthropological Theory Today*. Cambridge: Polity Press, 172-205.

Csordas (2011), Thomas J. Csordas, "Cultural Phenomenology. Embodiment: Agency, Sexual Difference, and Illness". In: Mascia-Lees, Frances E. (Hg.), *A Companion to the Anthropology of the Body and Embodiment*. Malden, Mass.: Wiley-Blackwell, 137-156.

Darnton (1984), Robert Darnton, The Great Cat Massacre and Other Episodes in French Cultural History. New York: Basic Books.

Douglas (1988): Mary Douglas, *Reinheit und Gefährdung. Eine Studie zu Vorstellungen von Verunreinigung und Tabu*. Frankfurt am Main: Suhrkamp. (1966)

Douglas (1986): Mary Douglas, *Ritual, Tabu und Körpersymbolik. Sozialanthropologische Studien in Industriegesellschaft und Stammeskultur*. Frankfurt am Main: Fischer. (1970)

Durkheim (1981): Émile Durkheim, *Die elementaren Formen des religiösen Lebens*. Frankfurt am Main: Suhrkamp. (1912)

Fabian (2001): Johannes Fabian, Im Tropenfieber. Wissenschaft und Wahn in der Erforschung Zentralafrikas. München: Beck.

Farnell (1999): Brenda Farnell, "Moving Bodies, Acting Selves". In: *Annual Review of Anthropology 28*, 341-373.

Foucault (1973): Michel Foucault, *Die Geburt der Klinik. Eine Archäologie des ärztlichen Blicks*. München: Carl Hanser. (1963)

Foucault (2002): Michel Foucault, „Nietzsche, die Genealogie, die Historie". In: Foucault, Michel, *Schriften in vier Bänden, Bd. 2*. Frankfurt am Main: Suhrkamp, 166-191. (1971)

Foucault (2008a): Michel Foucault, „Überwachen und Strafen. Die Geburt des Gefängnisses". In: Ders.: *Hauptwerke*. Frankfurt am Main: Suhrkamp, 166-191. (1975)

Foucault (2008b): Michel Foucault, „Der Wille zum Wissen. (Sexualität undWahrheit 1)". In: Ders.: *Hauptwerke*. Frankfurt am Main: Suhrkamp, 166-191. (1976)

Foucault (2008c): Michel Foucault, „Der Gebrauch der Lüste. (Sexualität undWahrheit 2)". In: Ders.: *Hauptwerke*. Frankfurt am Main: Suhrkamp, 166-191. (1984)

Foucault (2008d): Michel Foucault: „Die Sorge um sich. (Sexualität undWahrheit 3)". In: Ders.: *Hauptwerke*. Frankfurt am Main: Suhrkamp, 166-191. (1984)

Frazer (1989): James George Frazer, *Der Goldene Zweig. Das Geheimnis von Glauben und Sitten der Völker*. Reinbek bei Hamburg: Rowohlt. (1890)

Geertz (1987): Clifford Geertz, „Dichte Beschreibung. Bemerkungen zu einer deutenden Theorie von Kultur". In: Geertz, Clifford: *Dichte Beschreibung. Beiträge zum Verstehen kultureller Systeme*. Frankfurt am Main: Suhrkamp, 7-43. (1973)

Gell (1993): Alfred Gell, *Wrapping in Images: Tattooing in Polynesia*. Oxford: Clarendon, Oxford University Press.

Gennep (1986): Arnold van Gennep, *Übergangsriten*. Frankfurt am Main / New York: Campus. (1909)

Goffman (1971): Erving Goffman, *Interaktionsrituale. Über Verhalten in direkter Kommunikation*. (1967)

Hall (1968): Edward T. Hall, "Proxemics". In: *Current Anthropology 9/2-3*, 83-95.

Hertz (1973): Robert Hertz, "The Pre-eminence of the Right Hand. A Study in Religious Polarity". In: Rodney Needham (Hg.), *Right and Left. Essays on Dual Symbolic Classification*. Chicago: Chicago University Press, 3-31.

Hirschauer, (2004): Stefan Hirschauer, "Praktiken und ihre Körper. Über materielle Partizipanden des Tuns." In: Hörning, Karl H. / Reuter, Julia (Hg.), Doing Culture. Bielefeld: Transcript, 73-91.

Jackson (1983): Michael Jackson, "Knowledge of the Body". In: *Man* (N.S.) 18, 327-45.

Jackson (1989): Michael Jackson, *Paths Toward a Clearing. Radical Empiricism and Ethnographic Inquiry*. Bloomington and Indianapolis: Indiana University Press.

Kafka (2007): Franz Kafka, *In der Strafkolonie. In: Die Erzählungen und andere ausgewählte Prosa*. Frankfurt am Main: Fischer, 164-198. (1914)

Kasten (2006): Erich Kasten, *Body-Modification. Psychologische und medizinische Aspekte von Piercing, Tattoo, Selbstverletzung und anderen Körperveränderungen*. München, Basel: E. Reinhardt.

Krais / Gebauer (2002): Beate Krais / Gunter Gebauer, *Habitus*. Bielefeld: Transcript.

Lévi-Strauss (1967b): Claude Lévi-Strauss, „Die Zweiteilung der Darstellung in der Kunst Asiens und Amerikas". In: *Strukturale Anthropologie 1*. Frankfurt am Main: Suhrkamp, 267-291.

Lock (1993): Margaret Lock, Cultivating the Body: Anthropology and Epistemologies of Bodily Practice and Knowledge". In: *Annual Review of Anthropology 22*, 133-155.

Mauss (1989a): Marcel Mauss, „Die Techniken des Körpers". In: *Soziologie und Anthropologie 2*. Frankfurt am Main: Fischer, 199-220. (1935)

Mauss (1989b): Marcel Mauss, „Eine Kategorie des menschlichen Geistes: Der Begriff der Person und des ‚Ich'". In: *Soziologie und Anthropologie 2*. Frankfurt am Main: Fischer, 223-252. (1938)

Merleau-Ponty (1966): Maurice Merleau-Ponty, *Phänomenologie der Wahrnehmung*. Berlin: de Gruyter. (1945)

Mesenhöller (2006): Peter Mesenhöller, „O Le Tatau Samoa. Samoanische Tatauierungen". In: *Journal Ethnologie: Hautzeichen – Körperbilder*. Internet-Journal des Museum der Weltkulturen, Frankfurt am Main. <http://www.journal-ethnologie.de/Deutsch/ Schwerpunktthemen/Schwerpunktthemen_2006/Hautzeichen_-_Koerperbilder/Editorial/ index.phtml>

Müller (2010): Klaus E. Müller, *Die Siedlungsgemeinschaft. Grundriss der essentialistischen Ethnologie*. Göttingen: V&R unipress.

Nietzsche (1887): Friedrich Nietzsche, Zur Genealogie der Moral. Leipzig: Verlag C.G. Neumann.

Platz (2006): Teresa Platz, *Anthropologie des Körpers. Vom Körper als Objekt zum Leib als Subjekt von Kultur*. (Berliner Beiträge zur Ethnologie 10). Berlin: Weißensee Verlag.

Polit (2013): Karin Polit, „Verkörperung". In: Brosius, Christiane / Michaels, Axel / Schrode, Paula (Hg.), *Ritual und Ritualdynamik. Schlüsselbegriffe, Theorien, Diskussionen*. Göttingen: Vandenhoeck & Ruprecht, 215-221.

Raabe (2006): Eva Raabe, „Die Verwandtschaft mit dem Krokodil. Initiation und Narbentatauierung bei den Iatmul in Papua-Neuguinea". In: *Journal Ethnologie: Hautzeichen – Körperbilder*. Internet-Journal des Museum der Weltkulturen, Frankfurt am Main. <http:// www.journal-ethnologie.de/Deutsch/Schwerpunktthemen/Schwerpunktthemen_2006/ Hautzeichen_-_Koerperbilder/Editorial/index.phtml>Journal Ethnologie.de, Hautzeichen – Körperbilder.

Rein (2006): Anette Rein, „Von Reinheit, Zähnen und anderen Leidenschaften auf Bali". In: *Journal Ethnologie: Hautzeichen – Körperbilder*. Internet-Journal des Museum der Weltkulturen, Frankfurt am Main. <http://www.journal-ethnologie.de/Deutsch/ Schwerpunktthemen/Schwerpunktthemen_2006/Hautzeichen_-_Koerperbilder/Editorial/ index.phtml>

Schildkrout (2004): Enid Schildkrout, "Inscribing the body". In: *Annual Review of Anthropology 33*, 319-344.

Schlehe (1996): Judith Schlehe, „Die Leibhaftigkeit in der ethnologischen Feldforschung". In: Historische Anthropologie 4/3, 451-460.

Smith (1967): William Robertson Smith, *Die Religion der Semiten*. Darmstadt. (1889)

Spittler (2001): Gerd Spittler, „Teilnehmende Beobachtung als Dichte Teilnahme". In: *Zeitschrift für Ethnologie 126*, 1-25.

Stirn (2006): Aglaja Stirn, „Vom Pin-Up-Girl zur Kunst. Zum geschichtlichen Hintergrund der aktuellen Tattoo- und Piercingbewegung bei Jugendlichen". In: *Journal Ethnologie: Hautzeichen – Körperbilder*. Internet-Journal des Museum der Weltkulturen, Frankfurt am Main. <http://www.journal-ethnologie.de/Deutsch/Schwerpunktthemen/ Schwerpunktthemen_2006/Hautzeichen_-_Koerperbilder/Editorial/index.phtml>

Stoller (1989): Paul Stoller, *The Taste of Ethnographic Things. The Senses in Anthroplogy*. Philadelphia: The University of Pennsylvania Press.

Streck (2000): Bernhard Streck, „Körperveränderung". In: Ders. (Hg.), *Wörterbuch der Ethnologie*. 2. Auflage, Wuppertal: Peter Hammer, 129-133.

Taylor (2003): Diane Taylor, *The Archive and the Repertoire. Performing Cultural Memory in the Americas*. Durham: Duke University Press.

Turner (1997): Bryan Turner, „The Body in Western Society: Social Theory and Its Perspectives." In: Coakley, Sarah (Hg.), Religion and the Body. Cambridge: Cambridge University Press, 15-41.
Turner (2011): Terence Turner, "Bodiliness. The Body beyond the Body: Spatial, Material and Spiritual Dimensions of Bodiliness". In: Mascia-Lees, Frances E. (Hg.), *A Companion to the Anthropology of the Body and Embodiment*. Malden, Mass.: Wiley-Blackwell, 102-118.

Der Körper als Palimpsest: Erinnerungstopographien zwischen Schrift und Leiblichkeit

Thomas Reinhardt

I   Aufwachen

[W]enn ich mitten in der Nacht erwachte, wußte ich nicht, wo ich mich befand, ja, im ersten Augenblick nicht einmal, wer ich war: […] dann aber kam mir die Erinnerung […] gleichsam von oben her zu Hilfe, um mich aus dem Nichts zu ziehen, aus dem ich mir selbst nicht hätte heraushelfen können; in einer Sekunde durchlief ich Jahrhunderte der Zivilisation, und aus vagen Bildern von Petroleumlampen und Hemden mit offenen Kragen setzte sich allmählich mein Ich in seinen originalen Zügen wieder von neuem zusammen.

[…] Wenn ich […] in dieser Weise erwachte und mein Geist geschäftig und erfolglos zu ermitteln versuchte, wo ich war, kreiste in der Finsternis alles um mich her, die Dinge, die Länder, die Jahre. Noch zu steif, um sich zu rühren, suchte mein Körper je nach Art seiner Ermüdung sich die Lage seiner Glieder bewusst zu machen, um daraus die Richtung der Wand, die Stellung der Möbel abzuleiten und die Behausung, in der er sich befand, zu rekonstruieren und zu benennen. Sein Gedächtnis, das Gedächtnis seiner Seiten, seiner Knie und Schultern bot ihm nacheinander eine Reihe von Zimmern, in denen er schon geschlafen hatte, an, während rings um ihn die unsichtbaren Wände im Dunkel kreisten und ihren Platz je nach der Form des vorgestellten Raumes wechselten. Und bevor mein Denken, das an der Schwelle der Zeiten und Formen zögerte, die Wohnung durch ein Vergleichen der Umstände eindeutig festgestellt hatte, erinnerte er – mein Körper – sich von einem jeden an die Art des Bettes, die Lage der Türen, die Fensteröffnungen, das Vorhandensein eines Flurs, gleichzeitig mit dem Gedanken, den ich beim Einschlummern gehabt hatte und beim Erwachen wiederfand.[1]

Wenn wir so denken, wenn wir erkennen, wenn wir erinnern – was denkt, erkennt und erinnert da? Ein Geist? Ein Bewusstsein? Ein Ich? Ein Selbst? Und was wäre ein solches Ich? Ein solches Selbst? Ist es Masse? Materie? Struktur? Prozess?

---

1   Proust (1964), 12f.

Descartes hat uns im 17. Jahrhundert den Dualismus von Geist und Körper, Mens und Physis, *res cogitans* und *res extensa* eingebrockt, und bis heute sind wir nicht wesentlich darüber hinausgekommen. Nahezu alle Versuche, die große kartesianische Trennung zu überbrücken, haben am Ende nur neue Klüfte aufgetan und zu ihrer Re-Affirmation unter neuen Überschriften geführt. Wir mussten lernen, dass wir „Leib sind" und „Körper haben",[2] doch wurde damit lediglich der Körper wieder zum *corpus,* zum unbelebten, mechanischen Objekt, das den Raum der *res extensa* bewohnt und mit unserer Erfahrung der Welt wenig oder gar nichts zu tun hat.[3]

Die Phänomenologie des vergangenen Jahrhunderts hat versucht, den Dualismus aufzubrechen, indem sie unsere Welterfahrung über die Doppelfigur einer leiblich basierten Wahrnehmung auf ein sensorisches Fundament stellte. Der Körper wird hier gleichsam zur „Außenseite des Leibes", das Bewusstsein firmiert als „leiblich verankerte Zugangsweise zur Welt".[4] Wirklich gewonnen ist auch damit freilich wenig. Das Paradox des „lebendigen Körpers" lässt sich über terminologische Spitzfindigkeiten nicht lösen. Immer noch bliebe zu definieren, wo unser Selbst seinen Ort hat – und immer noch verfallen wir dabei wahlweise in einen kruden Naturalismus oder sein konstruktivistisches Gegenteil. Wir essentialisieren den Körper oder er entschlüpft uns vollständig in das Sprechen über Diskurse und ihre sprachlichen und medialen Regulierungen.

Vielleicht ist das aber eine sehr deutsche Grübelei. Im Englischen oder Französischen hat die Differenz von Leib und Körper keine Entsprechung, und (entsprechend) gestaltet sich auch die Problemlage anders. Foucault etwa beschreibt den Körper (*corps*) zunächst unbefangen als „gnadenlose Topie", als Bild, „das der Spiegel mir aufzwingt", als „Ort, von dem es kein Entrinnen gibt, an den ich verdammt bin".[5] Da ist sie wieder, die kartesianische Spaltung. Und sie scheint (!) unentrinnbar.

Doch wird sie im nächsten Augenblick aus einer anderen Richtung umso heftiger attackiert. Denn was, wenn nicht Körper wäre denn dieses Selbst? Eine körperlose, reine Seele? Ein ortloser Ort? Ein „glatter, kastrierter Körper, rund wie ein Stück Seife"?[6] Wie wäre es, fragt Foucault, wenn sich unsere Utopien nicht *gegen* den (topischen) Körper richteten? Wenn sie *nicht* den unvollkommenen Versuch

---

2   Plessner (1981), 303f.

3   Das homerische Griechisch kannte offenbar keinen Ausdruck, der unserem Wort ‚Körper' entsprochen hätte. Stattdessen spricht die Ilias von einzelnen Gliedern (*gyia, melea),* die sich zu einem pluralistischen Leib formieren, ohne dass sich daraus auf das Bewusstsein eines einheitlichen Organismus schließen ließe (vgl. Rappe [1995], 45).

4   Eglinger (2007), 14.

5   Foucault (2005), 25f.

6   Foucault (2005), 28.

bildeten, den Körper zum Verschwinden zu bringen wie eine Kerzenflamme, die man ausbläst? Wenn sie – im Gegenteil – aus ihm hervorgingen? Wenn der Körper der „große utopische Akteur" wäre, der mit „geheimen Mächten und unsichtbaren Kräften" „in Kommunikation" tritt?[7]

Der Körper wird Foucault so zum „Nullpunkt der Welt", zum Ort, „an dem Wege und Räume sich kreuzen", der selbst aber nirgendwo ist.[8] Elegant wird hier die Frage nach einer prädiskursiven Materialität des Körpers suspendiert und er wird geöffnet für kulturelle Zurichtungen und Einschreibungen aller Art.

II   Einschreiben?

Einschreibungen? Wer schreibt hier was in was ein? Die Antwort ist überraschend einfach: Das ist letztlich ganz egal. Wenn der Körper sich nie *jenseits von Diskursen* befindet, kann er zu keinem Zeitpunkt ein unbeschriebenes Blatt sein. Und ganz gleichgültig, ob sich Machtstrukturen oder Geschlechterrollen, vermeintliche oder echte Natürlichkeit oder Kultur in ihn einschreiben, jeder Schreibakt ist zugleich *Überschreibakt* und fügt dem Palimpsest des Körpers lediglich eine neue Bedeutungsschicht hinzu. Das klingt nach Beschränkung, ist aber in Wirklichkeit eine Befreiung. Der utopischen *tabula rasa* eines jungfräulichen Körpers brauchen wir damit nicht mehr nachzujagen. Stattdessen können wir unsere Energie auf die Betrachtung des Akts der Einschreibung selbst legen (im Sinne einer bestimmten Form der Performanz). Oder aber auf die Lektüre bereits (wie auch immer) eingeschriebener Inhalte.

Letzteres möchte ich nun versuchen, indem ich jenen Aspekt leiblicher Erkenntnisfähigkeit etwas näher betrachte, der sich in der Funktion des Gedächtnisses und des Erinnerns niederschlägt. Meist ist in diesen Zusammenhängen metaphorisch die Rede davon, dass sich bestimmte Inhalte in eine Gedächtnissubstanz „einschreiben". Mich interessiert jedoch eher, wie, um in der Metapher zu bleiben, die entsprechenden Inhalte wieder aus dem Gedächtnis herauskommen und dabei den Transfer vom Unbewussten ins Bewusstsein vollziehen. Leitmetapher meiner Ausführungen wird entsprechend nicht das Schreiben sein, sondern das Lesen. Wie *lesen* wir unser Gedächtnis? Welche Zugriffsmöglichkeiten bestehen auf die unbewussten Inhalte? Schließlich: Ist es denkbar, dass der Körper sich wenigstens gelegentlich gleichsam selbst liest? Dass sinnliche Erfahrungen ihre ganz eigenen

---

7   Foucault (2005), 31.
8   Foucault (2005), 34.

Bedeutungskonfigurationen zeichnen? Ohne regulierende Kontrolle durch ein zensierendes Bewusstsein?

III   Körper lesen!

Im Zentrum der folgenden Überlegungen steht also der *Körper* als Gegenstand einer deutenden Lektüre. Nicht der Körper als kartesianischer Antagonist zur *res cogitans,* nicht der Körper als „Außenseite des Leibes", sondern der Körper in einem quasi alltagssprachlichen Verständnis als Träger einer so ursprungs- wie autorlosen, „palimpsestuösen"[9] Hypertextualität.

Dass Körper Bedeutungsträger sind, braucht kaum ausgeführt werden. Sie erzählen von Geschlecht und Alter, sozialem Status und Gesundheitszustand, Gruppenzugehörigkeiten, Ernährungsgewohnheiten und psychischen Problemen. Auch *lügen* können sie – und erfüllen damit das wichtigste Kriterium für Phänomene, die in den Gegenstandsbereich der Semiotik fallen.[10]

Die Bedeutungsvarianz, die sich aus dieser Möglichkeit ergibt, erfordert Kontrollmechanismen. Mit anderen Worten: eine Hermeneutik, die die Rahmenbedingungen festlegt, nach denen Körper zu *lesen* sind, nach denen wir die Botschaften des Körpers interpretieren. Ein entsprechendes Instrumentarium hat die Medizin entwickelt, aber auch Theologie und Philosophie haben sich lange in physischer (respektive: physiognomischer) Hermeneutik geübt. „Physiognomik", schreibt Johann Caspar Lavater 1772, ist „die Wissenschaft, den Charakter (nicht die zufälligen Schicksale) des Menschen im weitläufigsten Verstande aus seinem Äußerlichen zu erkennen".[11] Und so freut sich der Zürcher Pfarrer schon bald, seinen Lesern mitteilen zu können, dass er den „Schlüssel zur Entzifferung der Dummheit" gefunden hat.[12] Das liest sich dann beispielsweise so:

> Jedes Gesicht ist dumm, dessen Unterteil, von der Nase an gerechnet, sich durch die Mittellinie des Mundes in zwei gleiche Teile teilet. (…) Jedes Gesicht ist von Natur dumm, dessen Stirne, mit einem weich-anliegenden Maße gemessen, beträchtlich kürzer ist, als die Nase, von dem Ende der Stirne auf dieselbe Weise gemessen, wenn auch das perpendikuläre Maß dieselbe Länge hätte. (…) Jedes Gesicht ist dumm,

---

9   Genette (1993), 533.

10   Eco (1991), 26.

11   Lavater (2013), 2.

12   Lavater (2013), 32.

was vom Augenwinkel an, bis mitten an den Nasenflügel, kürzer ist, als von dort
zur Mundspitze.[13]

Immerhin wegen Dummheit braucht sich keine Sorgen machen, wer ein „lang
hervorstehendes, nadelartiges, oder stark krauses, wildes, rohes, auf einem braunen Flecken gewurzeltes Haar an Kinn oder Halse" trägt.[14] Allerdings dürfte ein
solcher Mensch für derartige Sorgen ohnehin wenig empfänglich sein, spreche das
Merkmal doch „sehr entscheidend für großmächtige Voluptuosität, die selten ohne
großmächtigen Leichtsinn ist".[15] Und so fort.

Was sich heute als bizarre Petitesse liest, war seinerzeit durchaus ein Phänomen,
das das Interesse eines breiten Publikums zu fesseln verstand. Zu Lavaters Zuträgern
gehörten nicht nur religiöse Eiferer, sondern auch (wenigstens anfänglich) Männer
wie Goethe und Herder.[16] Es handelt sich also mitnichten um das wunderliche
Einzelprojekt eines kauzigen Schweizer Theologen, sondern um den durchaus
ernstzunehmenden (und ernstgenommenen) Versuch, der impliziten Aufforderung
gerecht zu werden, die aus dem biblischen „Gott schuf den Menschen sich zum
Bilde" (1. Mose 1,27) spricht. Wenn nämlich Gott den Menschen *als Bild* erschaffen
hat, zählt es selbstverständlich zu des Menschen edelsten Aufgaben, dieses Bild zu
entziffern und zu deuten.

Nichts anderes tun wir beständig im Alltag, wenn wir Andere nach ihrem körperlichen Erscheinungsbild beurteilen. Wir lesen – und wir werten. Was verraten
die Piercings im Gesicht? Was der tätowierte Steiß? Was die extrem geweiteten
Ohrlöcher oder die botoxstarre Stirn? Was sagen gestraffte Wangen, künstliche
Brüste oder Bauchplastiken über ihre Trägerinnen? Was der aufgepumpte Körper
des Bodybuilders? Mehrfarbige Frisuren, Blumenkohlohren, gebrochene Nasen, ein
Überbiss? Erste Urteile über Fremde – wie zutreffend oder nicht sie sein mögen –
verdanken sich fast immer einer kursorischen ‚Lektüre' ihrer äußeren Erscheinung.

---

13 Lavater (2013), 72-74.

14 Lavater (2013), 81.

15 Lavater (2013), 81.

16 Antje Stache weist zu Recht auf die zahlreichen Gemeinsamkeiten zwischen Lavaters
    Physiognomik und Goethes Morphologie hin (Stache [2010], 67). Derselbe Goethe, der
    1774 noch das „Lied des physiognomischen Zeichners" sang, sollte sich freilich schon
    fünf Jahre später spöttisch über „Gottes Spürhund" äußern (Goethe [1998a], 113). War
    Goethes Mitarbeit an den beiden ersten Bänden der *Physiognomischen Fragmente* noch
    sehr intensiv, nahm sie in Band 3 bereits deutlich ab und versiegte schließlich völlig.
    Irgendwann zwischen der Veröffentlichung der Bände 3 und 4 der *Fragmente* scheint
    auch der endgültige Bruch zwischen Goethe und Lavater erfolgt zu sein (vgl. Schögl
    [1997], 23). Vgl. dazu auch Goethes Ausführungen im dritten Teil von *Dichtung und
    Wahrheit* (Goethe [1998b], 15ff.).

Darüber aber ist das meiste längst geschrieben und braucht hier nicht wiederholt zu werden. Die weit spannendere Frage lautet: Wie ist es um die Wahrnehmung des *eigenen* Körpers bestellt? Lesen wir ihn anders als die Körper unserer Mitmenschen? Kann unser Körper uns selbst täuschen, wie wir offensichtlich Andere mit ihm täuschen können? Oder fehlt die der Täuschung unverzichtbare Differenz zwischen Täuscher und Getäuschtem?

IV Metaphern

Im Folgenden soll versucht werden, diese Fragen anhand einiger Überlegungen zu Problemen der Erinnerung und des Gedächtnisses zu beantworten. Das erste Problem, das sich dabei stellt, ist sprachlicher Natur: Gedächtnis und Erinnern entziehen sich der direkten Beschreibung. Sie scheinen sich nicht anders denken zu lassen als in Metaphern, die mal diesen, mal jenen Aspekt stärker gewichten. Bei aller Vielfalt der Beschreibungsversuche kristallisiert sich am Ende eine sehr überschaubare Zahl grundlegender Ideen heraus. Für Harald Weinrich sind es nicht mehr als zwei, was er der „Doppelheit des Phänomens Memoria" (als Gedächtnis und als Erinnerung) zuschreibt.[17] Beiden gemeinsam ist eine erstaunliche Persistenz.

Von Platons Taubenschlag[18] zu Augustinus' Schatzkammer,[19] von Edmund Spensers Bibliothek[20] zu Hegels „nächtlichem Schacht",[21] von Pierre Noras *„lieux de mémoire"*[22] zur poststrukturalistischen Vorstellung vom literarischen Text als „intertextuellem Gedächtnisraum"[23] – die von Weinrich als „Magazinmetaphern"[24] bezeichneten Veranschaulichungen lassen sich in allen historischen Epochen nachweisen. Was sie teilen, ist die Vorstellung einer Lokalisierbarkeit von Erinnerungen im Raum. Das Magazin, so Weinrichs Folgerung, ist daher vor allem der rhetorischen Mnemotechnik zuzuordnen. Es erlaubt das Wiederauffinden abgelegter Gedächtnisinhalte und ihre Reproduktion in einer vorab festgelegten Reihenfolge, zum Beispiel in der politischen Rede. Wie der Ort dieser topologi-

---

17 Weinrich (1976), 294.

18 Platon (1991), 197 b-e.

19 Augustinus (1888), X, 8.

20 Spenser (1596), 209, 59.

21 Hegel (2012), § 453.

22 Nora (1998).

23 Löbbermann (2002), 52.

24 Weinrich (1976), 291-294.

schen Erinnerung konkret aussieht, ist dabei unerheblich. Dachböden, Theater, Bibliotheken, Volieren, Häuser, Tempel, Schächte, Geldbeutel, Labyrinthe, Türme – die Gedächtnismetaphorik der vergangenen zweieinhalb Jahrtausende war nicht wählerisch. Es scheint, dass nahezu alles, was über eine räumliche Ausdehnung verfügt und als Aufbewahrungsort für *Etwas* dienen kann, bei irgendjemandem auch als Metapher Verwendung fand.

Gedächtnis*inhalte* sind freilich nur *ein* Aspekt der Erinnerung. Mindestens gleichberechtigt zur Seite steht ihnen das *Erinnern*, das aktive, prozesshafte, performative Aktualisieren dieser Inhalte. Dabei bleibt durchaus unklar, wie stark der eigentlich *kreative* Anteil an diesen Aktualisierungen ist. Jedenfalls aber wird die eher statische Auffassung vom Gedächtnis als Ort hier deutlich dynamischer gedacht. Schon Platons Wachstafel beschreibt die Gedächtnisleistung als eine des Einprägens, Abdrückens und Siegelns in eine Wachsmasse von individuell unterschiedlicher Qualität.[25] John Lockes mit Erfahrungen, Wahrnehmungen und Ideen zu beschreibende *Tabula rasa* zielt in eine ähnliche Richtung,[26] ebenso wie Freuds Wunderblock.[27] Weinrich hat diesen Veranschaulichungstypus unter dem Begriff „Tafelmetapher" zusammengefasst.[28] Stärker als in den Magazinmetaphern kommt in ihnen dem Aspekt des Lesens eine zentrale Rolle beim Erinnern zu. Die Erinnerungen selbst liegen quasi in schriftlicher Form vor. Nicht immer als vollständige Texte. Sie mögen überschrieben sein oder gelöscht, die Tilgungen sind aber nie so vollständig, dass es nicht möglich wäre, frühere Bedeutungsschichten wenigstens teilweise zu rekonstruieren.

Aleida Assmann hat Weinrichs Zweiteilung um einen dritten Typ ergänzt. Sie kontrastiert zunächst die Zweidimensionalität der Tafel mit der Dreidimensionalität des Magazins, um ihre Erinnerungstheorie dann mit „zeitlichen Gedächtnismetaphern" gleichsam in die vierte Dimension zu heben.[29] Augustinus' Rede vom Gedächtnis als „Magen der Seele"[30] zählt sie zu diesem Typus, aber auch Nietzsches Überlegungen zur Vergesslichkeit als „aktives, im strengen Sinne positives Hemmungsvermögen", das für den Geist leistet, was die Verdauung für die Ernährung tut.[31]

---

25 Platon (1991), 191c-d.

26 Locke (1690).

27 Freud (1924).

28 Weinrich (1976), 291.

29 Assmann (1999), 167.

30 Augustinus (1888) X, 14.

31 Nietzsche (1999), 291. Auch Umberto Ecos Ausführungen zum Vergessen „on account not of defect but of excess" (Eco [1988], 259) ließen sich dieser Kategorie zurechnen.

Ungeachtet aller Fragen nach der Dimensionalität von Gedächtnismodellen handelt es sich bei dem bisher Aufgezählten um sehr spezielle Formen metaphorischer Verdichtung: Wird gewöhnlich in einer Metapher Nicht-Zusammengehöriges zusammengebracht, um eine überraschende Wirkung zu erzielen und dadurch einen speziellen Aspekt des metaphorisierten Gegenstands hervorzuheben, konstituiert sich der Gegenstand der Gedächtnismetaphorik überhaupt erst durch die Metaphorisierung. „Wir können", schreibt Weinrich, „einen Gegenstand wie die Memoria nicht ohne Metaphern denken".[32] Wir haben es also nicht mit zwei sprachlich eindeutig bestimmten, konkreten Gegenstandsbereichen zu tun, die sich als Ergebnis einer metaphorischen Kombinatorik wechselseitig erhellen, sondern mit einem lexikalisch nicht gefassten Phänomen, das erst in der Metapher überhaupt zur Sprache findet. Gedächtnis und Erinnern scheinen mithin keinerlei ‚vor-metaphorische' Existenz zu besitzen. Jedenfalls keine, über die sich sprechen (oder auch nur nachdenken) ließe.

Kritik an diesen Gedächtnismodellen ist aus unterschiedlichen Richtungen möglich. Zwei dieser Kritiken möchte ich im Folgenden näher betrachten und versuchen, sie in eine Theorie des Gedächtnisses als (auch) einer Funktion des Körpers zu überführen.

V   Soma und Medium

Der erste Einwand ist gleichermaßen phänomenologischer wie epistemologischer Natur. Fast alle Gedächtnismetaphern setzen die erwähnte Trennung von Körper und Bewusstsein voraus. Das zu Erinnernde wird als etwas dem Bewusstsein Äußerliches gedacht, das diesem in irgendeiner Weise einverleibt werden muss. Das funktioniert über diverse Zugänge zum menschlichen Sinnesapparat.

Die Sinne – und damit der Körper – fungieren dabei lediglich als passiv vermittelnde Instanz. So überrascht es nicht, dass die Geschichte der Gedächtnismetaphern in weiten Teilen zugleich eine Geschichte der Medien ist. Es ist aber – und das wäre der zweite Einwand – letztlich so gut wie ausschließlich eine Mediengeschichte der *Schrift*, eine Geschichte *externer* Speicherung. Inhalte werden abgelegt und archiviert und bei Bedarf hervorgeholt. So funktionieren in der Tat das Buch und die Bibliothek, die Schatzkammer und das Monument. Mit Abstrichen auch Wachstafel und Wunderblock. Die hermeneutischen Anstrengungen bei der Entzifferung unterscheiden sich allenfalls graduell.

---

32 Weinrich (1976), 294.

Nicht auf diese Weise funktioniert hingegen das von Proust beschriebene schockhafte somatische Erinnern, das Marcel beim Verspeisen einer in Tee getauchten Madeleine ereilt. Die Stelle ist bekannt:

> [B]edrückt durch den trüben Tag und die Aussicht auf den traurigen folgenden [führte ich] einen Löffel Tee mit dem aufgeweichten kleinen Stück Madeleine darin an die Lippen. In der Sekunde nun, als dieser mit dem Kuchengeschmack gemischte Schluck Tee meinen Gaumen berührte, zuckte ich zusammen und war wie gebannt durch etwas ungewöhnliches, das sich in mir vollzog. Ein unerhörtes Glücksgefühl, das ganz für sich allein bestand und dessen Grund mir unbekannt blieb, hatte mich durchströmt. Mit einem Schlage waren mir die Wechselfälle des Lebens gleichgültig, seine Katastrophen zu harmlosen Mißgeschicken, seine Kürze zu einem bloßen Trug unsrer Sinne geworden; es vollzog sich damit in mir, was sonst die Liebe vermag, gleichzeitig aber fühlte ich mich von einer köstlichen Substanz erfüllt; oder war diese Substanz vielmehr nicht in mir, sondern ich war sie selbst. Ich hatte aufgehört, mich mittelmäßig, zufallsbedingt, sterblich zu fühlen.[33]

Constance Classen, David Howes und Anthony Synnott haben den Geruchssinn (und mit ihm den Geschmack) vor einiger Zeit zum „Sinn der Postmoderne" erklärt.[34] Wesentliches Merkmal dieser Postmodernität sind ihnen dabei die Ablösung des olfaktorischen Signifikanten von einem natürlichen Signifikat und die gleichzeitige De- und Re-Odorisierung der westlichen Gesellschaften. Duftlandschaften seien heute olfaktorische Inszenierungen ohne Referenten, Evokationen des Abwesenden, Präsenzen des Absenten.[35]

VI   Total Recall

Das genau ist es, was Prousts Marcel passiert, als ihn der Geschmack des in Lindenblütentee getauchten Gebäcks so unvermittelt wie vollständig in eine zunächst ganz und gar ungreifbare Kindheit zurückwirft. Die Erinnerung ist total. Sie ist Re-Präsentation im Wortsinne, eine „mystische Apokatastasis"[36], in der die Jahre zwischen ursprünglicher Erfahrung und Wiedererleben ausgelöscht sind und dem Bewusstsein nichts anderes zu tun bleibt, als die Differenz zwischen gefühlter und gewusster Gegenwart verwundert *festzustellen*. Marcel nimmt einen weiteren

---

33 Proust (1964), 63f.
34 Classen, Howes & Synnott (1994), 205.
35 Classen, Howes & Synnott (1994), 205.
36 Assmann (1999),163.

Schluck, dann noch einen. Und noch einen. Die Wirkung lässt nach. Er bemüht sich, sich des ersten Löffels voll Tee zu erinnern und des Gefühls, das sich dabei einstellte. Und ebenso plötzlich wie beim ersten schockhaften Erinnern wird irgendwann das Bild eines lange vergangenen Sonntagmorgens im Haus seiner Tante Léonie in Combray an die Oberfläche seines Bewusstsein gespült. Alles ist jetzt wieder da. „Das graue Haus mit seiner Straßenfront (…), die Stadt, der Platz (…), die Straßen (…) alle Blumen unseres Gartens (…) die Seerosen auf der Vivonne, die Leutchen aus Combray."[37] Im gleichen Augenblick aber, in dem das Erinnern *Erinnerung* wird, endet die magische Erfahrung einer zeitlosen Gegenwart.

In dieser kurzen Passage aus der *Recherche* begegnen uns offenbar zwei unterschiedliche Funktionsweisen des Gedächtnisses. Die eine möchte ich *diskursiv* nennen. Man könnte auch *rational* sagen oder *bewusst*. Es handelt sich um eine Funktion des episodische Gedächtnisses, das über Worte und Bilder operiert, die nach den Regeln der Magazin-Metapher irgendwo abgelegt und vergessen, dann wiedergefunden und aktualisiert werden. Interessanter finde ich die zweite Art von Erinnerung: Prousts *mémoire involontaire*. Erinnerung, die sich äußert als eine Art Substanz, „nicht in mir, sondern ich war sie selbst".[38] Denn, und damit komme ich auf meine Ausgangsfrage zurück, *wer* ist es, der erinnert, wenn doch die Erinnerung „ich selbst" ist? Die epistemologische Trennung zwischen erinnerndem Subjekt und erinnertem Gegenstand ist hier aufgehoben. Das funktioniert in der Regel nur für kurze – traumatische[39] – Momente. So stellt sich Marcels Gefühl der Mittelmäßigkeit, der Zufallsbedingtheit und der Sterblichkeit auch genau in dem Augenblick wieder ein, in dem die von der Geschmackskombination von Madeleine und Tee ausgelöste unfreiwillige Erinnerung in eine freiwillige, eine *bewusste* überführt wird.

Aleida Assmann ordnet auch diese Form von Erinnerung als „Ausgraben" dem Bereich der Raum- oder Magazin-Metaphern zu.[40] Allerdings wird diese Einschätzung nur *einem* Aspekt von Prousts *mémoire involontaire* gerecht – dem des An-die-Oberfläche-Kommens verborgener Erfahrungen. Unberücksichtigt bleibt hingegen die erwähnte Auflösung epistemischer Positionen in einer phänomenologischen Klammer. Der erinnernde Körper als Erinnerung und Erinnerndes unterläuft die medial vermittelte Zeichenhaftigkeit des bewussten Gedächtnisses. Er liest sich gleichsam selbst, und doch ist es eine Lektüre, der ein wesentliches Element

---

37 Proust (1964), 67.

38 Proust (1964), 64.

39 Ich gebrauche den Begriff mit Roland Barthes als „Suspendierung der Sprache" und „Blockierung der Signifikation" (vgl. Barthes [1982], 23).

40 Assmann (1999),163f.

herkömmlicher Textauslegung abgeht: die Möglichkeit der Lüge. Zu einem gewissen Grad schließt das auch die Möglichkeit der Täuschung ein. Denn das proustsche „Substanz-Sein" bewegt sich jenseits der Bewusstseinskategorien wahr und falsch. Wenn wir sagen wollen, dass der Körper hier als Medium oder Zeichen fungiert, so folgt er einer „eingliedrigen" Zeichenlogik. Er erzeugt eine „Verfahrenseinheit von Bezeichnung und konkreter Vergegenwärtigung des Bezeichneten".[41]

VII   Transzendenz?

Am Ende scheint sich somit im sich selbst lesenden Körper tatsächlich das Fenster zu einer milden Form der Transzendenz zu öffnen. Ein Fenster nur und nur eine milde Form, denn letztlich handelt es sich um eine Transzendenz, die in sinnlicher Erfahrung wurzelt. Dennoch überschreitet sie die Grenzen eines räumlich und zeitlich eindeutig lokalisierbaren Bewusstseins.

Das Gedächtnis, zumal das sogenannte kollektive, ist zweifellos in vielfacher Hinsicht eine soziale Funktion des Geistes. Maurice Halbwachs, von dem der Begriff stammt, ging so weit, die Existenz eines nicht sozial determinierten Gedächtnisses generell zu bestreiten. „Individuelles Gedächtnis", schreibt er, ist „nicht möglich ohne jene Instrumente, die durch die Worte und Vorstellungen gebildet werden, die das Individuum nicht erfunden und die es seinem Milieu entliehen hat."[42] In dieser Radikalität trifft das freilich nur auf unsere *bewussten* Erinnerungen zu. Tatsächlich ist die Topographie unseres Gedächtnisses weit komplexer, als es die Zuspitzung auf kollektive Formen der Erinnerung suggeriert. Außen vor bleiben in dieser Konzeption all jene leiblichen Aspekte des Erinnerns, die sich jenseits der Sprache vollziehen.

Der sich selbst lesende Körper bleibt damit ein Paradox. Wo Leser, Medium und Inhalt der Mitteilung zusammenfallen, wo kein Code die Botschaft medial fassbar macht, wo Ambivalenz und Mehrdeutigkeiten ausgeschlossen sind, lässt sich nur mit großer semantischer Toleranz von einer ‚Lektüre' sprechen. Die leibliche als totale Erinnerung entzieht notwendig allen hermeneutischen Bemühungen den Boden, indem der Versuch, sie zu entziffern, sie zum Verschwinden bringt. Sie ist entweder da – dann kann man sie bemerken. Oder sie ist wieder weg. Dann lässt sie sich (vielleicht) verstehen.

---

41 Landfester (2012), 31.
42 Halbwachs (1967), 55.

Wenn wir also so denken, wenn wir erkennen, wenn wir erinnern – was denkt, erkennt und erinnert da? Ein Geist? Ein Bewusstsein? Ein Ich? Ein Selbst? Und was wäre ein solches Ich? Ein solches Selbst? Ist es Masse? Materie? Struktur? Prozess? Ein wenig von allem, wie mir scheint. Körper, Sprache, Schrift und Schreiben sind so vielfältig miteinander verwoben, dass es wenig sinnvoll wäre, sie auseinanderdividieren zu wollen. Die Grenzen unseres Körpers werden ständig überschritten, wenn wir Wahrnehmungen filtern und mit unserem persönlichen Repertoire innerer Bilder kombinieren.[43]

Augenscheinlich wird das nicht zuletzt (wieder), wenn wir schlafen: „Der Schlafende spannt in einem Kreise um sich den Ablauf der Stunden, die Ordnung der Jahre und der Welten aus", schreibt Proust[44] und greift damit Foucaults Idee vom Körper als dem „Nullpunkt der Welt"[45] voraus:

> Manchmal entstand in meinem Schlaf aus einer falschen Lage wie Eva aus der Rippe Adams eine Frau. Während sie aus der Lust hervorgegangen war, die ich erlebte, bildete ich mir ein, daß diese mir erst durch sie zuteil geworden sei. Mein Leib verspürte in dem ihren seine eigene Wärme und drängte zu ihr, ich wachte auf.[46]

Literatur

Assmann (1999): Aleida Assmann, Erinnerungsräume: Formen und Wandlungen des kulturellen Gedächtnisses. München.

Augustinus (1888): Aurelius Augustinus, *Die Bekenntnisse des heiligen Augustinus*. Leipzig.

Barthes (1982). Roland Barthes. „Le message photographique". In: ders., *L'obvie et l'obtus. Essais critiques III*. Paris, 9-24.

Belting (2001): Hans Belting, *Bild-Anthropologie: Entwürfe für eine Bildwissenschaft*. München.

Classen, Howes & Synnott (1994): *Constance Classen, David Howes und Anthony Synnott, Aroma: The Cultural History of Smell*. London.

Eco (1988): Umberto Eco, „An Ars Oblivionalis? Forget It!" In: *Publications of the Modern Language Association of America (PMLA) vol. 103*, 1988:3, 254-261.

Eco (1991): Umberto Eco, *Semiotik: Entwurf einer Theorie der Zeichen*. 2. korrigierte Auflage. München.

Eglinger (2007): Hanna Eglinger, *Der Körper als Palimpsest: Die poetologische Dimension des menschlichen Körpers in der skandinavischen Gegenwartsliteratur*. Freiburg/Br.

---

43   Belting (2001), 70f.

44   Proust (1964), 11.

45   Foucault (2004), 34.

46   Proust (1964), 11.

Freud (1924): Sigmund Freud, „Notiz über den ‚Wunderblock'". In: *Internationale Zeitschrift für Psychoanalyse*, 10 (1), 1-5.

Genette (1993): Gerard Genette, *Palimpseste: Die Literatur auf zweiter Stufe*. Frankfurt/M.

Goethe (1998a): Johann Wolfgang von Goethe, „Gedichte und Epen I". In: ders., *Werke*. Hamburger Ausgabe, Bd. I. Frankfurt.

Goethe (1998b): Johann Wolfgang von Goethe, „Aus meinem Leben: Dichtung und Wahrheit". Autobiographische Schriften II. In: ders., *Werke*. Hamburger Ausgabe, Bd. X. Frankfurt/M.

Halbwachs (1967): Maurice Halbwachs, *Das kollektive Gedächtnis*. Stuttgart. (zuerst 1950)

Hegel (2012): Georg Wilhelm Friedrich Hegel, *Enzyklopädie der philosophischen Wissenschaften im Grundrisse*. Frankfurt/M. (zuerst 1830)

Landfester (2012): Ulrike Landfester, *Stichworte: Tätowierung und europäische Schriftkultur*. Berlin.

Lavater (2013): Johann Caspar Lavater, *Von der Physiognomik*. Paderborn. (zuerst 1772)

Locke (1690): John Locke, *An Essay Concerning Human Understanding*, London.

Löbbermann (2002). Dorothea Löbbermann, *Memories of Harlem: Literarische (Re)Konstruktionen eines Mythos der zwanziger Jahre*. Frankfurt/New York.

Nietzsche (1999): Friedrich Nietzsche, „Zur Genealogie der Moral: Eine Streitschrift". In: ders.: *Werke. V.* Hg. Von Giorgio Colli und Mazzino Montinari. München, 245-421.

Nora (1998): Pierre Nora, *Zwischen Geschichte und Gedächtnis*. Frankfurt/M.

Platon (1991): Platon, „Theaitetos". In: ders., *Sämtliche Werke VI*. Frankfurt/M., 151-367.

Plessner (1981): Helmuth Plessner, „Die Stufen des Organischen und der Mensch: Einleitung in die philosophische Anthropologie". In: ders., *Gesammelte Schriften, Bd IV*, Frankfurt/M. (zuerst 1928)

Proust (1964): Marcel Proust, „Auf der Suche nach der verlorenen Zeit". In *Swanns Welt I*. Frankfurt/M.

Rappe (1995): Guido Rappe, *Archaische Leiberfahrung: Der Leib in der frühgriechischen Philosophie und in außereuropäischen Kulturen*. Berlin.

Schögl (1997): Uwe Schögl, „Goethe, Lavater und die ‚Physiognomischen Fragmente'". In: Konrad Scheurmann und Ursula Bongaerts-Schomer (Hg.), *... endlich in dieser Hauptstadt der Welt angelangt! Goethe in Rom – Publikation zur Eröffnung der Casa di Goethe in Rom. Bd. 1*. Mainz, 20-24.

Spenser (1596): Edmund Spenser, The Faerie Queene. London.

Stache (2010): Antje Stache, *Der Körper als Mitte: Zur Dynamisierung des Körperbegriffs unter praktischem Anspruch*. Würzburg.

Weinrich (1976): Harald Weinrich, „Metaphora memoirae". In: ders., *Sprache in Texten*. Stuttgart, 291-294.

3

Die Leiblichkeit verkörperter Gründe

Gründe als Rechtfertigungen – verkörpert, eingebettet und freistehend

Matthias Jung

Moderne, pluralistische und demokratische Gesellschaften kultivieren eine tolerante Konzeption des öffentlichen Raums. Modernisierungs- und Säkularisierungprozesse[1] haben dazu geführt, dass soziale Integration nicht mehr durch eine einzige, dominante und alles überwölbende Weltanschauung garantiert, sondern in immer höherem Maß über vielfältige ‚social imaginaries' (Ch. Taylor) vermittelt und variablen Aushandlungen anheimgestellt wird. Pluralismus und Toleranz sind nun aber – durch fundamentalistische Religionen und Weltanschauungen immer wieder gefährdete – soziale Errungenschaften, denen auf der anderen Seite ein erhöhter Druck zur Rechtfertigung und Begründung der eigenen Positionen entspricht. Der Sozialphilosoph Rainer Forst spricht in diesem Zusammenhang vom „Recht auf Rechtfertigung"[2]: Alle Mitglieder demokratischer Gesellschaften schulden allen anderen *Gründe* für ihr Verhalten – nicht im privaten Raum, aber doch immer dann, wenn das fragliche Verhalten in einem unklaren oder gar problematischen Verhältnis zu weithin akzeptierten Normen steht. Die in Deutschland im Zusammenhang des sog. *Kölner Urteils*[3] vor einigen Jahren geführte Beschneidungsdebatte ist hierfür ein gutes Beispiel.

In vielen Konzeptionen aus dem Bereich der Sozial- und Moralphilosophie wird dieser Anspruch auf öffentliche Begründung recht rigoros an die begriffliche Unterscheidung zwischen dem *Guten* und dem *Richtigen* als zwei unterschiedlichen Sphären der Öffentlichkeit gekoppelt. Das Gute wird dabei konsequent pluralisiert

---

1   Unter dem programmatischen Titel „Spielräume der Modernisierung – Das Ende der Eindeutigkeit" liefert Wolfgang Knöbl (Knöbl [2001]) eine luzide Analyse des modernisierungstheoretischen Paradigmas.

2   So der Titel eines Buches von Forst: „Das Recht auf Rechtfertigung. Elemente einer konstruktivistischen Theorie der Gerechtigkeit" (Forst [2007]).

3   Vgl. Heil/Kramer (2012).

gedacht, im Sinne einer irreduziblen Vielfalt privater bzw. gruppenspezifischer Vorstellungen gelingender und sinnhafter Lebensführung, das Richtige hingegen als der universalistische, für alle verbindliche Rahmen konzeptualisiert, in dem Gründe zählen. Diese begriffliche Weichenstellung hat zur Konsequenz, dass alle Formen der Begründung, die unauflöslich mit historischen Narrativen oder anderen partikularen Traditionen, z. B. religiösen, verbunden sind, aus dem öffentlichen Raum weitgehend ausgeschlossen werden.

Für Jürgen Habermas beispielsweise schulden sich die „Bürger eines demokratischen Gemeinwesens [...] reziprok Gründe, weil nur dadurch politische Herrschaft ihren repressiven Charakter verlieren kann",[4] und zwar solche Gründe, die „auch unabhängig von ihrem religiösen Einbettungskontext ‚zählen‘"[5] müssen. Dabei stellt Habermas immer wieder die enorme Bedeutung des semantischen Potentials der Weltreligionen für die Entwicklung immer umfassenderer Konzeptionen der moralischen und juristischen Inklusion aller heraus – nur müssen diese Sinnressourcen eben, wie er das nennt, jeweils „übersetzt" werden, um in der Öffentlichkeit als Gründe zu zählen.[6] In diesem Übersetzungsprozess aber geht zwangsläufig jene Eingebettetheit bzw. Verkörpertheit verloren, die nicht nur für umfassende Weltanschauungen typisch ist, sondern ein generelles Merkmal des menschlichen Weltverhältnisses darstellt.

Natürlich hat Habermas einen guten Grund für sein Bestehen auf dem Ausschluss verkörperter Gründe aus dem öffentlichen Diskurs: Würden solche Gründe zugelassen, könnte dies die Geltung der hochabstrakten moralischen und juristischen Prinzipien gefährden, auf denen pluralistische Demokratien beruhen. Sich beispielsweise in Fragen der Gestaltung politischer Institutionen auf den geoffenbarten Willen Gottes zu berufen, wäre offensichtlich mit dem Gedeihen eines Gemeinwesens nicht vereinbar, in dem Menschen sehr verschiedener Religionen und

---

4   Habermas (2005), 127.

5   Ebd., 128.

6   Meine Darstellung ist hier vereinfacht. Habermas unterscheidet zwischen einer informellen Öffentlichkeit (in der auch „unübersetzte" Äußerungen zugelassen und sogar willkommen sind) und der institutionellen Öffentlichkeit von „Parlamenten, Gerichten, Ministerien und Verwaltungen" (Habermas 2005, 136), innerhalb derer nur „säkulare Beiträge" (ebd., 137) gehört werden dürfen. Die Beschränkung auf säkulare Gründe betrifft also nur den engeren Öffentlichkeitsbegriff. Grundlegend bleibt aber eine Vorstellung von Begründung, die erstens begrifflich einen scharfen Schnitt zwischen allgemein zugänglichen Gründen und partikularen Traditionen vornimmt und zweitens davon ausgeht, dass die öffentliche Identifikation von Redebeiträgen als zu der einen oder der anderen Gruppe gehörig jeweils möglich ist – und ohne Paternalismus auskommen kann.

Weltanschauungen friedlich miteinander leben können sollen. Habermas benennt also eine reale Gefahr, die eine sorgfältige Diskussion erfordert; dennoch meine ich, dass er mit seinem Kantianischen Versuch zur Reinigung der öffentlichen Vernunft von kontingenten Beimischungen verkörperter Traditionen das Kind mit dem Bade ausschüttet. Wenn nur freistehende, nicht mehr verkörperte Gründe zählen, wird der demokratische Prozess des wechselseitigen Begründens von Einstellungen, Handlungen und Institutionen in einer exklusivistischen Weise verkürzt.

Die menschliche Fähigkeit, die eigenen Einstellungen und Gewohnheiten zu reflektieren, ist so unauflöslich mit qualitativen Erlebnissen, mit sensomotorischen Aktivitäten und mit dem Verstehen des physischen und sozialen Kontextes verbunden, dass Habermas' ‚freistehendes' – also auf kontextunabhängige Argumente beschränktes – Verständnis moralischer Begründung nur eine begrenzte Reichweite beanspruchen kann. Unentbehrlich erscheint es mir für die Klärung und kritische Überprüfung, nicht aber für die Begründung des universalistischen Rahmens der Menschenrechte; für die Erhellung wirklicher moralischer Konflikte und Wertedebatten in pluralistischen Gesellschaften hingegen scheint mir sein Nutzen eher gering. Dieser Punkt ist schon häufig bemerkt worden. So betont etwa Michael Sandel immer wieder, wie wichtig es für die Vitalität demokratischer Gesellschaften ist, dass sie einen offenen und öffentlichen Austausch über verschiedene „dichte" und in Traditionen, Praktiken und Weltanschauungen verkörperte Vorstellungen des gemeinsamen Guts unterhalten.[7] Hans Joas hat diesen Punkt noch entscheidend präzisiert, indem er das (ursprünglich von Talcott Parsons stammende) Konzept der *Wertegeneralisierung* entwickelt hat,[8] das er keineswegs als Ersatz, wohl aber als unverzichtbare Ergänzung abstrakter moralischer und politischer Diskurse versteht.

Es kommt hinzu, dass ein argumentativer Universalismus, wenn man ihn wie Habermas in seiner *Theorie des kommunikativen Handelns* als normativen Kern der menschlichen Vernunft versteht,[9] mit den Erkenntnissen der Kognitionswissenschaft über die Verkörpertheit und Kontextualität kognitiver Prozesse nur schwer zu vereinbaren ist.[10] Nun ist Habermas selbst in seinen jüngsten Publikationen sehr

---

7   Vgl. Sandel (2012), 13f.

8   Vgl. Joas (2004), 47f.

9   Vgl. etwa Habermas (1981), 49: „Die grundlegende Intuition, die wir mit Argumenten verbinden, lässt sich [...] am ehesten durch die Absicht kennzeichnen, ein *universales Auditorium* zu überzeugen [...]." (kursiv im Original).

10 Damit meine ich keineswegs, dass kognitionswissenschaftliche Erkenntnisse universalistische Intuitionen in der Moralphilosophie obsolet machen. Es zeigt sich nur im Licht des „embodiment-turn", dass auch die menschliche Fähigkeit zur Dezentrierung und Dekontextualisierung eng mit leibgebundenen Vollzügen verbunden ist und nicht durch einen begrifflichen Schnitt von diesen abgetrennt werden kann.

weit darin gegangen, die Bedeutung von verkörperten Gründen zu betonen, wobei er zwischen kultureller, psychosozialer und materieller Verkörperung unterscheidet.[11] Schaut man jedoch genauer hin, so wird deutlich, dass die Verkörperungskonzeption des späten Habermas' paradigmatisch für die Art und Weise ist, in der prominente Theorien der Moderne die Bedeutung des Körpers herunterspielen. In dem Vortrag „Die Lebenswelt als Raum symbolischer verkörperter Gründe" stellt er Verkörperung als eine Einbahnstraße dar, durch die soziale Normen verinnerlicht und so in die Körper (und das Leibbewusstsein) der Gesellschaftsmitglieder eingeschrieben werden. „Auf dem Weg der Internalisierung", so schreibt Habermas, „werden die tragenden Werte und normativen Gründe einer Gesellschaft in den Persönlichkeitsstrukturen ihrer Mitglieder buchstäblich verkörpert."[12] Seine Darstellung von Verkörperung ist hier sehr einseitig. Sie betont die Internalisierung symbolisch artikulierter Normen und narrativ tradierter Werte, unterschätzt aber den Einfluss individuell und gruppenspezifisch verkörperter Gründe auf die Entstehung sozialer Werte und Normen. Für den späten Habermas ist im Unterschied zu der in der *Theorie des kommunikativen Handelns* vertretenen Position Begründung essentiell mit Verkörperung verbunden, jedoch einzig *top-down*: als die individuelle Inkorporation von kollektiven Werten und potentiell universellen Normen. Verkörperung *bottom-up*, verstanden als eine Quelle rationaler Kreativität, als *locus* von qualitativen Intensitäten und sensomotorischen Gewohnheiten, die den Lauf der diskursiven Vernunft inspirieren und formen, wird vollständig vernachlässigt. Der Grund dafür ist vermutlich der bereits erwähnte, nämlich die Befürchtung, durch ein *reziprokes*, nicht bloß *top-down* argumentierendes Verständnis von Verkörperung die Tür für partikularistische Wertkonzeptionen zu öffnen, die dem normativen Universalismus feindlich gegenüberstehen. Der Preis jedoch, der gezahlt werden muss, wenn man bei der Analyse und normativen Bewertung moderner Gesellschaften auf die Berücksichtigung von *bottom-up-embodiment* verzichtet, ist zu hoch.

Wir müssen daher nach einer Konzeption öffentlicher Deliberation Ausschau halten, die Verkörperung in beide Richtungen (Internalisierung sozialer Werte / Normen *und* Bedeutung erstpersonal-leiblicher Erfahrung für deren Entstehung) ernst nimmt und gleichzeitig in der Lage ist, einseitige Konzeptionen und ungerechtfertigte Geltungsansprüche verkörperter Vernunft zu kritisieren. Hierfür stütze ich mich auf das Denken John Deweys, in dem sich eine hochemphatische Konzeption partizipatorischer Demokratie mit hoher Sensibilität für die Grenzen abstrakter Reflexion und die Bedeutung körperlicher Erfahrungen verbindet. Für

---

11 Vgl. Habermas (2012), 72.
12 Ebd.

Dewey ist „Demokratie [...] ein Name für ein Leben in freier und bereichernder Kommunikation"[13] – Letztere wird aber, anders als bei Habermas, immer von der unmittelbaren, verkörperten face-to-face-Interaktion her gedacht (wobei der Begriff der Öffentlichkeit dann auch die zunehmende Bedeutung *indirekter* Interaktionsfolgen systematisch berücksichtigt). Für den Rest dieses Beitrags werde ich nun zunächst die Grundlinien der deweyschen Konzeption erläutern (I) und dann auf dieser Grundlage drei Typen öffentlichen Vernunftgebrauchs (freistehende, eingebettete und verkörperte Deliberation) unterscheiden und aufeinander beziehen (II). Abschließend wende ich mich einem konkreten Beispiel aus dem antiken Israel zu (III).

I   Verkörperte Vernunft nach John Dewey

Dewey zufolge müssen die Beziehungen zwischen verkörperter Erfahrung, sozial und historisch eingebetteten Praktiken sowie diskursiver, ‚freistehend' argumentierender Vernunft als reziprok gedacht werden.[14] Immer dann, wenn Menschen eine reflexive Einstellung einnehmen, sich also nicht mehr einfach kausal von ihren primären Antrieben und unbewussten Motivationen leiten lassen, sind alle drei rationalen Instanzen beteiligt. Dewey ist hier stark von Charles Sanders Peirce und dessen triadischer Ontologie sowie Semiotik beeinflusst.[15] Den Ausgangpunkt bildet der qualitative Charakter unmittelbarer Erfahrung, die immer durch eine dominierende, alles durchdringende Qualität bestimmt ist.[16] Diese unmittelbar

---

13 Dewey (1996), 155.

14 Die Differenz zwischen Motiven und Gründen wird damit keineswegs eingeebnet, wohl aber deutlich gemacht, dass erlebte Qualitäten und historische Erfahrungen notwendige, wenn auch natürlich keineswegs hinreichende Bedingungen realer Begründungsprozesse darstellen. Kantianer mögen sich diesen Sachverhalt daran verdeutlichen, dass Kant selbst schließlich dem kategorischen Imperativ (als Paradebeispiel freistehender Begründung und Dreh- und Angelpunkt des Universalismus) keineswegs inhaltliche Normbegründungen oder gar situationsangemessene Entscheidungen zugetraut hat, sondern seine Rolle ausschließlich in der Geltungsbegründung sah.

15 Für eine scharfsinnige Analyse der Weise, in der Peirce systematisch die drei Kategorien von Firstness, Secondness und Thirdness verschränkt und damit eine Betrachtungsweise etabliert, in der Allgemeinheit, Interaktion und Individualität unauflöslich verbunden sind, vgl. Dewey (1998).

16 Vgl. Dewey (2003).

leiblich erlebte Qualität bildet den intuitiven[17] Ausgangspunkt allen Denkens und den Hintergrund aller Objektivierungen, kann aber selbst nicht objektiviert werden: Im Erleben ist immer mehr präsent als das, was durch sprachliches oder nichtsprachliches Handeln zum Ausdruck gebracht werden kann. Leibliche Erfahrung ist daher primär nicht von Repräsentation, sondern von Präsenz geprägt. Hierin gründet die gedächtnisprägende Kraft von physischen Veränderungen des Körpers, also Einschnitten und Einschreibungen im wörtlichen Sinn. Bedeutsam sind z. B. Beschneidungspraktiken oder Tätowierungen vor allem deshalb, weil in ihnen die sekundäre, repräsentationale Bedeutung untilgbar mit primärer, leiblicher und erstpersonaler Erfahrung verkoppelt wird. Ein aus der Theologie übertragener Begriff, der *character indelebilis*, bietet sich hier zur Illustration an: Einschreibungen in den Körper erzeugen ein „unauslöschliches Merkmal", etwas, das nicht rückgängig gemacht werden kann, im Unterschied zu der jederzeit reversiblen, weil kontingenten Koppelung des repräsentierenden Zeichens an das durch es Repräsentierte.

Gefühlte Qualitäten, so wichtig sie für das Zustandekommen bewusster Kognitionen auch sind, lassen sich jedoch nur begrifflich aus dem größeren Ganzen heraustrennen, zu dem sie als unselbstständige Komponenten gehören. Sie konfrontieren uns mit intensiv anmutender Bedeutungshaftigkeit, die aber vage und unbestimmt bleibt – mit Möglichkeiten, die noch keine Wirklichkeiten sind. *Wovon* erlebte Qualitäten handeln, ist in ihnen nicht enthalten, es muss erst artikuliert werden. Der erste Schritt hierzu besteht darin, ihre Bedeutung innerhalb der Interaktionen zu bestimmen, die den menschlichen Organismus mit seiner Umwelt verbinden. Die qualitative Signatur des Erlebens gehört in den Kontext der philosophisch viel zu lange unterschätzten sensomotorischen Beziehungen[18] hinein, durch die wir mit unseren Mitmenschen verbunden sind, lange bevor wir ein Gespräch mit ihnen führen können. Solche sensomotorischen Verschränkungen der Bewegungsmuster von *ego* und *alter* (einschließlich ihrer Gesten) sind bereits früh in der Menschheitsgeschichte zu ritualisierten Formen ausgebaut worden.

An dieser Stelle bietet es sich an, eine Brücke zu der Theorie kognitiver Evolution zu schlagen, die der Kognitionspsychologe Merlin Donald entwickelt hat.[19] Er geht davon aus, dass die auf symbolischer Sprache basierende Kultur des modernen Men-

---

17 Vgl. ebd., 101: „Intuition bezeichnet die Erkenntnis einer durchdringenden Qualität, und zwar so, dass sie die Bestimmung relevanter Unterscheidungen oder all dessen reguliert, was zum akzeptierten Objekt des Denkens wird."
18 Vgl. dazu den Beitrag von Susan Stuart in diesem Band.
19 Vgl. Donald (1991) und Donald (2001).

schen einen evolutionären Vorgänger in Form der so genannten „mimetischen"[20] Kultur hatte, die auf ritualisierten Nachahmungen typischer Interaktionssequenzen und Reaktionen anderer, auf Körpersprache und -bemalung, Gesichtsausdrücken und der körperschematischen Verschränkung von Haltungen und Posen beruhte. Diese mimetisch-rituelle, körperbasierte Ausdrucksdimension ist nach Donald nicht nur fundamental für die Entstehung sozialer Kooperativität gewesen, sie stellt auch in der heutigen Kultur weiterhin einen zentralen Faktor dar – und beeinflusst, so ergänze ich, auch den Verlauf unserer reflexiven Kognitionen nachdrücklich. Mit der Semiotik von Peirce formuliert, kombiniert sie ikonische, also auf Ähnlichkeit basierende Elemente mit indexikalischen, die auf leiblicher Präsenz und erfahrenem Widerstand basieren. Peirce spricht hier treffend von dem „sense of acting and being acted upon", der gleichzeitig „our sense of the reality of things"[21] sei.

Wenn uns qualitative Situationen präsent sind und wir uns im sensomotorischen Austausch mit anderen Subjekten und der natürlichen Welt befinden, ist eine unmittelbare Erfahrung der Realität gegeben, und aus dieser Quelle entspringen die meisten unserer Gedankengänge und sprachlichen Bestimmungsversuche. Aber leiblich-qualitatives Erleben und Aktion / Reaktion im Umgang mit der physischen und sozialen Welt statten uns nur mit *prima-facie*-Gründen aus. Erst die dritte, irreduzible Komponente der Rückkoppelungsschleifen zwischen uns und unserer Umgebung, nämlich die symbolische Artikulation, ist in der Lage, explizit zu machen, was in unserem Erleben und in unserem Handeln implizit bleibt. Was uns z. B. im Erleben anzieht, kann sich der Reflexion als wertlos oder gar normativ verboten erweisen,[22] was wir zu glauben geneigt sind, nur als Wunschdenken. Nur durch den Gebrauch symbolischer, das Hier und Jetzt physischer Präsenz transzendierender Sprache können wir rational argumentieren und das Potential der Übereinstimmung über diejenigen hinaus erweitern, die mit uns durch ähnliche qualitative Gefühle und / oder die Partizipation an einer rituellen Gemeinschaft verbunden sind. Doch die Fähigkeit von symbolischer Sprache, sich von direkter Erfahrung zu entfernen (nicht: zu lösen!) und universelle Gründe zum Ausdruck zu bringen, gründet in der Unmittelbarkeit leiblicher Vertrautheit mit der Welt im Gefühl und Handeln. Aus diesem Grund identifiziert Deweys Konzeption des öffentlichen Austauschs Rationalität gerade nicht mit der Fähigkeit zur freistehenden, kontextlosen Argumentation, sondern hat zwei komplementäre Brennpunkte: Einerseits steht die Frage im Zentrum, wie aus konkreten Problemen Öffentlichkeiten entstehen,

---

20　Vgl. Donald (1991), chap. 6; Donald (2001), 260ff.

21　Peirce (1998), 4.

22　Dewey spricht hier prägnant vom Unterschied zwischen dem Erwünschten (*desired*) und dem tatsächlich Wünschenswerten (*desirable*): vgl. Dewey 2004, 324.

die auch wieder vergehen können, wenn diese Probleme hinreichend bearbeitet sind, andererseits geht es um das, was Hans Joas mit Parson Wertegeneralisierung genannt hat.[23] Wertegeneralisierung unterscheidet sich von anderen prominenten Konzeptionen des öffentlichen Vernunftgebrauchs wie etwa Rawls' *Theory of Justice* oder Habermas' *Theorie des kommunikativen Handelns* dadurch, dass sie Vernunft unaufhebbar als verkörpert versteht. *Mit* diesen Autoren geht das Konzept der Wertegeneralisierung davon aus, dass pluralistische und demokratische Gesellschaften ein Rahmenwerk an universalistischen Normen in Moral und Recht brauchen, um die Begegnung der unterschiedlichen Wertegemeinschaften zu moderieren und einzuhegen. *Entgegen* vor allem der Position von Habermas ist aber pragmatistische Sozialphilosophie nicht der Meinung, dass die Begründung der eigenen Werte und politischen Überzeugungen gegenüber den anderen Gesellschaftsmitgliedern auf verkörperte Gründe, die sich nicht in rein diskursives Sprachspiel übersetzen lassen, Verzicht leisten muss.

Wertegeneralisierungen sind Neuartikulationen partikularer Wertvorstellungen, die deren Herkunftsgeschichte und emotionale Intensität nicht aufheben, sie aber eben so neu fassen, dass sie als Spezifizierungen übergreifend geteilter, universeller Werte verständlich werden. Dafür gibt es historische Beispiele. Die Geschichte der Menschenrechte, vor allem die allgemeine Erklärung der Menschenrechte vom Dezember 1948 und der anschließende Aufstieg des Menschenrechtsdiskurses zu einer Art moralischer Globalsprache zeigen, dass Wertegeneralisierungen möglich und in prominenten Fällen auch wirklich geworden sind. Der springende Punkt ist hier, dass universelle Werte auch ohne einen harten Schnitt mit den partikularen Werttraditionen, nämlich durch eine Form der Kommunikation zwischen diesen entstehen können, die die verkörperten Gründe nicht transzendierend zurückstößt, sondern gerade auf ein abstrakteres Allgemeines bezieht. Auch das lässt sich an den Menschenrechten besonders gut zeigen:[24] So können beispielsweise Konfuzianer sie auf der Basis der von Konfuzius entwickelten Idee der *ren*, allgemeiner Menschlichkeit, bejahen, Christen sich auf die in der Gottesebenbildlichkeit gründende Würde des Menschen berufen und säkulare Kantianer in ihnen eine Konkretisierung des kategorischen Imperativs in seiner dritten Formulierung (der Selbstzweckformel) erblicken.

Dadurch wird klar, dass Wertegeneralisierung sich von dem unterscheidet, was Habermas als diskursive Verflüssigung bzw. als Übersetzung bezeichnet. Schon die Metapher des Übersetzens legt ja nahe, dass es sich hier um zwei getrennte semantische Systeme handelt. „Zweisprachige" Bürger sind dann solche, die sowohl die

---

23 Vgl. Fn 8.
24 Ich folge hier Anregungen von Hans Joas in Joas (2011), Kap. 6.

verkörperte Sprache einer Herkunftstradition (primär ist hier an die Weltreligionen gedacht) als auch die abstrakte, religiös-metaphysisch enthaltsame Sprache des Universalismus beherrschen und deshalb als Übersetzer dienen können (was ihnen im Übrigen einen seltsamen Vorsprung nicht nur gegenüber den „einsprachigen" Mitgliedern der eigenen Religion, sondern auch gegenüber den „einsprachig" säkularen Bürgern verleiht). Wertegeneralisierung unterscheidet sich also von der Übersetzung ins säkulare Idiom dadurch, dass mehrere jeweils autochthone, in Traditionen, Praktiken und Institutionen verkörperte Wertgemeinschaften einen Lernprozess durchlaufen, in dem sie ihr Selbstverständnis jeweils so reartikulieren, dass es einen inneren Bezug auf universelle Werte und Normen gewinnt, die sich *auch* in einer säkularen Sprache formulieren lassen. Verglichen mit der Vorstellung einer von Verkörperung ablösbaren, freistehenden Begründungssprache, ist das Konzept der Wertegeneralisierung daher zugleich anspruchsvoller und realistischer. Der Realismus gründet in der Tatsache, dass er der Art und Weise näher steht, in der verkörperte rationale Wesen tatsächlich ihre normativen (und auch epistemischen!) Überzeugungen formen. Doch ist das Konzept durch eben diese Einsicht auch äußerst anspruchsvoll und formuliert lediglich eine fragile *Möglichkeit* menschlicher Kommunikation über Werte, deren Verwirklichung von vielen kontingenten Bedingungen abhängt und sehr oft durch Machtinteressen oder dogmatische Selbstabschottung zunichte gemacht wird.

II Drei Arten des Begründens

Wir sind nun in der Lage, zwischen drei verschiedenen Spielarten des öffentlichen Begründens von Überzeugungen und Handlungen zu unterscheiden: nämlich zwischen der *freistehenden*, der *eingebetteten* und der *verkörperten*. Diese Unterscheidungen stehen natürlich nicht für getrennte Bereiche der menschlichen Vernunft, sie stellen eher Phasen oder Elemente eines dynamischen und prozessualen Ganzen dar. *Freistehende*, von jeder Einbindung abstrahierende Vernunft ist außerordentlich wichtig, aber in ihrer Reichweite sehr begrenzt. Ihr steht nur eine einzige Ressource zur Verfügung, nämlich solche Weisen des Überlegens, die mit der Zustimmung einer jeden denkenden Person rechnen dürfen. Für Kantianer bedeutet dies eine Beschränkung auf transzendentale Reflexion über die notwendige *Form* jedes Nachdenkens und Begründens. Je abstrakter und universalistischer jedoch Argumente formuliert werden – hier liegt eine ironische Pointe –, desto weniger sind sie für ‚gewöhnliche', in der Kunst des abstrakten Denkens nicht trainierte Bürger noch zugänglich. Freistehende und deshalb auf kognitive Eliten beschränkte Argumentationsfiguren, wie sie von Kant bis Rawls und Habermas entwickelt wurden, sind deshalb am wichtigsten, wenn es darum geht, idealisierte Konzeptionen von universellen Rechten und globaler Gerechtigkeit zu entwickeln sowie die Spannungen und Inkompatiblitäten zwischen konfligierenden Geltungsansprüchen aus einer unparteiischen Perspektive zu bewerten. Das ist ein wichtiger Punkt, und idealisierende Modellbildungen wie Rawls' *Theorie der Gerechtigkeit* haben einen bedeutenden Einfluss etwa in der Politikberatung entwickeln können. Doch ist die motivationale Kraft solcher Argumentationsfiguren – von Habermas prägnant als der „zwanglose[.] Zwang[.] des besseren Arguments"[25] charakterisiert – recht gering, weil das Abstrahieren von konkreter leiblicher und historischer Erfahrung den Argumentationsprozess von allen tiefer verankerten Überzeugungen abschirmt.

Eingebettete Begründungen hingegen erreichen diese Überzeugungen. Sie beziehen sich auf gemeinschaftlich geteilte Erfahrungen, und zwar auf der Ebene tatsächlicher (aktueller bzw. erinnerter) Handlungen oder sogar Rituale, und sind deshalb mit einem viel stärkeren Sinn von Realität, Verbindlichkeit und Dringlichkeit verbunden. Wenn die Dinge gut laufen, können auch kontextspezifische Begründungen zu universalistischen Ergebnissen führen, wie im Fall der Verabschiedung der *Universal Declaration of Human Rights* durch die Vollversammlung der Vereinten Nationen am 10.12.1948 in Paris. Dieses wohl bedeutendste Beispiel für öffentlichen Vernunftgebrauch im 20. Jahrhundert bediente sich einer eingebetteten, keineswegs abstrakt-universalistischen Begründungsstrategie, indem vor dem Hintergrund der nationalsozialistischen Schreckensherrschaft und des Zweiten Weltkriegs auf die weithin geteilte historische Erfahrung von „barbarous acts which have outraged the conscience of mankind"[26] verwiesen wurde. Kontextsensible, in geteilte Erfahrungen eingebettete Argumente können aber natürlich auch genau in die gegenteilige Richtung gehen. In ihren bösartigsten Formen universalisieren sie dann gewaltsam gerade die partikularistischsten Züge einer Religion oder Weltanschauung und erklären allen Abweichlern den Krieg. Ist es so weit gekommen, hat sich dann auch die Idee des Argumentierens selbst gleich miterledigt. Sofern das erfahrungsbezogene, eingebettete Denken sich jedoch für eine Fortsetzung des Gesprächs mit Andersdenkenden offen hält, ist es immer auch auf das abstrakte, universalistische Denken angewiesen, weil dieses in der Lage ist, konfligierende Geltungsansprüche zu klären und sie im Lichte höherstufiger, abstrakter Ideale zu bewerten.

Auf der Ebene *verkörperter* Gründe schließlich entfaltet die leibliche Präsenz des menschlichen Organismus ihre volle Kraft. Die hier gegebene Unmittelbarkeit

---

25 Habermas (1981), 52f.

26 http://www.un.org/en/documents/udhr/index.shtml; Zugriff vom 4.9.2014.

des Erlebens darf freilich nicht naiv verstanden werden. Denn genauso, wie eingebettete Gründe bereits Denkformen und -inhalte enthalten, die in freistehender rationaler Argumentation expliziert werden können, ist auch das verkörperte Denken von Individuen jeweils schon vielfältig durch die sozial eingebetteten Argumentationsmuster, Sinnbilder, Rollenvorgaben etc. präfiguriert, also von dem *Me*, aus dem gemäß der Sozialpsychologie von Georg Herbert Mead erst das *I* als kreativ-individuelle Instanz hervorgeht. Im Unterschied zu der Position von Habermas müssen die im leiblichen Weltverhältnis verankerten Gründe aber auch als eine wichtige und kreative Rationalitätsinstanz eigenen Rechts betrachtet werden. Was der lebendige Leib ausagiert und was ihm eingeschrieben wird, kann in aller Kürze durch den Hinweis auf die Relevanz von Präsenz als Möglichkeitsbedingung von Repräsentation beschrieben werden. Dasjenige, was dem Selbst in der Form gefühlter Bedeutung gegenwärtig ist, stellt nicht nur den Ausgangspunkt aller abstrakteren Kognitionsformen dar, es ist auch durchgängig mit einer Prägnanz und Dringlichkeit versehen, die der Reflexion oft fehlt. Mir scheint, dass eben diese Eigentümlichkeit all dessen, was in einem wörtlichen Sinn inkorporiert wird, auch den Grund für die Ubiquität entsprechender kultureller Praktiken darstellt. Es kann beispielsweise sicher viele Gründe dafür geben, sich den Namen einer geliebten Person eintätowieren zu lassen – und ebenso viele Gründe dafür, dies zu einem späteren Zeitpunkt von Herzen zu bereuen –, aber die semiotische Funktion dieses Handelns scheint immer dieselbe zu sein: die kontingente Natur der Repräsentation zu unterlaufen und einen *character indelebilis* zu schaffen, also etwas, das nicht rückgängig gemacht werden kann, weil der eigene Leib einen dauernd daran erinnert.

Friedrich Nietzsche und Franz Kafka haben, ohne die hier verwendete semiotische Terminologie zu benutzen, am Zusammenhang zwischen kultureller Verbindlichkeit und Einschreibung in den Körper keinen Zweifel gehabt. In seiner *Genealogie der Moral* bezeichnet Nietzsche den Schmerz als „das mächtigste Hülfsmittel der Mnemonik" und bezeichnet den Satz „„Man brennt etwas ein, damit es im Gedächtnis bleibt"" als den „Hauptsatz aus der allerältesten (leider auch allerlängsten) Psychologie auf Erden".[27] Und in Kafkas Erzählung *In der Strafkolonie* wird bekanntlich das Urteil wortwörtlich dem Körper des Delinquenten eingeschrieben. Sowohl Nietzsche als auch Kafka konzentrieren sich auf den schmerzenden Körper als ein düsteres Hilfsmittel der kulturellen Erinnerungstechnik, aber sie scheinen damit nur Teilaspekte dessen zu erfassen, was tatsächlich bei Einschreibungen in den Körper eine Rolle spielt. Im Anschluss an Überlegungen Roy Rappaports,[28] der

---

27 Nietzsche (1988), 295.
28 Vgl. Rappaport (1999), bes. 15.

rituelle Handlungen als Kompensationen für die kontingente und deshalb immer prekäre Natur sprachlicher Repräsentationen deutet, möchte ich auch Einschreibungen oder Einschnitte in den Körper semiotisch-anthropologisch deuten, und zwar an einem konkreten historischen Beispiel.

III   Der Bundesgedanke im alten Israel und die semiotische Verschränkung der Gründe

Ich möchte meine Überlegungen mit einem konkreten Beispiel abschließen und habe dafür den Gedanken eines Bundes zwischen Gott und den Menschen gewählt, wie er die Schriften der hebräischen Bibel prägt. Dabei bin ich mir schmerzlich dessen bewusst, dass meine rationalitäts- und verkörperungstheoretisch inspirierten Überlegungen jede religionswissenschaftliche oder exegetische Akribie und Expertise vermissen lassen. Es geht mir ausschließlich um die Frage, ob sich an einem konkreten historischen Exempel zumindest plausibel machen lässt, dass die hier entwickelten Differenzierungen tatsächlich in aufschlussreicher Weise auf kulturelle Phänomene angewandt werden können.

Der semiotisch entscheidende Gesichtspunkt, wenn es um den Bundesgedanken geht, ist nun die extreme Spannung zwischen der Partikularität eines Bundes, den Gott mit einem einzelnen Volk schließt, und der Betonung der Transzendenz dieses Gottes. In Dtn 4, 15 erinnert Gott Moses an seine Gestaltlosigkeit und begründet damit das Bilderverbot: „Nehmt euch um eures Lebens willen gut in acht! Denn eine Gestalt habt ihr an dem Tag, als der Herr am Horeb aus dem Feuer zu euch sprach, nicht gesehen. Lauft nicht in euer Verderben, und macht euch kein Gottesbildnis […]." Gott kann, so lässt sich das deuten, nicht ikonisch *präsentiert* werden: Es gibt kein Zeichen, das eine Ähnlichkeitsbeziehung zu ihm unterhält, weshalb er in symbolischer Sprache *re-präsentiert* werden muss – in einer Sprache also, die von kontingenten, arbiträren Beziehungen zwischen Zeichen und Bezeichnetem durchdrungen ist und eben darum über die Möglichkeit verfügt, diese Beziehung auch explizit zu machen. Zumindest scheint mir das eine plausible Lesart des Bilderverbots im Kontext der achsenzeitlichen Entwicklungen darzustellen. Mit dem Stichwort „Achsenzeit" verweise ich auf eine verzweigte und vielschichtige Debatte,[29] aus der hier nur ein zentraler Zug herausgehoben werden kann: die *Entdeckung der Transzendenz* sowohl in semiotischer (als die Transzendenz des Symbolisierten über das Symbol) als auch in inhaltlicher (metaphysischer oder religiöser) Hinsicht. Das,

---

29 Zum jüngsten Stand der Debatte vgl. Bellah (2011) und Bellah/Joas (2012).

was die Wirklichkeit im Kern ausmacht, wird nun als etwas verstanden, das sich radikal von allen Angelegenheiten dieser Welt unterscheidet, wobei die so benannte Denkbewegung sowohl theistische (wie im Judentum) als auch agnostische Formen (wie im ursprünglichen Konfuzianismus und Buddhismus) annehmen kann. Komplementär dazu gerät der sprachliche Prozess der Bezeichnung in eine Krise,[30] muss doch der Versuch, Transzendenz semiotisch zu fixieren, sich notwendig in Paradoxien verstricken. Symbolische Bedeutung kann deshalb nicht ‚re-ikonifiziert‘ werden, weil nur Worte, nicht aber Bilder in der Lage sind, die radikale Differenz offenzuhalten zwischen dem, worauf der Akt der Bezeichnung tatsächlich zielt, und dem, was im Sprechen oder Schreiben tatsächlich gesagt worden ist. Die Kehrseite der Sache: Der Gebrauch von symbolischer Sprache, wenn er nicht mehr in die leibhaftige, direkte Erfahrung von Individuen und Kollektiven eingebettet ist, verliert den Kontakt mit der Realität, unabhängig davon, ob diese nun weltlichen oder transzendenten Charakter trägt. Nicht zufällig ist unsere gewöhnliche Sprache mit deiktischen Ausdrücken durchsetzt, deren Funktion darin besteht, die indirekte Referenz symbolischer Ausdrücke auf die leibliche Gegenwart der Sprecher im Hier und Jetzt zu beziehen. Semiotisch formuliert, entsteht durch die Entdeckung der Transzendenz die Notwendigkeit, die radikale Andersheit des Göttlichen oder Metaphysischen und das damit einhergehende Gefühl von Kontingenz mit erlebter, direkter und erstpersönlicher Erfahrung zu verbinden. *Präsenz* und *Repräsentation* müssen in eine Balance gebracht werden.

Das bringt mich zurück zu meiner oben entwickelten Unterscheidung zwischen freistehenden, eingebetteten und verkörperten Gründen. Die außergewöhnliche kulturelle, religiöse und sozialintegrative Kraft des Bundesgedankens, so schlage ich vor, lässt sich aus seiner Fähigkeit heraus verstehen, alle drei Formen des Begründens miteinander zu integrieren. Vasallenbündnisse, häufig verbunden mit der Idee sakralen Königtums, sind im Nahen Osten zur Zeit der Antike weit verbreitet gewesen, und auf diesem Vorstellungskomplex baut auch die jüdische Bundesidee

---

30 Folgt man der Argumentation von Rappaport (1999), dann handelt es sich streng genommen bereits um die *zweite* Krise. Denn lange vor den kognitiven Umwälzungen, die wir mit dem Begriff der Achsenzeit verbinden, haben Menschen bereits symbolisch kommuniziert. Damit hatten sie auch die Möglichkeit, zu lügen und gezielt zu täuschen – etwas, das mit ikonischen und indexikalischen Zeichen gar nicht möglich ist, weil es hier eine direkte Beziehung von Zeichen und Bezeichnetem gibt. Die zentrale Rolle von Ritualen – also kollektiven, semiotisch gesprochen indexikalisch dominierten – physischen Performanzen in der menschlichen Kulturentwicklung deutet Rappaport nun als Reaktion auf diese mit den Symbolsprachen in die Welt gekommene Bedrohung des sozialen Zusammenhalts durch ‚Lug und Trug‘. Rituale bewirken nach dieser Deutung soziale Integration durch die Teilnahme an ihnen und dies unabhängig von einer evtl. vorhandenen *reservatio mentalis* auf Seiten einiger Teilnehmer.

auf. Doch die Idee der Transzendenz fügt ein neues, entscheidendes Ingredienz hinzu. Zunächst ist festzuhalten, dass symbolische, universalistisch-freistehende Argumentation mit Notwendigkeit die Grenzen jeder partikularistischen Bindung überschreitet. Dieser Aspekt kommt in Dtn 4 massiv zum Tragen. Dort wird die Idee des Bundes zusammen mit der des verheißenen Landes an die Befolgung des Gesetzes gekoppelt – eines Gesetzes, das in Textform, also in symbolischer Sprache, vorliegt. Und dieses Gesetz wird in einer Sprache artikuliert, die sich eines moralisch-universalistischen Vokabulars bedient, wie sich das auch bei vielen der israelischen Propheten beobachten lässt.[31] Damit bewegen wir uns auf der Ebene einer universalistischen, symbolischen Sprache, auf der die partikulare Relation zwischen einem Volk, seinem Heimatland und seinem Gott zumindest virtualisiert wird. Die Bibel wird so, wie Heinrich Heine das prägnant formuliert hat, zu einem „portative[n] Vaterland"[32], einem nicht mehr an ein physisches Terrain gekoppelten Zuhause.

Doch ist die Ortsunabhängigkeit, die mit der universalistischen Orientierung einhergeht, nur die eine Seite der Medaille. Der Zusammenhang mit partikularer, kontingenter Erfahrung bleibt nämlich sowohl auf der kollektiven wie auf der individuellen Ebene erhalten. Der Beweglichkeit (im Sinne von Adaptivität in verschiedenen Kontexten bei Bewahrung der Identität) einer Religion, die sich auf heilige Texte und damit auf Symbole, nicht auf direkte Erfahrung bezieht, entspricht nämlich auf der anderen Seite der starke Bezug auf das *gelobte Land*, das für die Sehnsüchte einsteht, die aus starken kollektiven Erfahrungen des Exils, des Nicht-beheimatet-Seins, erwachsen. In Dtn 34 wird das gelobte Land Moses vor seinem Tod durch Gott gezeigt. Hier handelt es sich um ein besonders subtiles Exempel einer kontextspezifischen, indexikalisch auf etwas verweisenden Begründung: Direkte Erfahrung wird eingeführt, präsentiert – denn Moses *sieht* ja tatsächlich das gelobte Land – aber gleichzeitig wird sie zurückgehalten, denn der direkte Erfahrungszugang wird auf das Sehen und damit auf den Distanzsinn beschränkt. Das Land wird nicht betreten, wohl aber geschaut, was schon auf eine Virtualisierung hinausläuft, denn „gelobt" ist das Land schließlich nicht als Objekt sinnlicher Anschauung aus der Ferne, sondern als zukünftiger Lebensraum eines Volkes. Hinzu kommt noch, dass die reale Gegenwart der unmittelbaren Erfahrung, die zudem aufs Schauen restringiert ist, gerade nicht performativ präsentiert, sondern

---

31 Besonders prägnant Amos 5, 22-24: „Wenn ihr mir Brandopfer darbietet, ich habe kein Gefallen an Euren Gaben, / und Eure fetten Heilsopfer will ich nicht sehen. Weg mit dem Lärm deiner Lieder! / Dein Harfenspiel will ich nicht hören, sondern das Recht ströme wie Wasser, / die Gerechtigkeit wie ein nie versiegender Bach."

32 Heine (1967), 357.

in der repräsentationalen, symbolischen Sprache einer Verheißung formuliert wird. Dtn 34 ist ein *Text*, der sich als Text auf etwas nicht Textförmiges, nämlich Präsenz bezieht. Und ein wirklich sichtbares, aber sich im selben Zug der Inbesitznahme noch verschließendes gelobtes Land ist natürlich auf der Ebene eingebetteter Gründe eine perfekte Analogie für den Bund eines Volkes mit einem transzendenten Gott.

In der geschilderten Weise wird der universalistische, moralische Kern des Gesetzes mit einer Kollektiverzählung integriert, die für die Mitglieder einer Erinnerungsgemeinschaft *eingebettete* Gründe liefert, ihre Identität zu bewahren. Aber die Geschichte ist hier noch nicht zu Ende. Der Beschneidungsritus nämlich verbindet die kollektiv-historische Erfahrung mit einer individuellen, die im wörtlichsten Sinn *verkörpert* ist. Er erzeugt einen *character indelebilis*, ein dauernd gegenwärtiges Erinnerungsmittel, das nicht aufgehoben und rückgängig gemacht werden kann. Der folgende Gedankengang ist vielleicht ein wenig spekulativ, aber mir erscheint es plausibel anzunehmen, dass dieser Einschnitt in den Körper eine Erfahrung des Andersseins erzeugt, die als Erinnerung an die Andersartigkeit eines transzendenten Gottes zu dienen vermag. Aus einer semiotischen Perspektive betrachtet, sind irreversible Veränderungen des Körpers mit Zeichencharakter eine effektive Methode, um Bedeutung und Präsenz zu verkoppeln. Diese Verkoppelung setzt naturgemäß voraus, dass direkte und indirekte Erfahrung integriert werden. Eine aus medizinischen Gründen vorgenommene Circumcision erzeugt keinerlei Bedeutung, denn sie hat weder präsentationalen noch repräsentationalen Charakter, sondern lediglich eine *Funktion*. Allein der semiotischen Einheit direkter (also leibgebundener) und indirekter (also symbolisch vermittelter) Erfahrung gelingt es, qualitativ erlebenden und durch physische Interaktion mit anderen verbundenen Individuen Gründe vorstellig zu machen, die auch tatsächlich motivationale Kraft zu entwickeln vermögen.

Ich fasse zusammen: In sozialen Begründungsdiskursen können freistehende, rationale Argumente nur eine begrenzte Rolle spielen. Wie oben schon angedeutet, sind solche Argumente dennoch ganz unverzichtbar und wichtig: immer dann nämlich, wenn es um die Rechtfertigung von Normen geht, die kontextgebundene Werte überfangen und das Verhältnis zwischen den Vertretern divergierender Werthaltungen regulieren.[33] In konkreten Argumentationen sind es in aller Regel semiotische Muster, die direkte Erfahrung auf der individuellen wie kollektiven Ebene ins Spiel bringen. Was ich mit Parsons und Joas Wertegeneralisierung genannt habe, steht dabei für die Möglichkeit der Öffnung partikularer, an verkörperte

---

33 Für ein differenzierte Darstellung des Verhältnisses zwischen universellen Normen und partikularen (nicht zwangsläufig auch partikularistischen) Werten vgl. Joas (1997), Kap. 10.

Gründe gebundener Werthaltungen für eine *von innen* kommende Universalisierung (im Unterschied zu der bei Habermas dominierenden Vorstellung eines harten begrifflichen Schnitts zwischen Werten und Normen).

Diese strukturellen Einsichten habe ich im letzten Abschnitt versuchsweise auf ein eminentes Beispiel aus der Religions- und Kulturgeschichte angewendet, nämlich auf den biblischen Themenkomplex *Bundesgedanke-Gelobtes Land-Beschneidung-Moralisierung des Gesetzes*. Die Stärke und Dauerhaftigkeit dieser kulturellen Tradition – das konnte zumindest plausibel gemacht werden – kann darauf zurückgeführt werden, dass ihr eine tiefe Integration verkörperter (Beschneidung), eingebetteter (Landverheißung) und freistehender (moralischer Universalismus des Gesetzes) Gründe gelungen ist. Genauso offensichtlich ist aber auch, dass hier ein unauflösbares Spannungsverhältnis vorliegt: Aufgrund der Universalität freistehender Gründe ist es niemals möglich, eingebettete und verkörperte Gründe isomorph auf diese zu projizieren. Diese Spannung ist *strukturell* unaufhebbar, und sie lässt sich entsprechend auch in den soziokulturellen Konflikten moderner Gesellschaften zwischen ihrem universalistischen Rahmenwerk (Gleichheit vor dem Gesetz, Menschenrechte etc.) und den nicht miteinander kompatiblen Ansprüchen einer Vielfalt von Religionen und Weltanschauungen wiederfinden. Die bundesdeutsche Beschneidungsdebatte ist hierfür ein gutes Beispiel. Die Gefahr, der verkörperte und eingebettete Begründungen immer ausgesetzt sind, liegt in der Verwechslung intensiver partikularer Erfahrungen mit universellen Geltungsansprüchen, während umgekehrt freistehend-universalistische Argumentationen immer Gefahr laufen, so abstrakt zu werden, dass sie mit der direkten Erfahrung leibhaftiger Menschen nicht mehr vermittelt werden können. Wenn wir aber einmal realisiert haben, dass auch freistehende Begründungen stets auf häufig unsichtbar bleibenden verkörperten und eingebetteten Gründen aufruhen, dann können wir auch besser verstehen, was für die Beteiligten zur Disposition steht, wenn Konflikte zwischen universalistischen Normen und partikularen Wertvorstellungen entstehen, die an verkörperte und eingebettete Gründe gebunden sind. Wir haben dann die naive Vorstellung hinter uns gelassen, alles Partikulare sei partikularistisch und wirkliche Rationalität könne ohne Verweise auf historisch eingebettete und erstpersönlich erlebte Erfahrungen auskommen. Nicht aufgegeben haben wir hingegen die Vorstellung, dass das soziale Geschehen des Gebens und Nehmens von Gründen eine transzendierende, Wertegeneralisierungen ermöglichende Kraft entwickeln kann. Doch auch diese Kraft bleibt verkörpert; sie kann nicht mehr nach dem Modell der diskursiven Selbstbefreiung des Geistes durch freistehende Argumentationen gedacht werden.

Literatur

Bellah (2011): Robert Bellah, Religion in Human Evolution. From the Paleolithic to the Axial Age. Cambridge, MA/London.

Bellah / Joas (2012): Robert Bellah / Hans Joas (Hg.), The Axial Age and its Consequences. Cambridge, MA/London.

Die Bibel (1982). Einheitsübersetzung der heiligen Schrift. Stuttgart.

Dewey (1996): John Dewey, Die Öffentlichkeit und ihre Probleme (The Public and its Problems, 1927). Bodenheim.

Dewey (1998): John Dewey, „Peirce's Theory of Quality". In: Larry A. Hickman, Thomas M. Alexander (Hg.), The Essential Dewey, vol. 2, Ethics, Logic, Psychology. Bloomington and Indianapolis, 371-376.

Dewey (2003): John Dewey, „Qualitatives Denken" (1930). In: ders., Philosophie und Zivilisation. Frankfurt/Main, 94-116.

Dewey (2004): „Theorie der Wertschätzung" (Theory of Valuation, 1939). In: ders., Erfahrung, Erkenntnis und Wert. Frankfurt/Main, 293-361.

Donald (1991): Merlin Donald, Origins of the modern mind. Three stages in the evolution of culture and cognition. Cambridge MA/London.

Donald (2001): Merlin Donald, A Mind so rare. The evolution of human consciousness. New York.

Habermas (1981): Jürgen Habermas, Theorie des kommunikativen Handelns, Bd. 1, Handlungsrationalität und gesellschaftliche Rationalisierung. Frankfurt/Main.

Habermas (2005): Jürgen Habermas, „Religion in der Öffentlichkeit. Kognitive Voraussetzungen für den ‚öffentlichen Vernunftgebrauch' religiöser und säkularer Bürger". In: ders., Zwischen Naturalismus und Religion. Philosophische Aufsätze. Frankfurt/Main, 199-154.

Habermas (2012): Jürgen Habermas, „Die Lebenswelt als Raum symbolisch verkörperter Gründe". In: ders., Nachmetaphysisches Denken II, Aufsätze und Repliken, Berlin, 54-76.

Heil / Kramer (2012): Johannes Heil / Stefan Kramer (Hg.), Beschneidung: Das Zeichen des Bundes in der Kritik. Zur Debatte um das Kölner Urteil. Berlin.

Heine (1967): Heinrich Heine, „Geständnisse". In: Heines Werke in fünf Bänden, fünfter Band. Berlin und Weimar.

Joas (2004): Hans Joas, Braucht der Mensch Religion? Über Erfahrungen der Selbsttranszendenz. Freiburg.

Joas (2011): Hans Joas, Die Sakralität der Person. Eine neue Genealogie der Menschenrechte. Berlin.

Knöbl (2001): Wolfgang Knöbl, Spielräume der Modernisierung. Das Ende der Eindeutigkeit. Weilerswist.

Nietzsche (1988): Friedrich Nietzsche, „Zur Genealogie der Moral". In: ders., Jenseits von Gut und Böse. Zur Genealogie der Moral (KSA 5, hg. von Giorgio Colli und Mazzino Montinari). München.

Peirce (1998): Charles S. Peirce, „What is a sign?". In: The Peirce Project (Hg.), The Essential Peirce vol. 2 (1893-1913). Bloomington and Indianapolis, 4-10.

Rappaport (1999): Roy Rappaport, Ritual and Religion in the Making of Humanity. Cambridge etc.

Sandel (2012): Michael J. Sandel, What Money can't buy. The Moral Limits of Markets. New York.

Wie wird Freiheit verkörpert?
Breaking Bad: Eine Fallstudie

Magnus Schlette

Die Vereinigten Staaten von Amerika besitzen nicht nur wie alle übrigen Staaten auf der Welt eine Nationalflagge und ein Staatswappen, sondern sie haben sich auch zwei nationale Motti gegeben, die man auf jeder Dollarnote nachlesen kann: „In God We Trust" steht auf der einen Seite, „E pluribus unum" auf der anderen. Gemeint ist die stets aufs Neue zu bewährende Hoffnung, dass die Angehörigen verschiedener Ethnien und Kulturen auf amerikanischem Territorium zu einer Nation zusammenfinden. Dass dies gelingen möge, ist nicht in menschlicher Hand und doch nur durch gemeinsame Anstrengung zu erreichen. So erinnert jeder Geldschein, der als solcher die Spielmarke des Besitzindividualismus ist, durch den Aufdruck der beiden Staatsmotti an die spannungsvolle Dialektik von Eigeninteresse und Gemeinwohlbindung. Und weil die USA eine föderale Republik sind, die sich aus fünfzig Bundesstaaten zusammensetzt, sollten auch diese jeweils ihren eigenen Wahlspruch haben, der sich zu dem des Gesamtstaates füge. Arkansas entschied sich zum Beispiel für „Regnat Populus", Wyoming hoch im Norden machte „Equal Rights" zu seinem Motto, Kalifornien sinnigerweise „Eureka" – gemeint ist wohl der Pazifik. Alle diese Wahlsprüche betonen auf die eine oder andere Weise zentrale Werte der amerikanischen Nation: Individualismus, Pioniersinn, Gemeinschaft und – geradezu martialisch im Motto des Neuengland-Staates New Hampshire – Freiheit. Die Einwohner dieses Bundeslandes stellen nämlich jeden Neuankömmling vor die Wahl: „Live Free Or Die".

Die Freiheit, welche die Bürger New Hampshires zum Gegenstand ihres Staatsmottos erkoren haben, begegnet uns im Gewande der adverbiellen Bestimmung eines Verbs – to live, leben –, das, wie uns die Grammatik belehrt, zur semantischen Klasse der Zustandsverben gehört. Frei zu leben kann daher nur heißen, sich in einem durch Freiheit qualifizierten Zustand zu befinden. Die Lebensführung ist gegebenenfalls zuständlich als frei qualifiziert. Der Staat New Hampshire stellt uns allerdings vor eine Alternative: Lebe entweder frei oder stirb! Offenbar unterstellt er uns die Fähigkeit, zwischen zwei Optionen zu wählen und daher auch, den angesonnenen Zustand herbeizuführen. Aber leben wir erst dann frei, wenn wir uns in dem erstrebten Zustand befinden oder auch schon dann, wenn wir uns noch bemühen, ihn herzustellen? Und wie müssen wir uns vorstellen, dass wir das Beabsichtigte erreicht haben? Sind wir dann einfach so frei, ohne fernerhin noch etwas dafür tun zu müssen? Ich kann lesen, während ich auf der Couch liege, also liegend etwas tun; ich kann aber nicht in der begrifflich gleichen Weise lebend tätig sein, zum Beispiel Fußball spielen, während ich lebe. Denn ich lebe in meinen und durch meine Tätigkeiten und nicht zusätzlich zu ihnen. Daher erfordert der Anspruch frei zu leben: in meiner und durch meine Lebensführung frei zu sein, mein Leben auf freie Weise zu führen, und das schließt sämtliche Anstrengungen ein, die ich aufbringen muss, um das erstrebte Ziel zu erreichen. Natürlich ist das Staatsmotto von New Hampshire, das den Bürgern die Alternative ansinnt, diesem Anspruch nachzukommen oder andernfalls besser zu sterben, eine Erinnerung an den Geist des amerikanischen Unabhängigkeitskriegs, wie jedes Schulkind in den USA lernt. Es ist ein Appell an das Engagement für politische Freiheit im Sinne der demokratischen Selbstorganisation des amerikanischen Gemeinwesens, dessen Gründungsstaaten in Neuengland sich von der Fremdherrschaft der englischen Kolonialmacht befreit hatten. Schon dieser konkrete historische Bedeutungshintergrund macht deutlich, dass die beanspruchte Freiheit nur durch die immer neue Bewährung eines bestimmten – eben freiheitlichen – Lebensvollzugs erfüllt und daher niemals in einen unbezweifelbaren Besitz überführt werden kann. Das Staatsmotto mahnt an, dass eine freiheitliche Lebensweise die Voraussetzung der Gewährleistung freiheitlicher staatlicher Institutionen ist und pointiert damit die Fragilität der politischen Freiheit. Damit deutet der Wahlspruch New Hampshires allerdings einen generischen Sinn des Freiheitsanspruchs an, der über die politische Freiheit auf die grundsätzliche Frage hinausweist, in welchem Sinne wir davon sprechen können, in und durch unsere Tätigkeiten frei zu sein.

Geläufigerweise orientiert sich der Begriff menschlicher Freiheit an der Unterscheidung zwischen Willensfreiheit und Handlungsfreiheit. Unter Handlungsfreiheit ist zu verstehen, dass wir frei sind zu tun, was wir tun wollen, unter Willensfreiheit, dass unsere Entscheidungen darüber, was wir wollen, ohne inneren oder äußeren Zwang zustande kommen. In unserem Handeln sind wir daher frei, wenn die subjektiven Handlungsbedingungen und die objektiven Umstände der Handlungssituation unsere Willensausübung nicht verhindern. Allerdings konzedieren wir, dass unsere Handlungsfreiheit durch die Widerstände qualifiziert wird, die wir überwinden müssen. Ganz ohne die Erfahrung von Widerständen würden wir wahrscheinlich über kurz oder lang an Realitätsverlust leiden, da sich unser Wille an keinen objektiven Bedingungen seiner Ausübung mehr reiben müsste. Wird er dagegen durch die objektiven Verhältnisse gebeugt, sind wir eben nicht mehr frei in unserem Handeln. Insofern ist unsere Handlungsfreiheit immer eine situational bedingte und durch ihre Bedingtheit qualifizierte Freiheit. Was unseren Willen frei macht, ist schwieriger zu sagen. Immerhin urteilen wir auch über die Willensfreiheit eines Subjekts immer relativ auf das ihm potentiell Mögliche; es wäre absurd, mit den Armen fliegen zu wollen, und daher ist das nichts, was wir billigerweise wollen könnten. Darüber hinaus dürfte unsere Willensbildung in vielfältiger Weise durch ihre Vorgeschichte bedingt sein, also dadurch, was für Menschen wir jeweils sind, was unsere Erfahrungen, unsere Präferenzen und vor allem unsere Abneigungen: Gebranntes Kind, sagt der Volksmund, scheut das Feuer. Aber die Bedingtheit der Willensbildung schließt ihre Freiheit nicht notwendig aus; sie ist nicht *eo ipso* ein Freiheitsproblem, ebenso wenig wie die Bedingtheit der Willensausübung, wenn auch aus anderen Gründen. Wie schon im Falle der Handlungsfreiheit stellt sich auch für unseren Willen das Freiheitsproblem erst, wenn wir näher bestimmt haben, wodurch der Wille jeweils bedingt ist.[1] Verdankt er sich einem Bildungsprozess der besonnenen Abwägung von Optionen oder bricht er in uns wie eine unaufhaltsame Macht hervor, die uns befremdet und vielleicht sogar ängstigt? Wenn der Wille durch das Resultat von Deliberationen bedingt ist, nennen wir ihn frei, werden wir dagegen von unserem Willen getrieben, ist er zwanghaft oder unbeherrschbar heftig, wird er erzwungen oder manipuliert, dann nennen wir ihn unfrei.[2]

Aber weder Handlungs- noch Willensfreiheit erfassen hinreichend, was in der Alternative „Live Free Or Die" gemeint ist. Das Freiheitsverständnis, das dem Staatsmotto von New Hampshire zugrunde liegt, enthält eine Pointe, die durch die beiden herkömmlichen Freiheitsbegriffe nicht hinreichend erfasst ist. Es gibt eine existentielle Dimension unseres Freiheitsverständnisses, die unsere Vorstellungen von einem guten Leben zutiefst prägt und nicht in Begriffen der Handlungs- und Willensfreiheit aufgeht. Zur Bezeichnung dieser Freiheitsdimension möchte ich von adverbieller Freiheit sprechen. Diese Freiheit, die Freiheit, auf freie Weise zu leben, qualifiziert die Ausübung der für mein Leben relevanten Tätigkeiten als frei. Über die freie Ausübung meiner Tätigkeiten entscheidet aber *nicht schon* die Willensfreiheit und *nicht erst* die Handlungsfreiheit. Denn die adverbielle Freiheit ist eine Eigenschaft, die wir allenfalls der Wechselbeziehung zwischen der Bildung unseres Willens und seiner Artikulation durch eine bestimmte Art des Tätigseins zuschreiben können. Ist diese Wechselbeziehung in einer bestimmten

---

1   Vgl. zu Gemeinsamkeit und Differenz von Willens- und Handlungsfreiheit Bieri (2003), 29-53.

2   Eine reiche und präzise Phänomenologie dieser Fälle von Unfreiheit bietet Bieri (2003), 84-126.

Weise beschaffen, dann ist es gerechtfertigt zu sagen, dass wir in und durch unsere Tätigkeiten frei sind. Die Freiheit, die unser Tätigsein in diesem Fall auszeichnet, ist die Freiheit eines gelungenen Passungsverhältnisses zwischen Willensbildung und -artikulation.[3]

Die folgenden Überlegungen bemühen sich um die Plausibilisierung dieser These. Dabei verfolge ich zunächst die Strategie einer phänomenologischen Sensibilisierung für meine These durch die Charakterschilderung des Protagonisten von *Breaking Bad,* einer US-amerikanischen Fernsehserie, die auf beiden Seiten des Atlantiks während ihrer Ausstrahlung Kultstatus erworben hat und für das Thema der adverbiellen Freiheit einschlägig ist (I). Darauf aufbauend umreiße ich die begrifflichen Implikationen des Freiheitsverständnisses, das in der Hauptfigur der Serie narrativ verdichtet worden ist (II). Nach einer kurzen Zusammenfassung der bisherigen Argumentation werde ich dann meine Hauptthese vorbereiten: Das Passungsverhältnis von Willensbildung und -artikulation *verkörpert* die Freiheit, die wir in und durch eine bestimmte Weise unseres Tätigseins erlangen können; adverbielle Freiheit ist wesentlich *verkörperte* Freiheit (III). Abschließend kehre ich zu meinem Fallbeispiel zurück und deute in einer Art Ausblick auf weiterführende systematische Erwägungen an, dass eine Orientierung des Lebens an adverbieller Freiheit in ein unauflösbares Spannungsverhältnis zur Orientierung am prudentiell Guten und moralisch Guten gelangen kann; dieses Spannungsverhältnis charakterisiert eine Spielart des Individualismus, die ich als *anarchischen* Individualismus bezeichnen möchte, insofern er die Verkörperung von Freiheit über die Moral und das Gute stellt (IV).

I

Noch einmal zurück zum Motto des US-Bundesstaates New Hampshire. Wer diesen Staat besucht, wird der Aufforderung „Live Free or Die" öfters begegnen, denn sie ziert jedes Nummernschild eines in New Hampshire zugelassenen Autos. In der

---

3   Ausgearbeitete Ansätze zu einer Theorie derjenigen Freiheit, die ich meine, finden sich im amerikanischen Pragmatismus, vor allem bei John Dewey. Vgl. Dewey (1969-1991a), 92-114, ein Aufsatz, in dem Dewey sich innerhalb einer von ihm entworfenen Typologie philosophischer Freiheitsbegriffe verortet; vgl. außerdem Dewey (1969-1991b), 193-203. Die folgenden Überlegungen sind von Dewey inspiriert, eine Interpretation seiner Freiheitstheorie im Sinne des hier von mir vorgestellten Konzepts der adverbiellen Freiheit ist aber einem anderen Aufsatz vorbehalten. Siehe Schlette (2015). Vgl. hierzu auch Stroud (2011); Pappas (2008), v. a. Kap. 10.

US-Fernsehserie *Breaking Bad* bricht ihr Protagonist, Walter White, in einem alten Volvo mit New Hampshire-Kennzeichen zum finalen *showdown* auf. Die emblematische Bedeutung des Nummernschildes wird bereits zu Beginn der ersten Folge der letzten Serienstaffel durch eine kurze erratische Sequenz angesonnen, die mit einem *close up* auf das Nummernschild des Wagens beginnt. Er fokussiert das New Hampshire-Motto auf dem Kennzeichen von Whites Volvo: „Live Free Or Die." Der volle Sinn der Sequenz, die hier zusammenhangslos und unverständlich aus dem Handlungsverlauf herausragt, erschließt sich erst ganz vom Ende der Serie her. In der letzten Folge, die an diese Sequenz narrativ anknüpft, verleiht das Staatsmotto den nachfolgenden Ereignissen, in denen sich der Plot der Serie dramaturgisch aufs Ende hin zuspitzt, ein unheilvolles Gewicht. Jetzt erfahren wir, dass White den Wagen auf dem weiträumigen Asphalt vor einem Schnellrestaurant geparkt hat, wo er von einem Waffendealer eine Maschinenpistole übernimmt. In letzter Zeit war ihm die Polizei auf den Fersen; er hatte sich darüber hinaus Ärger mit hinterwäldlerischen Drogenhändlern eingehandelt, die ihm das von ihm entwickelte Geschäftsmodell der Herstellung und Verbreitung von Methamphetamin entwendet haben und seinen ehemaligen Komplizen gefangen halten. White war derweil untergetaucht und kehrt zum Schluss noch einmal an seine frühere Wirkungsstätte zurück, um seinen Komplizen zu befreien und seine verwickelten Angelegenheiten zu regeln.

White war Chemielehrer an einer Highschool in Albuquerque, der seine Familie mehr schlecht als recht durch die wirtschaftlich instabilen Zeiten manövrierte; ein Mann mit einstmals großen wissenschaftlichen Ambitionen, der bescheiden geworden ist und nur mehr das Desinteresse einer Horde Halbwüchsiger pädagogisch verwaltet. Abends muss er in einer Autowaschanlage noch dazu verdienen, damit der Familie von den Banken nicht das bescheidene Eigenheim entzogen wird. So manövriert er sich durch die Zumutungen seines Lebens, bis er von einem tödlichen Tumor erfährt und keinen anderen Ausweg mehr sieht, die Zukunft der Familie zu sichern, als seine wissenschaftliche Expertise für die Entwicklung einer konkurrenzlos wirkungsvollen Droge zu nutzen. Sie findet rasch Abnehmer, bald schon kann White finanziell für die Zukunft der Restfamilie nach seinem Tod vorsorgen. *Breaking Bad*: Aus einem harmlosen und unter der Last des Lebens frustrierten Familienvater der amerikanischen Provinz wird auf einem langen und verzweigten Weg ein leidenschaftlicher Drogenhändler und kaltblütiger Mörder. Diese Entwicklung ist White aber nicht einfach widerfahren, sondern er hat sie sich mühevoll erarbeitet. Zunächst musste er die Umsetzung seines illegalen Vorsorgeplans für die Familie starken widerstrebenden Willensimpulsen abringen; aber die Art und Weise, wie er das tat: die zunächst kleinen und schließlich immer gewagteren Lügen, mit denen er sein heimliches Treiben vor der Familie zu verbergen versuchte, und die Bemühungen um Anpassung an ein Milieu, in dem er sich nur durch die

Kommunikation von Gewaltbereitschaft behaupten kann, hinterlassen zusehends Erfahrungsspuren in seinem Charakter, die den ehemaligen Lehrer schließlich in ein neues Leben einspuren. Punkten die ersten Folgen mit sarkastisch grundierter Komik, wenn der Pädagoge seine potenten Produkte unter finsteren Gesellen an den Mann bringen will, die auf verwahrlosten Hinterhöfen ihre Rottweiler ausführen, so lehren die späten Folgen die Zuschauer das Gruseln, wenn White sich mit zunehmender Sicherheit auf dem abschüssigen Parkett der Unterwelt bewegt.

Die Charakterisierung eines Menschen als ‚gehemmt‘ bezieht sich auf Situationen, in denen er sich, wie wir dann auch sagen, in ‚unfreier‘ Weise aufführt. Ein Beispiel ist der vereinzelte Partygast, der sich verlegen ans Büffet klammert und nur mühevoll in die Gespräche hineinfindet. Die Gastgeber werden nachher von ihm sagen, er habe nicht zu den anderen Gästen ‚gepasst‘. Der besagte Gast mag sich frei entschieden haben, die Geselligkeit der Party zu suchen und damit Ziele verfolgt haben, die er auch erreicht hat, zum Beispiel einen neuen Kontakt zu knüpfen, der seinem beruflichen Fortkommen förderlich ist. Und doch hat er sein Ziel auf eine ‚unfreie‘ Weise erreicht, er hat sich nicht in derselben Weise frei unter den anderen Gästen bewegt, wie – ein anderes Beispiel – der Fußballspieler, der nach langer Verletzung erstmals wieder ‚befreit‘ aufgespielt hat. Der Fußballer hätte vielleicht auch in ängstlicher Befangenheit durch seine gerade überwundene Bänderzerrung das Tor geschossen, das ausschlaggebend für das Ziel der Spieler war, den Klassenerhalt zu schaffen. Der Rekonvaleszent teilt dieses Ziel, und doch ist es etwas anderes für ihn, wenn er befreit aufzuspielen vermag und im Schwung der wiedergewonnenen Spielfreude das Tor erzielt. Die ersten Folgen von *Breaking Bad* zeigen Walter White als einen in nahezu allen seinen Tätigkeiten gehemmten Mann. Seine Hemmungen haben sich ihm zu der Erfahrung eines unfreien Lebens verdichtet. Diese Erfahrung hat sich in einem zaudernd-hektischen Verhaltensstil zum Ausdruck gebracht, der sich besonders in Krisensituationen bemerkbar macht, die aus der Routine des Alltäglichen herausfallen. White, so könnte man auch sagen, steht ganz ‚neben sich‘. Er hat das Gefühl, dass er in dem, was er tut, nicht gut ist, dass er es nicht auf eine gute Weise tun kann, dass er – mit anderen Worten – eine schlechte Figur macht in seinem Leben.[4] Das gelingt ihm später, als gefürchteter Unterweltler, viel besser, aber erst als Ergebnis eines langen Weges, auf dem sich die Entscheidung für ein neues Leben durch den Gestaltwandel der Lebensführung sukzessiv durchsetzen und bewähren musste.

Auf der letzten langen Fahrt mit dem Volvo aus seinem Unterschlupf vor der Polizei im winterlichen New Hampshire zurück in seine frühere Heimat in dem Wüstenstaat New Mexico stattet White auch seiner Frau einen Besuch ab, die längst

---

4   Vgl. dazu jetzt Schlette (2014), 117-130.

mit ihm gebrochen hat und mit ihren Kindern aus dem früheren gemeinsamen Domizil weggezogen ist. „I did this" – hebt er an, um die Motive der Verbrechen zu erläutern. Die Frau unterbricht ihn, denn was sie ihn zu sagen erwartet, hat sie schon oft gehört, seitdem er sein neues Tätigkeitsfeld vor ihr nicht mehr verbergen konnte. „Oh, don't tell me again, you did this for the family", faucht sie ihn an. Mit tonloser Stimme entgegnet Walter – wer sein Gesicht sieht, weiß sofort, dass er keine Mimikry an eine Lüge mehr betreibt, sondern zum ersten Mal die Wahrheit sagen wird: „I did it for me. […] I was good at it. I was feeling alive." Plötzlich ist heraus, was uns die Bildsprache dieses Serienmeisterwerks über dutzende von Folgen bereits gezeigt hat: dass nämlich Whites Metamorphose in einen szeneberüchtigten Drogenmagnaten nur deshalb gelingen konnte, weil der ehemalige Lehrer die damit verbundenen impliziten Verhaltensanforderungen der Kriminalität nicht nur als äußeren Druck erfahren hat, der von einer unwirtlichen Umwelt auf ihn ausgeübt wurde, sondern weil er im gestaltrichtigen Handeln nach den Maßstäben, die der erfolgreiche Drogenhandel aufstellt, seine Persönlichkeit artikulieren konnte. Das Faszinosum der Serie besteht nicht zuletzt darin, Whites Lehrzeit sinnfällig zu machen, in der er sich die Voraussetzungen erarbeitet, das Handwerk des Drogenhandels gut auszuüben. Die Serie spart nicht mit drastischen Beispielen, wie man das auch schlecht machen kann.

Man kann ein unbeholfener Schurke sein wie Walter White als Lehrling der Unterwelt, oder man kann auch alles im Griff haben, wie es bei seinen späteren Missetaten zunächst genialisch aufblitzt, bis die Kaltblütigkeit, mit der er das nur kaltblütig Machbare tut, ihm schließlich zur zweiten Natur seiner Gangstereien geworden ist. Die Glatze rasiert er sich noch mit der Absicht, sich den *air* von Hartgesottenheit zu geben, die Schärfe seiner Gesichtszüge wirkt zunächst markiert, gestellt; später bedarf es nicht mehr der Attitüde, ihr Ausdruckscharakter stellt sich schließlich unwillkürlich ein. Die Furchen im Gesicht – ein Lob auf den Schauspieler Bryan Cranston – zeugen von unablässiger Anspannung eines längst entschiedenen Willens. Das alte, fahrige Selbst, das in besonders kritischen Situationen dazu neigte, die Nerven zu verlieren, macht sich bis fast zuletzt immer wieder bemerkbar, aber es gewinnt nicht mehr die Oberhand, wenn es darauf ankommt. Die Zuschauer können nachvollziehen, wie White sich in seiner neuen Rolle immer besser zurecht findet und sie als gelungenen Ausdruck seiner veränderten Lebenseinstellung akzeptiert. So entdeckt der ehemalige Lehrer im weiteren Verlauf der Serie *in actu*, worin seine Stärken liegen; er entdeckt es an der wachsenden Verhaltenssicherheit, mit der er Herr der unmöglichsten Situationen bleibt, in die er sich hineinmanövriert hat. Sein Meisterstück absolviert er bei einem Überfall auf einen Güterzug, den er mit bewundernswertem Gefühl für das richtige *timing*, mit sachlichem Überblick bei der Koordination der Beteiligten und einer die atemberaubende Geschwindigkeit der Aktion entschleunigenden *coolness* geradezu zelebriert: „I was good at it. I was feeling alive."

II

Es ist überhaupt keine Frage, dass die Serie mit der herausragenden Bedeutung, die sie dem New Hampshire-Nummernschild in der emblematischen Szene zumisst, in der sich Walter White auf einem Shopping Mall-Parkplatz für das Ende rüstet, einen Zusammenhang zwischen der adverbiellen Freiheit des „Live Free Or Die" und der adverbiellen Güte von Handlungen behauptet, durch die wir eine bestimmte Tätigkeit in guter Weise ausüben. Die Tätigkeit des Drogenhändlers gut auszuüben besteht unter anderem darin, belastbare Kontakte zu qualitativ hochwertigen Produzenten aufzubauen, einen effizienten Vertrieb zu entwickeln, die Mitglieder dieser Unternehmung mit Bedacht auszuwählen und zu führen, sich in der Konkurrenz mit rivalisierenden Anbietern zu behaupten usw. „I was good at it. I was feeling alive": Vince Gilligans Regie in *Breaking Bad* macht plausibel, dass es ein Freiheitsverständnis gibt, in dem die Bewährung am adverbiell Guten zum Maßstab des prudentiell Guten wird. Das adverbiell Gute bezeichnet die gute Ausübung einer Tätigkeit, und diese betrifft ihre Anerkennungswürdigkeit nach Maßstäben, die den besagten Tätigkeiten immanent sind. Die Maßstäbe des Guten beschränken sich aber auch nicht auf die Funktionalität der Tätigkeit für ihr Resultat, auf welche die logisch attributive Verwendung von „gut" abhebt – etwa im Falle der Rede vom ‚guten Messer', das vom Können des Handwerkers ein Zeugnis abgibt, der es hergestellt hat. Denn die Vollzugsweise funktional gleichermaßen zielführender Tätigkeiten kann immer noch besser oder schlechter sein.[5]

Das den jeweiligen Tätigkeiten immanente Kriterium des adverbiell Guten ist eben jene Erfahrung, in seinem Leben eine gute Figur zu machen, die Walter White seiner Frau eingesteht: „I was good at it. I was feeling alive." Und der Kern dieser Erfahrung, so behaupte ich, ist das Passungsverhältnis zwischen Willensbildung und -artikulation. Die Pointe dieses Passungsverhältnisses ist, dass es auf einer Entwicklung beruht, in der Willensbildung und Willensartikulation (durch den Vollzug von willentlichen Handlungen) einander *wechselseitig stimulieren*. Demnach beruht das Passungsverhältnis darauf, dass das *feedback* von Tätigkeiten auf den Handelnden, der ihre Ausübung als individuell angemessen erfährt, ihrerseits Einfluss auf die Willensbildung hat. Daher ist die adverbielle Güte der Tätigkei-

---

5   Vgl. zum Begriff der adverbiellen Güte von Handlungen Tugendhat (2003), Kap. 4.

ten weder *schon* diejenige der – besseren oder schlechteren, mehr oder weniger gebundenen – Willensbildung noch *erst* diejenige des Handlungsresultats, in dem sich der Wille – mehr oder weniger abhängig von praktischen Bedingungen seiner Verwirklichung, mehr oder weniger erfolgreich – objektiviert hat. Die adverbielle Güte qualifiziert stattdessen den Handlungs*vollzug*, in dem sich Willensbildung und -artikulation aneinander schärfen und justieren. Gut ist dieser Vollzug dann, wenn er Erfahrungen ermöglicht wie diejenigen, die Walter White schließlich zu einer radikalen Neuorientierung seines ganzen Lebens bewegt haben.

White war nicht immer ein frustrierter Lehrer, bevor er sich, zunächst aus dem Gefühl der Not und Unausweichlichkeit heraus, später aus ganzem Herzen, für die Umschulung auf Drogenhandel entschied. Die Serie deutet verstreuterweise an, dass er vielmehr ein passionierter Lehrer war. Zu seiner Gehemmtheit, zum Selbstgefühl, eine schlechte Figur in seinem Leben zu machen, wird auch sein Eindruck beigetragen haben, dass er die eigene Begeisterung für die Chemie nicht an die Schüler vermitteln konnte. Unterstellen wir ihm eine freie Willensbildung in Bezug auf den Lebensplan, Lehrer zu werden, so entgeht dem Zuschauer andererseits nicht, dass White die für sein Leben maßgebliche Berufstätigkeit als Missverhältnis zwischen Willensbildung und -artikulation erfahren hat (abgesehen von der ganz handfesten Misslichkeit des zu geringen Verdienstes). Und die Angemessenheit dieser Erfahrung ist schlecht objektivierbar. Jedenfalls schien White allen Andeutungen zufolge im Kollegium als guter Lehrer zu gelten; es gab auch keine Andeutungen darüber, dass seine Schüler hinter den Leistungsansprüchen zurückblieben. Whites fundamentale Erfahrung der Gehemmtheit, der Unfreiheit ist keineswegs, dass sein Wille (Chemielehrer zu werden) nicht zu den Ansprüchen der Schulbehörde oder der Schüler oder der Familie passt, sondern dass er nicht zu *ihm selbst* passt, dass er, Walter White, während er chemische Experimente vorführt und Formeln lehrt, zugleich ‚neben sich steht‘. Es reicht auch nicht hin zu sagen, er sei an seinem eigenen Anspruch gescheitert, die Schüler für Chemie zu enthusiasmieren. Allenfalls wäre ihre Begeisterung eine Bestätigung dafür gewesen, in seiner Tätigkeit als Lehrer eine gute Figur zu machen.

Whites Existenzkrise hatte begonnen, bevor die Ärzte einen bösartigen Tumor diagnostizierten; sie besteht in einer fundamentalen Verunsicherung darüber, welchen Willen er, White, in einer ihn bezeugenden Weise artikulieren könnte. Erst auf seinem neuen Betätigungsfeld gewinnt er diese Sicherheit schrittweise zurück, bis hin zu dem Eingeständnis: „I did it for me. [...] I was good at it. I was feeling alive.“ Die Formulierung, er habe sich in der Ausübung seiner illegalen Tätigkeiten lebendig gefühlt, muss als Hinweis darauf verstanden werden, dass sich das Passungsverhältnis zwischen Willensbildung und -artikulation dem Handelnden als ein affirmatives Selbstgewahrsein bezeugt; in diesem Fall bin ich meiner selbst in gesteigerter Weise *zuständlich* als Handelnder bewusst. Das Passungsverhältnis wird erfahren als ein Lebensgefühl, durch das sich der Handelnde unbestimmt als derjenige erfährt, auf den es hier und jetzt ankommt.[6] Es ist ein Aktivitätsbewusstsein nicht von dieser oder jener Handlung, sondern – so könnte man vielleicht in Anlehnung an Kant sagen – von ‚Tätigkeit überhaupt' und damit zugleich auch davon, dass ich die Handlungssituation durch mein Handeln gestalte, dass die Situation, in der ich mich finde, zugleich auch von mir ‚gemacht' wird, dass ich Herr des Geschehens bin.[7]

Tätigkeiten, bei deren Ausübung sich uns in eindrücklicher Weise bezeugt, dass ihr Gelingen von unserem persönlichen Engagement abhängt, können wir als ‚Ich'-Handlungen bezeichnen.[8] Von solchen Handlungen sagen wir häufig auch unter grammatisch redundanter Verwendung des entsprechenden Pronomens, wir hätten sie *selbst* ausgeführt. Der pronominale Verwendungssinn von ‚selbst' fängt präzise das Lebensgefühl ein, das ‚Ich'-Handlungen begleitet. Sie sind dadurch qualifiziert, dass sie sich „nur als Gegenstand eines Sich-Ansprechens in einem Selbstimperativ verstehen" lassen,[9] weil sie die Anmutung haben, dass jeweils unvertretbar *ich* es bin, der sie ausübt und dass es von mir abhängt, wie sie ausgehen. Entscheidend für das zuständliche Bewusstsein der ‚Selbstbeteiligung' an meinen ‚Ich'-Handlungen ist, dass sie der Art und Weise des Handlungs*vollzugs* gegenüber dem Handlungsziel besonderes Gewicht verleiht. Dass das Gelingen oder Misslingen der ‚Ich'-Handlungen von mir abhängt, heißt nämlich nichts anderes, als dass es davon abhängt, *wie* sie vollzogen werden. Im *Wie* ihres gelungenen Vollzugs verkörpert sich ihr ‚Ich'-Charakter, durch die Art und Weise ihrer gelungenen Ausübung werden sie als ‚Ich'-Handlungen individuiert. Umgekehrt wird ihr Misslingen als ein Verlust von ‚Selbstbeteiligung' erfahren, als eine Art Verselbständigung des Handlungsverlaufs, an dem ich nur mehr als ausführendes Organ der Handlung teilhabe.

---

6   Zum Begriff der Zuständlichkeit vgl. Fellmann (1996), 213-226.

7   Kant spricht im Rahmen seiner Kritik der ästhetischen Urteilskraft von einer „Erkenntnis überhaupt" (Kant, *KdU* A 28f.). Die Kant-Interpretin Brigitte Scheer erkennt in diesem Begriff zu Recht einen Schlüsselbegriff der Kantischen Ästhetik. Bei der „Erkenntnis überhaupt" kann der Verstand als „Inbegriff der Konzeptualisierung tätig sein und muß sich nicht auf *ein* Konzept oder *eine* Hinsicht auf die Anschauungsmannigfaltigkeit einschränken. Das ästhetisch reflektierende Subjekt erfährt sich so in der Vollständigkeit seiner Erkenntniskräfte und als lebendiger Ursprung aller Deutungs- und Sinnbildungsprozesse am anschaulich Gegebenen" (Scheer [1997], 91). Analog soll unter der ‚Tätigkeit überhaupt' eine Art des subjektiv belebenden Tätigkeitsbewusstseins verstanden werden.

8   Zum Begriff der ‚Ich'-Handlung vgl. Tugendhat (2003), Kap. 3.

9   Tugendhat (2003), 62.

Stellt sich im Gelingensfall das Passungsverhältnis zwischen Willensbildung und Artikulation ein, das Passungsverhältnis zwischen dem, was ich will, und dem, was ich vermag, so wird es im Misslingensfall verfehlt und die Handlung verliert dabei für den Handelnden den Charakter einer Artikulation, sinkt zurück in die als losgelöst vom Willen erfahrene Faktizität bloßer Tätigkeit, wie sie von jedem beliebigen anderen Handelnden auch ausgeführt werden könnte.

Zwar hängt es wesentlich von mir ab, ob meine ‚Ich'-Handlungen gelingen oder nicht, aber es gibt Kriterien, die zwingend erfüllt sein müssen, um ihnen ihr Gelingen oder Misslingen zu attestieren, und diese Kriterien stehen nicht in meiner Verfügungsgewalt, sondern sind durch den Typus von Tätigkeiten vorgegeben, die ich ausübe. Wie ich auf unbefestigtem Terrain gut Fahrrad fahre, was ich tun muss, um mit einer Segeljolle unter starken Böen eine Wende zu fahren, was meine Gäste erwarten können, wenn ich ein mehrgängiges Menü koche, hängt nicht allein von meinen Gelungenheitsanmutungen, sondern von objektiven Standards ab, die *cum grano salis* für alle Radfahrer, Segler und Köche gelten. Ihnen entnehmen wir, was als spezifische Widerstände dieser Handlungen gilt, an denen sich das Passungsverhältnis zwischen Willensbildung und -artikulation, zwischen dem, was ich will, und dem, was ich vermag, zu bewähren hat. Die Widerständigkeit der Welt gegenüber meinen Willensaspirationen beschränkt sich nicht nur auf die natürlichen Bedingungen, unter denen ich handeln muss, sondern sie schließt die intersubjektiven *constraints* ein, denen ich mich zu stellen habe. Deswegen individuieren gelingende ‚Ich'-Handlungen immer auch persönliche Interpretationen der Standards, an denen sie sich als solche bewähren. Tastfunker können einander am Rhythmus ebenso erkennen, wie die Musikliebhaber den Pianisten am Anschlag: die individuelle Nuancierung des Standards bezeugt den Handelnden.

Von dem Lebensgefühl her, das sich mit dem Passungsverhältnis zwischen Willensbildung und -artikulation einstellt, das also die adverbielle Güte von ‚Ich'-Handlungen qualifiziert, erschließt sich nun auch der Begriff der adverbiellen Freiheit im Unterschied zur Willens- und Handlungsfreiheit und damit diejenige Freiheit, die wir meinen, wenn wir uns dem Anspruch stellen, frei, auf freie Weise zu leben. Verstehen wir unter Handlungsfreiheit die Freiheit, unseren Willen auszuüben und unter Willensfreiheit die Freiheit, diesen Willen in zwangloser Deliberation der Alternativen zu bilden, so bedeutet adverbielle Freiheit die Freiheit eines gelungenen Passungsverhältnisses zwischen dem, was ich will, und dem, was ich vermag, eines Passungsverhältnisses, dessen ich mir in gesteigerter Weise in meinen ‚Ich'-Handlungen gewahr bin. Niemals ist die Erfahrung der adverbiellen Freiheit befreiender, als wenn eine oftmals erprobte und unter Widrigkeiten ausgeübte Tätigkeit erstmals eine gelungene Gestaltschließung gespurt hat. Je widerständiger die Umstände sind, woraufhin ich handele, desto deutlicher kann mein Handeln auch meine Signatur tragen, das heißt durch die Gelungenheit ihrer Gestaltschließung bezeugen, dass *ich* es war, der gehandelt hat: „I was good at it. I was feeling alive." Diese Freiheit ist nun aber auf ein bestimmtes Maß an Willens- und Handlungsfreiheit angewiesen, und zwar genau auf dasjenige Maß, das nötig ist, um die intrinsische Wechselbeziehung zwischen Willensbildung und -artikulation zu ermöglichen. Grundsätzlich muss das *feedback* der Handlungsvollzüge Einfluss auf den Willensbildungsprozess nehmen können und dessen Verlauf offen für die motivationalen Konsequenzen sein, die sich aus den Erfahrungen der Willensartikulation ergeben. Umgekehrt ist zumindest soviel Handlungsfreiheit dafür nötig, auf freie Weise zu leben, dass Handlungen tatsächlich grundsätzlich als *Artikulationen* jeweils meines Willens vollzogen werden können.

Fahradfahrer, Tastfunker und Pianisten, so ließe sich einwenden, mögen vielleicht illustrieren, was es heißt, sich in der gelingenden Ausübung einer Tätigkeit in gesteigerter Weise als lebendig zu erfahren, aber sie passen nicht zu dem Anspruch, auf freie Weise zu leben, den Walter Whites Kriminellenkarriere veranschaulichen sollte. Wo White sich seiner Frau gegenüber zu seiner Unterweltexistenz mit den Worten bekennt: „I was good at it. I was feeling alive", bezieht er sich nicht nur auf einzelne Tätigkeiten, sondern auf Praxis im Sinne eines übergreifenden und integrierenden Zusammenhangs von Tätigkeiten, zu denen in seinem Fall die Herstellung und Distribution von Methamphetamin, die Abschreckung rivalisierender Anbieter, die Täuschung der *Drug Enforcement Agency* usw. gehörte. Aber dieser Einwand verpasst die Pointe des Begriffs der adverbiellen Freiheit, die darin besteht, auf dem Wege einer Exploration des Passungsverhältnisses zwischen dem, was ich will, und dem, was ich vermag, langsam in einen Zusammenhang von zueinander passenden Tätigkeiten hineinzuwachsen, durch den sich allererst klärt, was ich als gut für mein Leben im Ganzen erfahre. Ich hatte meine Diskussion von *Breaking Bad* mit der Behauptung eingeleitet, der Regisseur demonstriere uns in der Figur Walter Whites ein Freiheitsverständnis, das die Bewährung am adverbiell Guten zum Maßstab der Exploration des prudentiell Guten eines individuellen Lebens macht. Das ist ungewöhnlich, eher würde man in Orientierung an der philosophischen Grundfrage, wie es gut ist zu leben,[10] das Umgekehrte erwarten. Aber White geht gerade nicht von einem allgemeinen Begriff des guten Lebens aus, in dessen Licht sich diese oder jene Praxis als lebenswürdig erschließt. An einem solchen allgemeinen Begriff hatte er sich ja bereits orientiert, als er sein jetzt verworfenes bürgerliches Leben danach einrichtete. Seine Krebsdiagnose und eine Reihe weiterer Zufälle hatten ihn dann, wie wir sahen, auf eine andere Bahn gestoßen. Die Pointe seines zweiten Lebens ist, dass es ihm aus der Perspektive des

---

10 Vgl. Platon, *Gorgias*, 492d 5, 500d 4, 512a 7.

prudentiell Guten und des moralisch Guten als verwerflich erscheint, er aber aus der Perspektive des adverbiell Guten in einer Vielzahl von Bewährungsepisoden das Selbstvertrauen entwickelt, in dem, was er tut, eine gute Figur zu machen. Die Basis dieses Selbstvertrauens sind konkrete Erfahrungen eines gelingenden – und wachsenden – Passungsverhältnisses zwischen Willensbildung und -artikulation, zwischen dem, was er will, und dem, was er vermag.

III

Ich fasse, wie angekündigt, meine bisherige Argumentation kurz zusammen. Die Überlegungen nahmen ihren Ausgang von der Intuition, dass der Begriff der politischen Freiheit, den der im US-amerikanischen Unabhängigkeitskrieg verwurzelte Wahlspruch des Neuenglandstaates New Hampshire pointiert, auf einem Verständnis individueller Freiheit aufruht, das in den philosophischen Konzepten der Willens- und Handlungsfreiheit nicht aufgeht. Frei zu leben bedeutet noch etwas anderes als seinen Willen in freier Deliberation bilden und ohne Restriktionen ausüben zu können. Zur Plausibilisierung dieser Intuition habe ich mich auf die fiktive Biographie des Helden einer Fernsehserie bezogen, der die Verwirklichung individueller Freiheit an die Erfahrung eines intensiven Lebensgefühls der Selbstmächtigkeit bindet, die er immer dann macht, wenn er Tätigkeiten, die für seine Lebensführung im Ganzen relevant sind, auf gute Weise ausübt. Die Güte dieser Tätigkeiten habe ich als adverbielle Güte bezeichnet und (in Abgrenzung von den Begriffen der prudentiellen und moralischen Güte) dadurch bestimmt, dass sie auf Kriterien beruht, die der Ausübung dieser Tätigkeiten immanent sind und ihnen nicht (wie im Falle der prudentiellen oder moralischen Güte) von außen vorgegeben werden. Die Pointe meines Fallbeispiels bestand darin, dass die individuelle Freiheit, die mit dem Anspruch frei zu leben gemeint ist, in eben diesen Erfahrungen der adverbiellen Güte subjektiv lebensrelevanter Tätigkeiten (etwa derjenigen der Berufsausübung, der sportlichen Betätigung oder des familiären Handelns etc.) fundiert ist. Für diese Freiheit gilt, dass sie Willens- und Handlungsfreiheit in dem Maße voraussetzt, wie es sie zur Verwirklichung des Passungsverhältnisses von Willensbildung und -artikulation benötigt.

Habe ich die Freiheit, die sich in der guten Ausübung lebensrelevanter Tätigkeiten erschließt, bisher als adverbielle Freiheit bezeichnet, so schlage ich nun vor, sie näherhin als verkörperte Freiheit zu bestimmen. Denn wenn es stimmt, dass der grundlegende Sinn des Anspruchs, frei zu leben, darin besteht, in und durch meine Tätigkeiten das Passungsverhältnis zwischen Willensbildung und

-artikulation zu verwirklichen, durch das es mir gelingt, in allen Handlungszusammenhängen, die mir wirklich wichtig im Leben sind, eine gute Figur zu machen, und wenn ferner diese Freiheit durch das zuständliche Bewusstsein des besagten Passungsverhältnisses, durch ein Bewusstsein mit der Qualität des Lebensgefühls der Selbstmächtigkeit authentifiziert wird, in dem sich der Handelnde unbestimmt als derjenige erfährt, auf den es hier und jetzt ankommt – dann ist die Freiheit, die ich meine, wesentlich an die Empfindungs- und Ausdrucksfähigkeit des Körpers gebunden.[11] Das Passungsverhältnis zwischen Willensbildung und -artikulation verwirklicht sich in der verkörperten, das heißt wesentlich körperlichen Wechselbeziehung zwischen Wille und Welt, es verwirklicht sich in der *performance* des Handelns, das den Willen in der Welt situiert. Zuständliches Bewusstsein des Passungsverhältnisses zwischen Willensbildung und -artikulation ist wesentlich ganzheitliches, unbestimmtes Situationsbewusstsein[12] von der Gestaltprägnanz der sich im Handeln ereignenden Wechselwirkung zwischen Wille und Welt, und es resultiert in „eine[r] permanente[n] Bereitschaft", ein durch praktische Bewährung erworbenes „Können zu aktivieren".[13]

Adverbielle Freiheit kann also in dem präzisen Sinn als verkörpert gelten, dass sie in den sensomotorischen Dispositionen des Handelnden fundiert ist, diejenigen *affordances* in der Erlebnismannigfaltigkeit einer gegebenen Situation zu fokussieren, auf die er nach Kriterien, die den jeweils aktualisierten Handlungsmustern inhärieren, folge- und gestaltrichtig reagieren wird.[14] Die Einstellung des Handels

---

11   Da sich im Zusammenhang der internationalen Diskussion der Begriff der Verkörperung durchgesetzt hat und die englische Sprache die Unterscheidung von Körper und Leib nicht explizit vornehmen kann, verwende ich im Folgenden ausschließlich den Begriff des Körpers. Die seit der Philosophischen Anthropologie der Zwanziger Jahre des letzten Jahrhunderts in Deutschland gebräuchliche Unterscheidung von Leib und Körper gibt deshalb Anlass zu Missverständnissen, weil der Begriff des Körpers einerseits als Gattungsbegriff verwendet wird, der nach dem Schema von genus proximum und differentia specifica den Begriff des Leibes unter sich befasst, andererseits als Artbegriff auf derselben begriffslogischen Ebene wie der Leibbegriff und in strikter Entgegensetzung zu ihm fungiert. Im Folgenden beschränke ich mich auf die generische Verwendung des Körperbegriffs im Sinne des genus proximum. Vgl. zur Geschichte des Leibbegriffs Alloa/Bedorf/Grüny/Klass (Hg.) (2012), *Leiblichkeit. Geschichte und Aktualität eines Konzepts.*

12   Vgl. Fellmann (1996), a..a. O., 214, 220.

13   Fellmann (1996), 221.

14   Den Begriff *affordance* borge ich von James Gibson, der ihn folgendermaßen einführt: „The affordances of the environment are what it offers to the animal, what it provides or furnishes, either for good or ill […] I mean by it something that refers to both the environment and the animal […] It implies the complementarity of the animal and the

und der zu ihr passende Erfolg resonieren in dem Gefühl der Selbstmächtigkeit, gleichsam ‚Herr der Lage‘ zu sein und sich in der Welt ‚zurecht zu finden‘. Dieses Gefühl hat eine proleptische Ausrichtung, es ist ein lebendiges Gewahrsein des Handelnden im Horizont von Erfolgsmöglichkeiten seines Handelns. Die Dispositionen, situationsgebundende *affordances* wahrzunehmen, qualifizieren das Lebensgefühl als die dem Handeln stets vorlaufende Zuversicht, die sich im Austausch mit der Welt stellenden Herausforderungen zu meistern. Die gebräuchliche Rede davon, in diesen oder jenen Tätigkeiten eine gute Figur zu machen, erweist sich auch in diesem Zusammenhang als eine nützliche Anschauungshilfe: Um eine gute Figur zu machen, bedarf es des Vorschusses von Selbstvertrauen in die Selbstmächtigkeit als Handelnder, und dieses Selbstvertrauen hat nicht in erster Linie den Charakter eines inneren Zuspruchs, mit dem der Handelnde sich seiner Fähigkeiten gedanklich versichert, sondern einer intuitiven antizipatorischen Gewissheit, den Aufgaben auch gewachsen zu sein, denen er sich stellen wird, einer Gewissheit, die sich in der unwillkürlichen Verhaltenssicherheit manifestiert, mit der er sich auf seine Tätigkeiten versteht.

Die Auffassung des Passungsverhältnisses zwischen Willensbildung und -artikulation als eine wesentlich körperliche Wechselbeziehung zwischen dem Handelnden und seiner Umwelt möchte ich nun als körperliche Vermittlung von Wille und Welt pointieren. Sich artikulierend, greift der Wille in die Befindlichkeiten und Bewandtnisse der Welt ein, den Willen bildend werden seine Bedürfnisse und Absichten von der Welt bestimmt; Willensbildung ist Anpassung des Willens an die Welt, Willensartikulation Anpassung der Welt an den Willen. Die wechselseitige Anpassung wird durch den Körper verwirklicht, dessen Responsivität auf die Ansprüche des Willens und die Widerständigkeit der Welt die Grundlage für das zuständliche Situationsverständnis des Menschen ist.[15] Im Bewusstsein der Selbstmächtigkeit des Handelnden in den für ihn relevanten Tätigkeiten, auf

---

environment“ (Gibson (1986), *The Ecological Approach to Visual Perception*). Gibsons Begriff der *affordances* hat sich in der Philosophie mittlerweile als Grundbegriff zur Bezeichnung der Verkörperung oder Einbettung des Mentalen in die Wechselbeziehung zwischen Körper und Umwelt etabliert. Vgl. dazu folgende beiden klassischen Beiträge zur verkörperungstheoretischen Philosophie des Geistes: Haugeland (1998), „Mind Embodied and Embedded“, in: ders., *Having Thought. Essays in the Metaphysics of Mind*; Varela/Thompson/Rosch (1993), *The Embodied Mind. Cognitive Science and Human Experience*.

15 Der Begriff körperlicher Responsivität spielt eine zentrale Rolle in der Leibphänomenologie von Bernhard Waldenfels, zu dessen Konzeption des ‚leiblichen Responsoriums‘ sicher Berührungspunkte bestehen, ohne dass ein ausdrücklicher Bezug auf Waldenfels’ komplexe Theorie hier beabsichtigt wäre. Vgl. Waldenfels (2000), *Das leibliche Selbst, Vorlesungen zur Phänomenologie des Leibes*, v. a. Kap. VIII; vgl. dazu Sternagel (2012),

die er sich unter den wechselnden Umständen praktischer Bewährung versteht, resoniert die Responsivität des Körpers auf die Ansprüche des Willens und die Widerständigkeit der Welt. Gleichzeitig ist es die phänomenale Resonanz körperlicher Responsivität im zuständlichen Situationsbewusstsein, die den Handelnden für jeden künftigen Erfolg in der wechselseitigen Anpassung von Wille und Welt disponiert. Die Verhaltenssicherheit, in der die Fähigkeit gründet, eine gute Figur zu machen, besteht wesentlich in der Flüssigkeit und Gestaltrichtigkeit, die der Körper in der Vermittlung zwischen Wille und Welt aufbringt. Es ist die Flüssigkeit und Gestaltrichtigkeit nicht *irgendeines* Körpers, in dem sich die Vermittlung von Wille und Welt beliebigerweise instanziieren könnte, sondern die Signatur jeweils einer individuellen und unvertretbaren Physis, in der die Vorgeschichte eines – mehr oder weniger – erfolgreichen Passungsverhältnisses zwischen Willensbildung und -artikulation aufbewahrt ist; das besagte Passungsverhältnis *individuiert* sich im Körper, in seiner Empfindungs- und Ausdrucksfähigkeit – und die spezifische Empfindungs- und Ausdrucksfähigkeit des Körpers *individualisiert* ihn als Träger der spezifischen Individuation des Passungsverhältnisses zwischen Willensbildung und -artikulation in der jeweiligen Lebensgeschichte des Handelnden.

Hubert Dreyfus, der Spiritus rector der gegenwärtigen philosophischen Forschung zur Verkörperung, hat die eminente Bedeutung des individualisierten Körpers für die Wechselbeziehung zwischen dem Handelnden und seiner Umgebung folgendermaßen hervorgehoben: „Im allgemeinen gilt: Wenn wir eine Fertigkeit erwerben – z.B. Autofahren, Tanzen oder eine Fremdsprache erlernen –, müssen wir zuerst langsam, mühselig und bewußt den Regeln folgen. Dann aber kommt der Augenblick, von dem an wir die Tätigkeit automatisch ausführen können. Nun scheint es aber nicht so zu sein, dass wir zu diesem Zeitpunkt diese starren Regeln einfach ins Unterbewußtsein fallenlassen. Eher verhält es sich so, daß wir sozusagen die Muskelstruktur des Erlernten einverleibt haben, die unserem Verhalten eine neue Flexibilität und Geläufigkeit gibt.“[16] Die von Dreyfus angesprochene „Flexibilität und Geläufigkeit“ verdankt sich dem Passungsverhältnis zwischen der Artikulation eines Willens in der Welt und seiner Bildung durch deren Widerständigkeit. Der Tänzer mag von Gestaltantizipationen einer gelungenen Walzerdrehung geleitet werden, die er auf dem Parkett umsetzen will, aber die für den Anfänger schlechterdings unvorhersehbare Wirkmächtigkeit der Handlungsbedingungen auf die Artikulation seines Willens – zu der neben dem glatten Tanzboden nicht zuletzt die Erwiderungen der Partnerin auf seinen Anspruch zählen, sie zu ‚führen' – bändigen

---

„Bernhard Waldenfels – Responsivität des Leibes“, in: Alloa/ Bedorf/Grüny/ Klass (Hg.) (2012), *Leiblichkeit*, a. a. O.

16 Dreyfus (1989), *Was Computer nicht können. Die Grenzen künstlicher Intelligenz.*

den ersten Schwung und zwingen ihn zu einer Zergliederung der antizipierten Tanzgestalt in ein ernüchterndes Klein-Klein von Bewegungsansätzen und ihrer sofortigen Korrektur. An die Stelle der Nonchalance, mit der er die Partnerin zum Tanz aufgefordert hatte, tritt eine Mischung aus Schüchternheit und Eifer, mit der die beiden durch den Saal rucken. Das sich betrüblicherweise sogleich einstellende Bewusstsein von der kläglichen Figur, die sie machen, grundiert die Bewegungen durch die Bemühung, ihre fast zwangsläufige Unbedarftheit zu vertuschen. Es bietet sich hier zwanglos an, die schier unversöhnlich anmutenden Gegensätze von Wille und Welt auf die mangelhaft nuancierte Responsivität des Körpers zurückzuführen. Flexibilität und Geläufigkeit werden sich erst im Verlauf der weiteren Lebensgeschichte einstellen, in der sich ein zunehmend feinkörnigeres Passungsverhältnis zwischen Willensbildung und -artikulation in die „Muskelstruktur des Erlernten" einschreibt und den Körper zum Ausdrucksorgan dieses spezifischen Passungsverhältnisses – zu dem vielleicht immer dieselbe Tanzpartnerin gehört – individualisiert.

Die Verhaltenssicherheit der Flexibilität und Geläufigkeit, der Flüssigkeit und Gestaltrichtigkeit reicht von der „Muskelstruktur des Erlernten" in einzelnen Bewegungssequenzen über höherstufige Verhaltensmuster, die sich durch die Kombination von Bewegungsmustern etwa der Gestik und Mimik expressiv in einem Gesamthabitus der Persönlichkeit ausprägen, bis zur Individualisierung von Handlungstypen in der habitusbasierten Ausübung von Tätigkeiten, die ihrerseits wiederum in einem komplexen, symbolisch vermittelten Sinnzusammenhang stehen. Wenn Walter White bei seinen kriminellen Konkurrenten auftritt, dann muss sich dieser Sinnzusammenhang einerseits pointiert in der einzelnen Geste verdichten, die gestaltprägnant zum Ausdruck bringt, dass er sich auf das, was er tut, versteht, andererseits hat sich dieser Sinnzusammenhang als individuell relevanter Sinnzusammenhang bottom up aus den Verhaltenssicherheiten heraus entwickelt, die dem ehemaligen Lehrer sukzessiv immer deutlicher haben werden lassen, was es ist, worin er gut ist, worin er sich lebendig fühlt. Die Verhaltenssicherheit, mit der es White gelingt, in einer Vielzahl von Handlungssituationen das ihm eigentümliche Passungsverhältnis von Willensbildung und -artikulation zu verwirklichen, möchte ich seinen persönlichen *Stil* nennen. Denn im Stil wird die Selbstmächtigkeit anschaulich, in den für das Leben relevanten Tätigkeiten eine gute Figur zu machen. Im Stil artikuliert sich die persönliche Art und Weise, im Angesicht der praktischen Bewährungsaufgaben, die sich dem Handelnden stellen, frei zu leben. Das Lebensgefühl einer ‚Tätigkeit überhaupt', durch das man sich in gesteigerter Weise seiner Selbstmächtigkeit in dem gelungenen Passungsverhältnis von Willensbildung und -artikulation gewahr wird, ist Stilbewusstsein. Man könnte hier auch von Geschmack sprechen, allerdings nicht im Sinne der Sicherheit gestaltrichtigen Urteilens über die ästhetischen Qualitäten von Gegenständen, sondern eben im Sinne der gestaltrichtigen Ausübung von Handlungen, durch die man in der Wechselbeziehung zwischen Willensbildung und -artikulation eine gute Figur macht. Von der Entwicklung dieses Stilbewusstseins gilt, dass man es nicht lehren kann. Was es heißt, in der Bewältigung der Widerstände, mit denen wir in der Ausübung unseres Willens konfroniert werden, einen eigenen Stil zu bilden, in dem ich, mich artikulierend, frei lebe, das kann man nur zeigen und irgendwie nachzumachen versuchen.[17]

Es ist die Rolle von Vorbildern, in uns eine Sensibilisierung für die Verkörperung von Freiheit anzuregen. Vorbilder müssen zugleich konkret und allgemein sein: konkret in der gestaltrichtigen Verkörperung ihrer individuellen Freiheit, allgemein in der Sinnfälligkeit, mit der die konkreten Anschauungsbeispiele verkörperter Freiheit zur Exploration eigener Freiheit *verführen*. Vor allem im Bildmedium des Films lässt sich diese Einheit von Individualität und Allgemeinheit zur vorbildhaften Figurendarstellung verdichten. Die großen Stars in ihren großen Rollen punkten bei uns durch die gestaltantizipatorische Suggestivität, mit der sie uns die Verkörperung von Freiheit andeuten. Der Film kann in der Tat nur andeuten: Mimik und Gestik, die Art sich zu bewegen und die Intonation der Sprache stehen für Haltungen, die dem Zuschauer das gelungene Passungsverhältnis von Willensbildung und -artikulation suggerieren.[18] Die tatsächliche Bewährung dieses Passungsverhältnisses kann nicht gezeigt werden, aber die Anziehungskraft der Filmstars besteht in ihrer Fähigkeit der symbolischen Verdichtung einer

---

17 In diesem Zusammenhang sei darauf hingewiesen, dass die Entwicklung stabiler Beziehungsmuster in der Interaktion mit anderen darauf angewiesen ist, dass sich in einem Frühstadium der Ontogenese in der Mutter-Kind-Dyade ein Vertrauensverhältnis entwickelt, indem das kleine Kind durch Imitation des Ausdrucksverhaltens der Mutter *schemes-of-being-with* einübt, in denen erste Formen der Verhaltenssicherheit in übersichtlichen Situationen erfahren werden können (vgl. zusammenfassend zur entwicklungspsychologischen Forschung in diesem Zusammenhang Fuchs (2008), *Das Gehirn – ein Beziehungsorgan. Eine phänomenologisch-ökologische Konzeption.* Imitationsverhalten bleibt von der Frühen Kindheit an eine wesentliche Dimension der Handlungsorientierung und der Einübung von Verhaltenssicherheit.

18 Darin besteht die strukturelle Verwandtschaft zwischen der Filmschauspielerei und der Hochstapelei. Filmschauspieler und Hochstapler haben gemeinsam, dass sie durch die gekonnte Selbstdarstellung ihre Fähigkeit suggerieren, stets eine gute Figur zu machen, ohne dass sie, wenn auch aus ganz unterschiedlichen Gründen, die erweckten Erwartungen einlösen können: Im Falle des Films ist die Zukunft des Handelns durch die Regie immer schon festgelegt, weshalb nur der *persona*, nicht aber dem Schauspieler, der sie verkörpert, Unerwartetes passieren kann. Und im Falle des Hochstaplers verspricht die Haltung, was die Praxis am Ende gar nicht einzulösen beabsichtigt.

Handlungssequenz zum Zeichen für die Selbstmächtigkeit ihrer *persona*, in den vielfältigen Bewandtnissen ihres Lebens eine gute Figur zu machen. Vince Gilligan weiß darum. Der Regisseur von *Breaking Bad* bestätigt die Leitfunktion des Mediums, indem er sie konterkariert. In der herkömmlichen filmischen Star-Feier begegnen wir den Virtuosen des Stilbewusstseins, denn was die Helden vor allem auszeichnet, die Guten wie die Bösen, ist ihre Fähigkeit, eine gute Figur zu machen. *Breaking Bad* dagegen schildert die Arbeit am Subjekt, den schlingernden Kurs, auf dem Walter White darum kämpft, endlich ein Idiom seines Handelns zu finden, das zu ihm passt. In Anlehnung an Dreyfus' Beispiel könnte man vielleicht sagen, dass White sich aufmacht, Tanzen zu lernen, einen anderen Tanz als das bürgerliche Lebenslaufschema für ihn vorgesehen hatte. Das Skandalon der Serie: Ihr ambivalenter Held ist ein Mörder und Drogenhändler.

IV

Die Pointe meines Fallbeispiels war, dass sie ein Freiheitsverständnis demonstriert, das die Bewährung am adverbiell Guten zum Maßstab der Exploration individueller Freiheit macht. Es ist kein Zufall, dass Vince Gilligan diese Freiheit von einer Figur verkörpern lässt, deren Lebensführung allen Vorstellungen moralischer und prudentieller Güte widerspricht. Denn er hat den anarchischen Impuls des amerikanischen Individualismus im Blick, der darin besteht, Freiheit zu verkörpern, indem man sich an den immanenten Maßstäben der Tätigkeiten misst, in denen und durch die zu leben es gelingt, eine gute Figur zu machen, welche auch immer das sein mögen. Meine Überlegungen haben diesem Individualismus eine verkörperungstheoretische Pointe abgewonnen, die meines Erachtens dazu beiträgt, plausibel zu machen, warum dieser Individualismus so attraktiv ist. Whites Lebensgeschichte mindert diese Attraktivität nicht, sondern verdeutlicht allenfalls die Konflikte, die sich aus ihr ergeben können. Denn der anarchische Individualismus bringt die Gefahr mit sich, die Frage des prudentiell Guten für erledigt zu halten, wenn es rund zu laufen scheint mit den Tätigkeiten, in denen wir eine gute Figur machen wollen. Aber die Frage bleibt, ob es für mich gut ist, diese oder jene Tätigkeit gut auszuüben. In dieser Frage geht dem anarchischen Individualismus – das macht seinen Charme aus – das Probieren über das Studieren. Er hat aber keine Antwort darauf, wie das, was wir gut tun können, zu anderen Tätigkeitsfeldern passt, in denen wir adverbielle Güte anstreben. Walter White log nicht, als er vorgab, er sei für die Familie kriminell geworden. Das war sein ursprüngliches und aufrichtiges Motiv, nur wurde es später von den Erfahrungen in den Hintergrund gedrängt, die ihn seinem bürgerlichen Leben immer mehr entfremden mussten. Sollten meine Thesen plausibel sein, dass es erstens ein Freiheitsverständnis gibt, das in einen scharfen Konflikt mit prudentiellen, klugheitsethischen Konzepten der Lebensführung treten kann, und dass zweitens die Attraktivität dieses Freiheitsverständnisses auf seiner körperlichen Dimension beruht, dann könnte aus den hier angestellten Überlegungen folgen, dass es für die praktische Philosophie attraktiv wäre, eine verkörperte Basis für das prudentiell Gute zu suchen. Am Ende verliert Walter White seine Familie, sein Leben endet im Desaster. Aber er folgt bis zum Schluss dem Gesetz der Konsequenz derjenigen Lebensführung, durch die er zu dem Menschen geworden ist, von dem er abschließend sich und seiner Frau eingesteht: „I did it for me. […] I was good at it. I was feeling alive." Das ist die verstörende Bilanz dieses Fallbeispiels verkörperter Freiheit.

Literatur

Alloa/Bedorf/Grüny/Klass (2012): Emmanuel Alloa, Thomas Bedorf, Christian Grüny, Tobias Klass (Hg.), *Leiblichkeit. Geschichte und Aktualität eines Konzepts*, Tübingen 2012.

Bieri (2003): Peter Bieri, *Das Handwerk der Freiheit. Über die Entdeckung des eigenen Willens*. Frankfurt/M. 2003.

Dewey (1969-1991a): John Dewey, "Philosophies of Freedom". In: ders., *Collected Works, The Later Works*, Bd. 3, hg. v. Jo Ann Boydston, Carbondale and Edwardsville, 92-114.

Dewey (1969-1991b): John Dewey, "Human Nature and Conduct". In: ders., *Collected Works – The Middle Works*, Bd. 14, hg. v. Jo Ann Boydston, Carbondale and Edwardsville, 193-203.

Dreyfus (1989): Hubert Dreyfus, *Was Computer nicht können. Die Grenzen künstlicher Intelligenz*, Frankfurt/M. 1989,

Fellmann (1996): Ferdinand Fellmann, „Intentionalität und zuständliches Bewußtsein". In: Sybille Krämer (Hg.), *Bewußtsein. Philosophische Beiträge*. Frankfurt/M., 213-226.

Fuchs (2008): Thomas Fuchs, *Das Gehirn – ein Beziehungsorgan. Eine phänomenologisch-ökologische Konzeption*, Stuttgart 2008.

Gibson (1986): James Gibson, *The Ecological Approach to Visual Perception*, Lawrence Erlbaum Associates: Hilldale, NJ, 1986.

Haugeland (1998): John Haugeland, „Mind Embodied and Embedded", in: ders., *Having Thought. Essays in the Metaphysics of Mind*, Havard University Press: Cambridge/MA, 1998

Immanuel Kant, *Kritik der Urteilskraft* (Werkausgabe Band X, hrsg. W. Weischedel), Frankfurt am Main 1974.

Pappas (2008): Gregory Fernando Pappas, *John Dewey's Ethics. Democracy as Experience.* Indiana UP, Indiana & Bloomington, 2008.

Platon, *Gorgias* (in: Sämtliche Werke 1), Reinbeck bei Hamburg 1957.

Scheer (1997): Brigitte Scheer, *Einführung in die philosophische Ästhetik*. Darmstadt.

Schlette (2014): Magnus Schlette, „Über die Gewissheit, eine gute Figur zu machen, und was sie uns darüber verrät, wie wir uns selbst verstehen". In: Susanne Schulte (Hg.), *So verstehen wir. Texte über das Verstehen.* Münster. 117-130.

Schlette (2015): Magnus Schlette, "Verkörperte Freiheit. Praktische Philosophie zwischen Kognitionswissenschaft und Pragmatismus". In: *Ethik und Gesellschaft* 1/2015: Pragmatismus und Sozialethik.

Sternagel (2012): Jörg Sternagel, „Bernhard Waldenfels – Responsivität des Leibes", in: Emmanuel Alloa, Thomas Bedorf, Christian Grüny, Tobias Nikolaus Klass (Hg.), *Leiblichkeit*, Tübingen, 116-129.

Stroud (2011): Scott R. Stroud, *John Dewey and the Artful Life. Pragmatism, Aesthetics, and Morality.* Penn State UP, University Park, PA, 2011.

Tugendhat (2003): Ernst Tugendhat, Egozentrizität und Mystik. Eine anthropologische Studie. München.

Varela/Thompson/Rosch (1993): Francisco Varela, Evan Thompson, Eleanor Rosch, *The Embodied Mind. Cognitive Science and Human Experience*, MIT Press: Cambridge/MA 1993

Waldenfels (2000): Bernhard Waldenfels, *Das leibliche Selbst. Vorlesungen zur Phänomenologie des Leibes*, Frankfurt/M. 2000.

4

Zur historischen Anthropologie von Körperritualen

Zur kulturellen Funktion des Leibes bei Herodot

Martin F. Meyer

Ὡς δὲ καὶ ἀνθρώπου σῶμα ἓν οὐδὲν αὔταρκές ἐστι·
Hdt. II 32

In den *Historien* behauptet der Athenische Gesetzgeber Solon (eine Schlüsselfigur des Werks), *der menschliche Körper sei nicht autark* (II 32). An diesem Gedanken sind die folgenden Überlegungen orientiert: Herodots Aussagen über den Leib und über die Körperriten lassen sich im Sinne einer Soma-Semantik lesen: Der Leib; genauer: *das, was der Mensch mit seinem Leib tut* und *mit ihm ausdrückt*, läßt sich als Zeichen von sozio-kulturellen Funktionen deuten. Herodot beschreibt den Leib in aller Regel als Teil eines sozio-kulturellen Ganzen. In Herodots Text ist der Leib ein Signifikant, seine kulturelle Funktion ein Signifikat, das es zu dechiffrieren gilt. Diese Lesart sieht die *Historien* als Teil eines Diskurses, dem es (grob vereinfacht) um die Differenz von Kultur und Kulturlosigkeit geht. Im Anschluß an Claude Lévi-Strauss (1971, 90) bin ich überzeugt, daß es in allen Kulturen solche Grenzziehungen gibt. Kulturen lassen sich daran unterscheiden, wo und wie sie diese Grenzen ziehen. Herodots Aussagen über den menschlichen Leib markieren Punkte auf einer von zwei Extremen begrenzten Skala: Das Attribut ‚wild‘ bildet den einen, das Attribut ‚zivilisiert‘ den andern Pol.

Einige Vorbemerkungen zu Herodot und den *Historien*. Über Herodots Leben gibt es kaum veritable Nachrichten: Das meiste ist den *Historien* selbst zu entnehmen. Herodot ist um 480 v. Chr. in der kleinasiatischen Küstenstadt Halikarnassos (dem heutigen Bodrum) geboren. Er unternahm (wie sein milesischer ‚Vorgänger‘

Hekaitos)[1], was noch eine Generation vor ihm unmöglich war,[2] weite Forschungs-reisen. Als gesichert gelten Aufenthalte in Delphi, Dodona, am Hellespont, der Schwarzmeerküste, in Ägypten (ca. vier Monate), Phönizien, Babylonien, Skythien, Thrakien, Makedonien und Unteritalien (vgl. Jacoby 1913, 247-281); für weitere Reisen liefern die *Historien* indirekte Indizien. Herodot hat vermutlich in Athen vor großem Publikum mündliche Vorträge über diese Reisen gehalten Diese *Logoi* hat er später in Athens Kolonie Thurioi (Süditalien) zu dem schriftlichen Gesamtwerk der *Historien* zusammengefügt (vgl. Jacoby 1913; Cobet 1977). Als dramaturgisches Grundmotiv (als roter Faden) gilt seit Otto Regenbogen[3] die sog. ‚Perserlinie': Herodot behandelt die Ursachen der sog. Perserkriege. Den Aufstieg des persischen Großreiches und die Unterwerfung der gegen Hellas versammelten Völker verbindet er mit ethnologischen Schilderungen. Dabei stützt er sich auf seine Reisenotizen bzw. darauf, „was er dort gehört hat". Geo- und ethnographi-sche Notizen bilden gleichsam als hyletische Basis die ältere Textschicht (Jacoby ebd.). Er verknüpft historiographische und ethnographische Elemente zu einer Einheit, indem er das synchrone Material in eine diachrone Ordnung überführt. Der Titel dieses bis dato umfangreichsten Prosawerks stammt nicht von Hero-dot selbst; er wurde im 2. Jh. v. Chr. dem ersten Satz des Werkes entlehnt.[4] Der rezeptionsgeschichtlich wegweisende[5] Ausdruck ‚Historia' kommt etymologisch

---

1 Vgl. Jacoby (1912); Heilen (2000), 38-54.

2 Vgl. Jacoby (1912); Heilen (2000), 43: Das Perserreich war noch zu Zeiten Anaximanders »unbetretbares Feindesland«. Erst ab 520 v. Chr. bot sich die Gelegenheit, »die riesigen Gebiete des Ostens aus eigener Anschauung kennenzulernen«. Dies ist ein gutes Beispiel für die Wirkung von politischen Faktoren auf die Forschung.

3 Vgl. Regenbogen (1965), 75. Der Terminus ‚Perserlinie' geht zurück auf G. F. Creuzers Übersetzung Περσικὸς λόγος.

4 Vgl. Jacoby (1913): Aristarch von Samothrake (ca. 216-155 v. Chr.) hat dieses umfang-reichste Werk des 5. Jahrhunderts in neun (den Musen geweihte) Bücher eingeteilt. Von Aristarch stammt auch der Titel des Werkes. Der Titel hat seinen Anhalt im Eingangs-satz des Werkes (dem sog. Proömion), der den Inhalt als eine Apodeixis der Historie qualifiziert; vgl. Erbse (1956), Hommel (1981) zum Proömion.

5 Vgl. Oser-Grote (2005), 254-255 zu den Bedeutungsvarianten des Begriffs bei Aristoteles. Der Ausdruck ›Historia‹ hat sich dann weit über die griechischen Sprachgrenzen hinaus in stets neuer Färbung tief in das kollektive Gedächtnis der Menschheit eingeprägt. Die Römer übernahmen das griechische Wort eins zu eins in ihre Sprache. ›Historia‹ wurde zum Signum für zahllose Klassiker der europäischen Naturwissenschaft. Exemplarisch: Theophrast, *Historia plantarum*; Plinius d. Ä. *Naturalis historia*; Leonhart Fuchs, *De historia stirpium commentarii insignes* (1542); Pierre Belon, *L'histoire de la nature des oyseaux* (1555); Conrad Gessner *Historia animalium* (1551-1558), *Historia plantarum* (1555-1565); John Ray, *History of plants* (1686-1704); Georges-Louis Leclerc de Buffon, *Histoire naturelle générale et particulière* (1716-1799); Georges Cuvier, *Tableau élémen-*

von ‚Zeuge sein'.[6] Herodot begreift sich als Zeuge, der den Griechen durch sein Zeugnis Unbekanntes kund tut.[7] Das Publikum sollte exotische Nachrichten hören und konnte die Berichte mit der vertrauten Nahwelt vergleichen. Dies ist wichtig zum prinzipiellen Verständnis seiner Methode: Herodot akzentuiert das unbekannte Fremde, das Exotische und Kuriose. Kulturelle Differenzen werden bewußt übertrieben. Diese auf Unterhaltungseffekte berechnete Absicht ist nicht zu vernachlässigen. Es wäre aber verfehlt, die *Historien* hierauf zu reduzieren.

Kulturwissenschaftlich bedeutsam ist die Differenz von ‚wild' und ‚zivilisiert'. Nachstehend werden die Aussagen über den Leib unter dieser Rücksicht interpretiert. Das griechische Substantiv σῶμα für Leib begegnet in den Historien an 48 Stellen. Der Ausdruck σῶμα meint in der älteren Tradition den toten Leib (die Leiche) und erst später (auch hier war Herodot ein Pionier) den lebendigen Leib. Im Text begegnen insg. ca. 60-70 differente ‚Ethnien' und ca. 30 ethnische Subspezies. Im Einzelfall ist nur schwer entscheidbar, was als ‚Ethnie', was als ‚Stamm' oder was als ‚Landmannsschaft' gilt.[8] In seiner Geschichte der antiken Ethnologie zeigt Klaus E. Müller, daß Herodot differenziert in Völker der Hochkultur, Bauern, Viehzüchter und Wildbeuter (1997, 119-122). Dieses Kategoriensystem korreliert mit der

---

taire de l'histoire naturelle des animaux (1798); Jean-Baptiste de Lamarck, *Philosophie zoologique, ou, Exposition des considérations relative à l'histoire naturelle des animaux* (1809); *Histoire naturelle des animaux sans vertèbres présentant les caractères généraux et particuliers de ces animaux, leur distribution, leurs genres, et la citation des principales espèces qui s'y rapportent: précédée d'une introduction offrant la détermination des caractères essentiels de l'animal, sa distinction du végétal et des autres corps naturels : enfin, l'exposition des principes fondamentaux de la zoologie* (1815-1822); Frédéric Georges Cuvier, *Histoire naturelle des mammifères avec des figures originales coloriées, dessinées d'après les animaux vivants* (1819-1842), *De l'Histoire naturelle des cétacés* (1836). Dazu grundlegend: Lepenies 1976.

6 Vgl. Snell (1924), 56-71. Snell gibt die wichtigsten Stellen in der vorplatonischen Literatur an. Er zeigt, daß sich mit der Funktion des Zeugen recht bald auch die Funktion des ›Schiedsrichters‹ verband. Diese Entwicklung hat ihre Parallelen in der deutschen Sprache [ahd. *forscunga*; mhd. *vorschunge*; nhd. *forschung*]. Belegt ist das Wort zunächst v. a. in Eidesformeln.

7 Vgl. Hdt. II 33: Als Methode gibt Hdt. an, „vom Unbekannten auf das Bekannte zu schließen".

8 Vgl. Müller (1997), 124: „Ein Volk wird [von Herodot] durch die Summe all derer bestimmt, die derselben *Abstammung* – und damit auch *Physis* – sind, dieselbe *Sprache* besitzen und sich zu derselben *Lebensordnung* und *Religion* bekennen" [kursiv: Müller]. Müller (ebd.) bemerkt zugleich, es sei Herodot „durchaus bewußt", daß die Völker unmöglich aus einer gemeinsamen genealogischen Wurzel stammen könnten, sondern vielmehr aus der Durchdringung und Verschmelzung ganz unterschiedlicher ethnischer Komponenten hervorgegangen sind.

räumlichen Gliederung der bekannten Welt: Die Hochkulturen (Hellas, Ägypten) stehen im Zentrum. Um diese Mitte gruppiert Herodot die seßhaften Bauern, die

nomadischen Viehzüchter und Wildbeuter. An den Rändern der Ökumene franst die Zivilisation dann aus (vgl. nebenstehende Abbildung ebd. 120). Wie Reinhold Bichler in dem verdienstvollen Buch Herodots Welt (2001) mit Blick auf diese sog. Randvölker zeigt, verlieren sich Herodots Schilderungen über den Rand der Ökumene in zunehmend wilderen Phantasien. Anders als bei diesen fachwissenschaftlichen Ansätzen (die v. a. die räumliche bzw. die zeitliche Dimension in den Vordergrund stellen) problematisiert die historische Epistemologie die logischen Strukturen, die ein solches Kategoriensystem konstituieren. Mich interessiert also weniger die sog. ‚historische Wahrheit‘ als Herodots Beobachterperspektive, sein Blick aufs Fremde. Bei den Kriterien, die Herodots ethnologischen Berichten zugrunde liegen, begegnen neben eher ‚objektiven‘ Beschreibungsmustern (Topographie, Populationsgröße, Klima, Naturraum) Kategorien wie Sprache, Götterglaube, Totenkult, Rechtssystem, ökonomische Merkmale, Kriegsrituale, Opferbräuche, historische Vorstellungen, technische Errungenschaften und somatische Aspekte.[9] Die Häufigkeit dieser Kategorien ist kontextabhängig und nicht symmetrisch verteilt: Die Ägypter werden in einem eigenen Buch (dem sog. ‚Ägyptischen Logos‘) bei weitem am ausführlichsten behandelt. Umgekehrt berichtet Herodot über viele Ethnien in nur wenigen Zeilen. Kategorien wie Recht, historische Vorstellungen oder technische Erfindungen kommen nur im Kontext der sog. Hochkulturen zur Sprache. Die Analyse der somatischen Aspekte hat insofern den methodischen Vorteil, daß dieses Deskriptionsmuster bei Herodot durchgehend vorkommt.

I   Der Leib als Zeichen sozio-kultureller Funktionen

Die *Historien* enthalten eine Vielzahl von ethnographischen Beobachtungen, die es mit somatischen Themen zu tun haben. Herodot akzentuiert v. a. Phänomene, die den Griechen fremd und exotisch vorkamen; Phänomene, bei denen ‚kulturelle

---

9   Vgl. Müller (1997), 113-115: Detaillierte Liste der ethnographischen Beschreibungsmuster.

Differenzen' besonders scharf hervortraten.[10] Um die Breite des Spektrums zu verdeutlichen, beginne ich mit einigen allgemeinen Bemerkungen zu den äußeren Erscheinungsformen des Leibes (I). Ich wende mich dann (der Ambivalenz des griechischen Ausdrucks σῶμα folgend) erst dem lebendigen (II), dann dem toten Leib zu (III). Herodot beschreibt oft das Aussehen der nicht-griechischen Menschen. Dies ist aber nur nötig, wenn ihr Äußeres von dem der Griechen markant abweicht. Der Leib fungiert meist als Signum kultureller Funktionen. In den *Historien* begegnen teils aberwitzige Behauptungen: so die von den Libyern erzählte Geschichte, es gäbe Menschen ohne Kopf, die ihre Augen auf dem Bauch trügen (IV 191); oder (die schon von Aristoteles kritisierte Bemerkung), der Samen der äthiopischen Männer sei schwarz. Anthropologisch virulent ist ein Passus, der offen läßt, ob es sich bei den genannten „wilden Männern und wilden Frauen" um Menschen oder Menschenaffen handelt.[11] Einige Textbeispiele:

> [Die Inder] haben alle dieselbe Hautfarbe, der der Äthiopier sehr ähnlich. Ihr Same, den sie an die Frauen abgeben, ist nicht wie bei anderen Menschen weiß, sondern schwarz wie ihre Haut. Einen solchen Samen haben auch die Äthiopier. (III 104)
>
> Die Budiner sind ein großes zahlreiches Volk und haben sämtlich ganz helle Augen und rötliche Haare. [...] Teils sprechen sie skythisch, teils griechisch. (IV 108)
>
> Die Babylonier [...] haben langes Haupthaar und binden es zu einem Schopf hoch, am ganzen Körper sind sie mit Myrrhe gesalbt. Jeder trägt einen Siegelring und ein handgearbeitetes Szepter. (I 195)
>
> Die Argiver aber, die seit dieser Zeit den Kopf kahlgeschoren tragen, während sie früher verpflichtet gewesen waren, langes Haar zu tragen, setzten durch ein Gesetz fest und sprachen einen Fluch aus, dass kein Argiver sein Haar wachsen lassen und keine Frau Goldschmuck tragen dürfe, [...]. Die Lakedaimonier aber bestimmten durch ein Gesetz das Gegenteil, nämlich von diesem Zeitpunkt an lange Haare zu tragen, während sie vorher keine langen Haare gehabt hatten. (I 82)

---

10 Vgl. Müller (1997), 107. Demnach berichtet Herodot nur über eine bestimmte Auswahl an Sitten und Gebräuchen, die er für besonders repräsentativ und typisch für das Ethos des Volkes hält.

11 Vgl. Hdt. IV 191. Die Stelle könnte inspiriert sein von einem ca. 470 v. Chr. verfaßten [von Plin. Hist. nat. 5.8 bezeugten] Reisebericht des Karthagischen Seefahrers Hanno über eine Gegend in Westafrika. Dort heißt es: »Und in diesem See lag eine weitere Insel, voll von wilden Menschen. Es waren überwiegend Weiber, die am ganzen Körper dichtbehaart waren; die Dolmetscher nannten sie goríllai. Wir verfolgten sie, konnten aber keine Männer fangen; sie entwischten alle, weil sie ausgezeichnete Kletterer waren und sich mit Felsbrocken zur Wehr setzten; Weiber aber fingen wir drei ein; sie bissen und kratzten und wollten denen, die sie führten, nicht folgen. Daher töteten wir sie, zogen ihnen die Haut ab und brachten die Bälge nach Karthago mit« (Übers. Bayer, in Ders. [1993], 353). Ferner heißt es in dem Bericht, »Waldmenschen in Tierfellen« hätten die Reisenden mit Steinen beworfen; vgl. dazu: Wuketits (2005), 14.

Anderswo tragen die Priester der Götter langes Haar; in Ägypten scheren sie es ab.
Bei Trauerfällen haben die anderen Völker die Sitte, daß die nächsten Angehörigen
sich das Haar abschneiden; in Ägypten läßt man, wenn jemand stirbt, Haupthaar
und Bart wachsen, während man sich sonst schert. (II 36)
Die libyschen Frauen tragen ein ledernes Gewand an dem mehrere Riemen hängen.
Über das Kleid werfen sie ein kahles, rot gefärbtes Ziegenfell. (II 189)[12]
Die Aithiopen hatten Panther- und Löwenfelle umgetan. [Es folgt die Beschreibung
ihrer Kriegstracht[13]] Für die Schlacht färbten sie ihren Körper zur Hälfte mit Kreide,
zur anderen Hälfte mit [rotem] Mennig. (Hdt. VII 69)
Die Maxyer [nordafrikanisches Volk, westl. des Triton-Flusses] lassen auf der rechten
Seite des Kopfes das Haar wachsen, die linke scheren sie ab, den Körper reiben sie
mit rotem Ocker ein. (IV 191)

Die Beispiele lassen drei allgemeine Folgerungen zu: (a) bei Herodot gibt es (wie
auch sonst in der klassischen Antike) keine ethnozentrischen oder gar rassistischen
Wertungen: Sog. ‚rassische' Merkmale werden vereinzelt zwar erwähnt, ziehen aber
keine normativen Wertungen nach sich; sie sind kulturwissenschaftlich belanglos;
(b) die Beschreibung *individueller Leiblichkeit* ist eine Ausnahme und ebenfalls
weitgehend irrelevant; (c) der *Leib als solcher* ist fast nie Thema, vielmehr verweisen Herodots Notizen (Kleidung,[14] Haartracht, Kriegsbemalung, Schmuck[15]) fast
ausnahmslos auf den Leib als Ausdruck sozio-kultureller Funktionen.

---

12 Vgl. Hdt. IV 189: Die Griechen haben diese Tracht übernommen und verwenden sie
    auch für die Abbildung der Athene. Aus dem Wort *aiges* für Ziegenfell stammt der
    griechische Ausdruck *aigis*, mit dem das Ziegenfell als mythologischer Gegenstand
    gemeint ist; vgl. Hdt II 50 zu weiteren Einflüssen der Libyer auf die Griechen (Erfindung
    des Viergespanns, Übernahme des Poseidon-Kultes).

13 Vgl. Hdt. VII 69: „Ihre Bogen waren aus Palmstreifen gemacht und nicht weniger als
    vier Ellen lang; dazu hatten sie kurze Rohrpfeile, deren Spitze nicht Eisen, sondern
    Stein war, wie man ihn auch zum Schneiden der Siegelringe benutzt. Ferner hatten
    sie Lanzen, denen ein zugespitztes Antilopenhorn als Spitze diente. Auch beschlagene
    Keulen führten sie."

14 Vgl. Zoepffel (2006), 322: Demnach spielt die „Kleidung […] in weniger informellen Ge-
    sellschaften als der heutigen eine bedeutende Rolle für die gesellschaftliche Einordnung
    einer Person. Deshalb gehörten Kleiderordnungen auch in den Rahmen der öffentlichen
    Regelungen".

15 Vgl. Zoepffel (2006), 324: Demnach führen insb. die Forschungen von Kenneth J. Dover zu
    dem Schluß „daß die Aufwendungen für Kleidung und Schmuck bei den [griechischen]
    Frauen derjenigen von Ausgaben für Hetären und andere sexuelle Vergnügungen beim
    Mann entsprechen. Die Komödie läßt aber erkennen, daß Männer ebenso gerne mit
    prächtiger Kleidung, Schmuck und Waffen oder mit Pferden prunkten".

II Der lebendige Leib

[*Der sexuelle Gebrauch des Leibes*] Herodots Berichte über den „Gebrauch der Lüste" (Foucault) sind ein guter Indikator für seine Verortung der ‚Ethnien' auf einer Skala von ‚zivilisiert' und ‚wild' (so auch Bichler [2001], 71). Auch hier wieder einige Textbeispiele:

> Geschlechtsverkehr pflegen alle Inder [...] wie das Weidevieh, in aller Öffentlichkeit. (III 101)
>
> Der Geschlechtsverkehr dieser Völker [am Kaukasus] geschieht in aller Öffentlichkeit wie bei Tieren. (I 203)
>
> Die Auseer [nordafrikanisches Volk am Triton-See] leben in Weibergemeinschaften und kennen kein eheliches Zusammenleben. Auch sie „begatten sich wie das Vieh". Solange ein Kind des Auseer-Volkes nicht erwachsen ist, ist der Vater unbekannt. Erst wenn ein bestimmtes Alter erreicht ist, sammeln sich innerhalb von drei Monaten alle möglichen Väter und entscheiden mit der Mutter, welchem Mann das Kind am ähnlichsten sieht. (IV 180)[16]
>
> [Sex in heiligen Bezirken] Fast alle anderen Menschen nämlich, ausgenommen Ägypter und Griechen, vereinigen sich mit Frauen in Heiligtümern und kommen, wenn sie von einer Frau aufstehen, ungewaschen ins Heiligtum, da sie meinen, daß die Menschen wie die übrigen Tiere sind. (II 64)
>
> [*ius primae noctis*] Das Volk der Adyrmachiden pflegt dagegen eine andere außergewöhnliche Sitte. Als einziges Volk Libyens haben sie den Brauch, dem König alle heiratsfähigen Mädchen vorzuführen. Das, welches ihm am besten gefällt, entjungfert er. (IV 168)
>
> [Polygamie] Die Leute nördlich der Krestonaier [in Thrakien] tun folgendes: Jeder Mann hat viele Frauen. Wenn nun einer von ihnen stirbt, gibt es einen großen Wettstreit unter den Frauen; auch ihre Freunde bemühen sich eifrig um die Frage, welche von ihnen von dem Mann am meisten geliebt worden ist, diejenige aber auf die die ehrenvolle Wahl fiel, wird gepriesen von Männern und Frauen, getötet über dem Grabe von dem, der ihr am nächsten steht; und die Getötete wird zusammen mit dem Mann bestattet. Die anderen Frauen aber tragen schwer daran, nicht gewählt worden zu sein, denn das gilt bei ihnen als größter Schimpf. (V 5)
>
> [Polygamie] Bei den [nordafrikanischen] Nasamonen wird der Brauch gepflegt, dass ein Mann viele Frauen hat, die gemeinsamer Besitz sind. Wenn eine Frau von einem Mann besucht wird, stellt dieser einen Stab vor die Hütte auf um ein Zeichen zu setzen. Zudem haben sie die Sitte, dass eine Frau in der Nacht vor ihrer Heirat mit sämtlichen Hochzeitsgästen schlafen muss und dafür Geschenke bekommt. (IV 172)

---

16 Vgl. Hdt. IV 180: An einem jährlichen Tag feiern sie das Fest der Athena. Die Jungfrauen dieses Volkes teilen sich in zwei Gruppen und kämpfen mit Steinen und Keulen gegeneinander. Die Jungfrauen, die an ihren Wunden sterben, gelten als falsche Jungfrauen. „Die Jungfrau, die am tapfersten gekämpft hat, wird mit einem korinthischen Helm und einer griechischen Kriegsausrüstung geschmückt und auf einem Wagen um den See herumgefahren".

[Frauengemeinschaft] Das skythische Volk der Agathyrsen lebt in einer Frauengemeinschaft, in der alle miteinander verwandt sind und somit kein Neid oder Zwietracht aufkommen kann. (IV 104)

[Verkauf der Kinder] Bei den anderen Thrakern gibt es folgenden Brauch: Sie verkaufen die Kinder zur Ausfuhr in die Fremde. Auf die Mädchen geben sie nicht acht, sondern lassen sie Umgang haben mit den Männern, die sie wünschen; die Frauen aber bewachen sie streng und kaufen die Frauen von den Eltern für teures Geld. (V 6)

[Prostitution für die Mitgift] Alle Töchter des niederen lydischen Volkes gehen der Prostitution nach und sammeln damit ihre Mitgift ein, bis sie heiraten. Sie arrangieren selbst ihre Verheiratung. (I 93)

[Heiratsmarkt mit Sozialausgleich] Die vernünftigste Sitte [der Babylonier] ist meiner Meinung nach diejenige, die (wie ich erfahren habe) auch die Eneter in Illyrien kennen: In jedem Dorf wurde einmal im Jahr folgendes gemacht: Sobald die jungen Frauen ins heiratsfähige Alter kamen, führte man sie alle dichtgedrängt auf einen Platz zusammen; um sie herum aber stellte sich eine Schar Männer auf. Ein Herold ließ dann jede einzeln aufstehen und bot sie feil. Zunächst die schönste von allen, dann, sobald diese für viel Geld verkauft war, bot er eine andere an, die nach jener die schönste war. Die begüterten Freier der Babylonier kauften die schönsten, indem sie die anderen überboten. Die Freier aus dem Volk aber verlangten nicht nach gutem Aussehen, sondern erhielten Geld und häßlichere junge Frauen. Sobald der Herold mit dem Verkauf der schönsten Frauen fertig war, ließ er die häßlichste oder eine verkrüppelte aufstehen und rief diese aus, wer sie gegen Auszahlung der geringsten Summe heiraten wollte, bis sie demjenigen zugeschlagen wurde, der die geringste Summe entgegennahm. Das Geld aber kam von den schönen Frauen – und so verhalfen die schönen den häßlichen und verkrüppelten zur Heirat. Man durfte aber nicht seine Tochter an jeden Beliebigen verheiraten; und es war auch nicht möglich, die Frau ohne Bürgen wegzuführen, wenn man sie gekauft hatte; man mußte vielmehr Bürgen dafür stellen, daß man tatsächlich gewillt war, die Ehe mit ihr einzugehen, und dann konnte man sie fortführen. Wenn man nicht miteinander auskam, war es üblich, das Geld wieder zurückzugeben. Auch aus einem anderen Dorf durfte der Kaufwillige kommen. Diese Sitte war die beste bei ihnen; sie bestand freilich nicht mehr in unserer Zeit, sondern neuerdings haben sie ein anderes Verfahren erfunden: [Prostitution der Töchter] Seit die Babylonier nämlich [den Persern] unterworfen sind [seit 539 v. Chr.], schickt jeder einfache Mann, weil er zu wenig zum Leben hat, seine Töchter zur Prostitution. (I 196)

[Tempelprostitution] Die verabscheuungswürdigste Sitte bei den Babyloniern ist folgende: Im Gegensatz zum Heiratsmarkt, den Herodot als vernünftige Sitte beschreibt, hält er die Tempelprostitution in Babylonien für verabscheuungswürdig: Jede einheimische Frau muß sich einmal in ihrem Leben in den Tempel der Aphrodite setzen und Geschlechtsverkehr mit einem fremden Mann haben (I 199) [...] Die Frauen dürfen den Tempel nicht verlassen, bevor ein Mann ihnen Geld in den Schoß geworfen und mit ihm außerhalb des Tempels verkehrt hat. Häßliche Frauen, die nicht begehrt sind, bleiben sogar drei bis vier Jahre in dem Tempel und warten auf einen Mann. (ebd.) [17]

---

17 Vgl. Bichler (2001), 123: „Die hübschen jungen Frauen sind das Objekt der Begierde, die häßlichen und alten Objekt des Spotts". Die Bloßstellung der Frauen steht im Gegensatz zu dem sonst positiven Eindruck über die babylonischen Sitten und das Verhalten.

öffentliche Promiskuität aller	sexuelle Handlungen in Tempeln und heiligen Bezirken	allgemeine Polygamie	Weiber- und Kindergemeinschaft	allgemeine Prostitution aller Jungfrauen	partielle Prostitution (zum Zweck der Mitgift für die Ehe)	‚Zuweisung' von Frauen an den Machthaber	keine sexuellen Handlungen in Tempeln und heiligen Bezirken	Eheverträge; keine vorehelichen heterosexuellen Sexualpraktiken

ANIMALISCH/WILD		ZIVILISIERT/HOCHKULTUR
Randvölker	semi-zivilisierte Kulturen	Ägypten/Griechenland

Die Texte belegen, daß Herodot (als Vertreter einer monogamistisch-patrilinearen Kultur) ein weites Spektrum sexueller Sitten vorführt.[18] In Griechenland gab es zu seiner Zeit zwar Diskussionen über die Nacktheit der Männer (etwa in den Ringschulen) und ebenfalls über den Status der männlichen Homosexualität. Die Nacktheit von Frauen, ihr vorehelicher Geschlechtsverkehr und weibliche Homosexualität waren indes strikte gesellschaftliche Tabus; da diese Vorschriften (aus feministischer Sicht) zur Unterdrückung der Frau oder (eher nüchtern betrachtet) v. a. zur Stabilisierung des in Hellas zentralen Instituts der privatvertraglich geregelten Ehe diente.[19] Die von Herodot geschilderten Sexualpraktiken haben den primären

---

18　Vgl. Bichler (2001), 123-131: ausführlich zu Sexualität und Frauenrollen; vgl. Foley (1981), Dewald (1981), Dewald (2007): Frauen und Heirat bei Herodot; Zoepffel (2006), 311-358 allgemein zur Rolle der Frau in der Antike.

19　Vgl. Zoepffel (2006), 333: „Im allgemeinen wird eine Ehe zwischen dem Vormund (*kyrios*) der Braut und dem Bräutigam ausgemacht [...]; es gab aber immer auch Ausnahmen“. Bei Herodot sei überliefert, daß ein Athener seinen Töchtern eine großzügige Mitgift

Sinn, die genannten Ethnien auf einer Skala von ‚tierisch/wild' bis ‚zivilisiert' zu verorten. Die hemmungslose Publizität von Sex nennt Herodot „tierisch"; d. h. Wildheit kennzeichnet einen tierähnlichen Zustand. Ebenfalls die Promiskuität von Jungfrauen, die Verkehrung der in Hellas festgeschriebenen Hochzeitsbräuche, die Zuweisung von Frauen an (demokratisch nicht legitimierte) Herrscher, die allgemeine Anerkennung der Prostitution von allen Frauen einer Kultur[20] bzw. (im Falle Babyloniens) Tempelprostitution, sexuelle Handlungen in heiligen Bezirken gelten aus griechischer Sicht als skandalös. Veranschaulicht man Herodots Aussagen schematisch, so ergibt sich die nebenstehende Tabelle.

An einigen Stellen informiert Herodot auch über postsexuelle Bräuche. Von den Frauen der (libyschen) Gindanen heißt es, sie legten nach jedem Geschlechtsverkehr ein neues Lederband um den Knöchel; diejenige mit den meisten Ringen werde für die Beste gehalten, da sie von den meisten Männern geliebt wurde (IV 176). Ähnlich wie bei den nördlichen Thrakern unterstreicht Herodot die allgemeine Anerkennung der Promiskuität durch sämtliche Mitglieder der Gemeinschaft. Im Falle der sonst hochzivilisierten Babylonier wird das Skandalon der Tempelprostitution durch die Notiz gemildert, daß die Babylonier (wie übrigens auch die Araber) nach dem Geschlechtsverkehr baden (I 198). Dies zeigt, daß Herodots Verortung der Kulturen auf der wild-kultiviert-Skala regelmäßig (da, wo umfangreichere Beschreibungen vorliegen) *nicht nur von einem einzigen Prädikat* abhängt, sondern auf komplexen Zuschreibungen basiert: Die negative Wertung durch ein Attribut wie Tempelprostitution wird durch die Zuschreibung eines positiven Merkmals wie Reinheit entschärft. Am Beispiel der Babylonier ließe sich zudem zeigen, daß die Nähe zu der ägyptischen Kultur eine positive Wirkung hat; Kultur ist für Herodot gewissermaßen ansteckend (vgl. Bichler 2001, 58 et passim).

[*Der reinliche Leib*] Die Reinheit des Leibes ist ein großes, an vielen Stellen der *Historien* wiederkehrendes Thema. Herodot unterscheidet differente Motive für Reinlichkeit. Neben dem (insb. bei den Ägyptern relevanten) Ziel der Gesundheit, spielen Motive der religiösen Reinigung (bzw. Reinhaltung)[21], die Säuberung vor

---

gegeben und ihnen freie Wahl des Ehemannes gewährt habe. Ein Gegenbeispiel findet sich bei Demosthenes. Zum Ehevertrag resümierend Zoepffel (334-5): „Aus früherer Zeit sind keine Eheverträge überliefert, und es fragt sich, worin eine ‚rechtmäßige Ehe' überhaupt bestand". Der Beginn der Ehe sei gleichsam durch konkludentes Handeln dokumentiert, wenn die Braut öffentlich von ihrem alten in den neuen Oikos überführte.

20 Vgl. Zoepffel (2006), 346 zur Prostitution in Griechenland: Demnach waren Prostituierte in aller Regel so teuer, daß sich nur Männer aus der Oberschicht entsprechende Liebesdienste leisten konnten.

21 Vgl. Hdt. II 123: „Als Erste auch waren die Ägypter diejenigen, die sagen, dass des Menschen Seele unsterblich ist; wenn der Körper vergehe, ziehe sie in ein anderes Lebewesen

oder nach dem Geschlechtsverkehr, teils auch soziale Privilegierung (Zugriff auf besondere Ressourcen von Reinigungsmedien)[22] eine Rolle. Die gute Gesundheit der „langlebigen Äthiopen" begründet Herodot mit Bädern in sehr sauberen Quellen.[23] Die exzellente Gesundheit und Langlebigkeit der Libyer und Ägypter führt er (neben klimatischen Faktoren und der in Ägypten überragenden medizinischen Versorgung) auf besondere Reinlichkeitsvorschriften zurück. Zu diesem Hygeia-Komplex gehören auch die Ausführungen über die Beschneidung der Ägypter:

> Wie der Himmel in Ägypten anders ist, so sind auch die Sitten und Gebräuche der Ägypter fast in allen Stücken denen der übrigen Völker entgegengesetzt. [...] Die Frauen lassen ihr Wasser im Stehen, die Männer im Sitzen. Die Entleerung macht man im Hause ab, essen tut man auf der Straße. Sie geben als Grund dafür an, daß man natürliche Bedürfnisse, soweit sie häßlich sind, im geheimen, soweit sie nicht häßlich sind, öffentlich befriedigen müsse. [...] *Die Geschlechtsteile lassen die anderen Völker, wie sie sind; nur die Ägypter und die es von ihnen angenommen haben, beschneiden sie.* [...] Sie trinken aus Bronzebechern und spülen sie jeden Tag aus, und zwar tut das jeder ohne Ausnahme. [...] Sie tragen Kleider aus Leinen, die stets frisch gewaschen sind; darauf achten sie genau. *Die Beschneidung der Geschlechtsteile geschieht aus Reinlichkeitsgründen; Reinlichkeit steht ihnen höher als Schönheit.* Die Priester schneiden alle zwei Tage ihre sämtlichen Körperhaare ab, *damit* sich keine Laus oder anderes Ungeziefer festsetzen kann. [...] Zweimal am Tage und zweimal des Nachts baden sie in kaltem Wasser und halten noch andere, geradezu unzählig viele Gebräuche inne. [...] Bohnen sähen Ägypter in ihrem Lande überhaupt nicht, und die von selber wachsenden ißt man nicht, weder geröstet, noch gekocht. (II 35-37)

Für Herodot stellt die Hygiene das eigentliche Motiv der ägyptischen Beschneidung dar. In der Forschung wird kontrovers diskutiert, inwiefern Herodots Angaben der ‚historischen Wahrheit' entsprechen bzw. wie er zu seinen Ansichten gekommen ist. Namentlich ist strittig, ob der (übrigens fast weltweit verbreitete)[24] Beschnei-

---

um, das gerade entsteht; wenn sie aber durch alle gezogen sei, die auf dem Land, im Meer und in der Luft leben, ziehe sie wieder in den Körper des Menschen, der gerade geboren werde."

22 Vgl. Hdt. I 188: Der Perserkönig kann es sich leisten, gekochtes Wasser aus dem eigens für ihn reservierten Fluß Choaspes zu trinken.

23 Vgl. Hdt. III 20 u. 23: Die Bewohner Äthiopiens sind die größten und schönsten aller Menschen. Die haben eine Lebenszeit von ca. 120 Jahren. Der Grund dafür liegt u.a. in ihrer einfachen Ernährung (gekochtes Fleisch und Milch), einer besonderen Quelle, durch welche die Haut wie Öl glänzt; das Quellwasser ist sehr wohlriechend und so leicht, daß nicht einmal Holz darauf schwimmt.

24 Vgl. Lloyd (1976), 159 (mit Verweis auf *Foucart* [1910], 659 ff.): Demnach ist die Beschneidung ein weltweit verbreiteter Brauch; Ausnahmen: Europa und das nicht-semitische Asien.

dungsritus seine Wurzeln eher in rituellen[25], sozialen[26] oder hygienisch-medizinischen[27] Motiven hatte. Offenbar aber ließ die Bindungskraft dieser Gebote im Laufe der Zeit nach und es setzte eine gewisse Beschneidungsmüdigkeit ein. Herodots Zeugnis ist vermutlich von der Beobachtung inspiriert, daß zu seiner Zeit die Beschneidung nicht *mehr an allen männlichen Ägyptern* vorgenommen wurde.[28]

---

25 Vgl. How/Wells (1928) [1912], 183; Sigerist (1963), 224: „Dies enthüllt auch deutlich den religiösen Charakter des Ritus. Herodot glaubte, daß er aus hygienischen Gründen ausgeführt wurde, als er schrieb: ‚Sie führten die Beschneidung aus Gründen der Reinlichkeit durch (II, 37); denn sie stellten die Reinlichkeit über die Schicklichkeit‘, eine Feststellung, die sicher falsch ist. Die Operation wurde in Tempeln vorgenommen, nicht von Ärzten, sondern von Priestern, mit einem Steinmesser und nicht mit einem Bronzemesser, ein Detail, das zeigt, wie alt die Sitte war. Vielleicht war es auch [...] eine Einweihung, besonders in den Kult von Re, denn eine alte Überlieferung, berichtete, daß Blut von Phallus des Gottes fiel, als er sich selbst verstümmelte. Dann wäre die Beschneidung eine Nachahmung der verstümmelten Operation des Gottes. Wie dem auch sein mag, über ihren religiösen Charakter gibt es wohl keinen Zweifel.“

26 Vgl. Lloyd (1976), 165: (a) (mit Verweis auf Wendland APF [1902]) Vorbereitung auf Fruchtbarkeit und Heirat; (b) (mit Verweis auf Evans-Pritchard [1954], 92) als Initiationsritus entweder (i) für den Eintritt in die erwachsene männliche Gemeinschaft oder [mit Hinweis auf Foucart] im besonderen Fall der Ägypter ursprünglich als Initial für den Eintritt in den Re-Kultus; so auch Ghalioungui 1963, 95.

27 Vgl. Lloyd (1976), 157-159 [zu Hdt II 36...] und 165 [zu Hdt. II 37] mit ausführlicher Diskussion des historischen Materials (Dokumente, Statuen, Mumien) und spezieller Forschungsliteratur. Lloyd weist darauf hin, daß *alle Leichen* der prädynastischen Grabstätten beschnitten waren; die archäologischen Befunde ergeben, daß die Beschneidungen meist zur Zeit der Pubertät vorgenommen wurden. Andererseits aber sei klar, daß die Beschneidung im *Alten Reich* und im *Mittleren Reich* keine durchgehende Praxis war („not a universal practice“). In der Forschung werde (obwohl es dafür keine positiven Beweise gebe) meist angenommen, daß in der Pharaonenzeit die Beschneidung vornehmlich unter den Priestern und Königen verbreitet war; in diesen Fällen lag das Beschneidungsalter vom ersten Lebensjahr bis zur Pubertät. (Für die Römische Zeit sei gesichert, daß Beschneidung für alle Priester Pflicht war: Hadrian hat sie per Gesetz verboten.) Üblicherweise wurde die Beschneidung mit einem Feuer-Steinmesser vorgenommen, wobei der Eingriff entweder (i) (bloß) einen Einschnitt in die Vorhaut oder (ii) (wie bei den Hebräern) die komplette Entfernung bedeuten konnte. Als hygienische Begründung kommt insb. die sog. Vorhautverengung [Phimose] in Betracht, die bei 96 v. H. aller männlichen Kleinkinder vorkommt. Die von Lloyd ebd 165 aufgeführten Gründe für ‚purity‘ (das Insistieren der Priester auf Beschneidung und die Abscheu vor den Nicht-Beschnittenen) sind indes wenig überzeugend, da sie m. E. eindeutig auf religiöse Motive hindeuten.

28 Vgl. Brandenburg (1976), 93: Demnach läßt sich anhand von Mumienfunden feststellen, daß nicht alle Ägypter beschnitten waren. Hätte die Beschneidung aus religiösen Gründen stattgefunden, so hätten sie bei allen Ägyptern durchgeführt werden müssen; vgl. Brandenburg (1976), 93 (mit Verweis auf Sigerist) auch zur *Beschneidung von Mäd-*

Die These, daß andere Völker die Beschneidung von den Ägyptern übernommen haben, wird in II 104 im Kern wiederholt; wobei Herodot an dieser Stelle einige Ethnien nennt (Kolcher, Phöniker, Äthiopier, Syrer[29] in Palästina), aber zugleich einschränkend bemerkt, er sei sich im Falle der Äthiopier nicht sicher, ob sie die Beschneidung von den Ägyptern übernommen hätten oder umgekehrt. Jedenfalls sei die Beschneidung „uralt" (II 104). Der Passus II 35-36 über die Beschneidung fällt in den Kontext einer scharfen Zeichnung der ägyptischen Hygiene: Die Priester rasieren wegen dieser Reinheitsgebote alle zwei Tage sämtliche Körperhaare. Sie waschen sich viermal täglich. Sie halten sich vom Verzehr bestimmter Tiere fern, sie säubern ihre Bestecke und Teller und „essen nicht einmal das Fleisch eines reinen Stieres, wenn es mit einem *griechischen Messer* geschnitten ist" (II 125). Sie würden nie einen Griechen zur Begrüßung auf den Mund küssen (II 41; 125). Aus ägyptischer Perspektive gelten die Griechen also teils als unrein. Überhaupt nehmen die Ägypter (wie die Skythen) keine Sitten von fremden Völkern an (II 91). Für die Perser bezeugt Herodot ähnliche Vorschriften, die aber hier in einem ganz anderen Licht erscheinen:

> Nie urinieren sie in einen Fluß oder speien hinein, waschen auch nicht ihre Hände darin oder dulden, daß es ein anderer tut. Vielmehr erweisen sie den Flüssen tiefste Ehrfurcht. Noch gibt es etwas Merkwürdiges bei den Persern, was sie selbst freilich nicht empfinden, wohl aber wir. (I 138)
> Der Rang eines Persers in der Gesellschaft ist an der Begrüßung mit einem anderen zu erkennen. Der hohe Rang küßt sich gegenseitig auf den Mund, der geringere auf die Wangen. Wenn der andere allerdings viel höher steht, muss er ihm zu Füßen fallen und ihm seine Verehrung deutlich machen. (I 134)

Die Ehrfurcht vor den Flüssen ist auch für Griechenland belegt: Hesiod (der eine Vielzahl an Flußgöttern anführt) rät, nicht in die Flüsse oder gegen die Sonne zu urinieren (Op. 727, 757-759).[30] Diese Vorschriften gehören in einen alten (der Scham vor den wachsamen Göttern geschuldeten) Sittenkanon. Hygienische Motive werden (wie oft in der Kulturgeschichte) erst später nachgeliefert und (wenn die ursprünglichen Riten nicht mehr verstanden werden) ex post konstruiert. Der

---

chen: Demnach ist über die Resektion der Klitoris und der kleinen Schamlippen nur bei Strabon eine kleine Notiz überliefert.

29 Vgl. Hdt. II 104: „Dafür aber, daß die Phöniker es im Verkehr mit den Ägyptern gelernt haben, ist auch Folgendes ein wichtiges Indiz: Diejenigen unter den Phönikern, die mit den Griechen Verkehr haben, ahmen nicht mehr die Ägypter nach, sondern beschneiden die Penisse ihrer Nachkommen nicht".

30 Vgl. Jamblichos, Protr. 21 [DK 58 C 6: Liste der pythagoreischen Akousmata; Nr. 15]: „Uriniere nicht der Sonne zugewandt!".

persische Begrüßungsritus ist keine *kulturelle*, sondern eine *soziale* Norm. Insb. der Kniefall (die sog. Proskynese) gehört zu den von den Griechen strikt abgelehnten Unterwerfungsgesten.[31] Über fremde Badesitten handelt Herodot u. a. in seinem skythischen Logos in Buch IV. Demnach waschen die Skythen sich nie mit Wasser. Sie favorisieren Dampfbäder. Die skythischen Frauen verwenden dabei besondere Kosmetika:

> Die Frauen [der Skythen] reiben auf einem rauhen Stein Zypressen- und Zedern- und Weihrauchholz und gießen Wasser dazu. Und dann bestreichen sie sich damit, das nun ein dicker Brei geworden, den ganzen Leib und das Gesicht. Dadurch nun erlangen sie einen lieblichen Geruch, und werden, wenn sie am nächsten Tag die Paste abnehmen, rein und glänzend. (IV 75)[32]

[*Eingriffe in den Körper*] Freilich sind nicht alle Eingriffe in den Körper hygienisch oder religiös motiviert. Zu den zweifellos brutalsten Interventionen gehört die Kastration, die die Sieger an den Besiegten teils zum Zweck der Demütigung, teils zum Zweck der künftigen Dezimierung der Feinde vornahmen. Herodot spielt darauf an zwei Stellen an. Deutlich wird, daß die Kastration keine kultische, hygienische oder religiöse Funktion hatte, sondern entweder (II 130) zur Ausmalung des brutalen persischen Herrschaftssystems oder (VIII 104-6) einer scharfen Kritik der Praktiken des Sklavenhandels dient (weiterführend: Brandenburg [1976], 122-128). Zu den eher freiwilligen Eingriffen in den Körper gehört die Tätowierung. Herodot bemerkt, bei den Thrakern gelten Tätowierungen als vornehm, das Fehlen der Tätowierung indes als unvornehm. (V 6)[33] Die Tätowierung ist hier (zumindest auch) Mittel der sozialen Differenzierung innerhalb einer Gruppe. Im Zusammenhang mit der besonderen Vertragstreue der Araber notiert Herodot, solche Bündnisse würden u. a. durch *Schwurbrüderschaft* besiegelt:

> Den Bund [Vertrag] schließen sie [die Araber] auf folgende Weise ab. Wenn zwei miteinander einen Bund schließen wollen, so stellt sich ein anderer zwischen beide in die Mitte und macht mit einem scharfen Steine einen Einschnitt in die Mitte der Hand neben dem Daumen der beiden, die den Bund miteinander eingehen; dann nimmt er von dem Mantel eines jeden der beiden einen Flocken und bestreicht mit dem Blut sieben vor ihnen liegende Steine: während er dies tut, ruft er Dionysos [MM: Baal] und Urania [MM: Aphrodite] an. Und wenn er dies vollbracht hat, so stellt derjenige, der

---

31 Vgl. Hdt. I 134; VII 136; dazu Brodersen (2007a), 269 (Anmerkung 150); vgl. auch Arist. Rhet.

32 Vgl. Brandenburg (1976), 135-139 hierzu und zu weiteren antiken Kosmetika.

33 Vgl. Bichler (2001), 67: Die Thraker nehmen eine ambivalente Position zwischen Wildheit und bekannter Zivilisation ein.

den Bund geschlossen hat, seine Freunde als Bürgen dem Fremdling oder auch dem Mitbürger vor, wenn er mit einem solchen den Bund eingegangen hat; die Freunde glauben dann ebenfalls von ihrer Seite den Bund heilig halten zu müssen. (III 8)

Das Blut der Vertragspartner bzw. die blutbestrichenen, den Göttern offerierten Steine dienen als Vertragsunterschrift, der die Gültigkeit des Vertrags sichtbar und gleichsam amtlich vor dritten Personen und Zeugen besiegelt. Herodot berichtet ebenfalls von einem medizinischen Eingriff in den Körper bei den nomadischen Völkern Nordafrikas; hier ist wieder ein gesundheitliches Motiv ausschlaggebend:

Diese libyschen Nomaden nämlich (ob alle, kann ich nicht mit Sicherheit sagen, jedenfalls viele) tun folgendes mit ihren Kindern. Wenn dieselben vier Jahre alt sind, brennen sie ihnen mit dem fettigen Schmutz der Schafswolle die Adern oben auf dem Kopf auf, einige auch die [Adern] an den Schläfen; und zwar deswegen, daß ihnen nicht die ganze Zeit über das Phlegma schadet, das aus dem Kopf in den Körper strömt. Darum seien sie so sehr gesund, behaupten sie. (IV 187)[34]

III   Der Leib der Toten und Trauernden

Auch der Körper der Toten wird dem kulturellen Ritus unterworfen – oder genauer gesagt: Im Ritus spiegelt sich der *soziale Leib* des Toten. Die *Historien* liefern eine ganze Reihe von Informationen zu diesem Thema. Ausführliche Nachrichten erhalten wir von den Trauerriten der Ägypter:

Totenklage und Begräbnis [bei den Ägyptern] gehen folgendermaßen vor sich. Wenn in einem Hause ein angesehener Hausgenosse stirbt, bestreichen sich sämtliche weiblichen Hausbewohner den Kopf oder auch das Gesicht mit Kot, lassen die Leiche im Hause liegen und laufen mit entblößter Brust, sich schlagend, durch die Stadt; alle weiblichen Verwandten schließen sich ihnen an. Auch die Männer schlagen sich und haben ihr Gewand unter der Brust festgebunden. (II 85)

Berühmt sind die hier folgenden Schilderungen der ägyptischen Praktiken zur Einbalsamierung und Mumifizierung der Leichen. Herodot unterscheidet sozial gestaffelt drei verschiedene Verfahren:

---

34   Vgl. Brandenburg (1976), 113-114. Die Vorstellung von dem pathogenen Phlegma im Kopf der jungen Kinder ist in der klassischen griechischen Medizin aufgenommen worden. In der hippokratischen Schrift *De morbo sacro* (5.1-5) wird das Nicht-Abfließen dieses Phlegmas als eine Ursache für die Entstehung der Epilepsie genannt.

Es gibt besondere Leute, die dies [die Einbalsamierung der Leiche] berufsmäßig ausüben. Zu ihnen wird die Leiche gebracht, und sie zeigen nun hölzerne, auf verschiedene Art bemalte Leichname zur Auswahl vor. Wonach man die vornehmste der Einbalsamierungsarten benennt, scheue ich mich zu sagen. Sie zeigen dann weiter eine geringere und wohlfeilere und eine dritte, die am wohlfeilsten ist. Sie fragen dann, auf welche der drei Arten man den Leichnam behandelt sehen möchte. Ist der Preis vereinbart, so kehren die Angehörigen heim, und jene machen sich an die Einbalsamierung. Die *vornehmste Art* ist folgende. Zunächst wird mittels eines eisernen Hakens das Gehirn durch die Nasenlöcher herausgeleitet, teils auch mittels eingegossener Flüssigkeiten. Dann macht man mit einem scharfen aithiopischen Stein einen Schnitt in die Weiche und nimmt die ganzen Eingeweide heraus. Sie werden gereinigt, mit Palmwein und dann mit geriebenen Spezereien durchspült. Dann wird der Magen mit reiner geriebener Myrrhe, mit Kasia und anderem Räucherwerk, jedoch nicht mit Weihrauch, gefüllt und zugenäht. Nun legen sie die Leiche ganz in Natronlauge, 70 Tage lang. Länger als 70 Tage darf es nicht dauern. Sind sie vorüber, so wird die Leiche gewaschen, der ganze Körper mit Binden ans Byssosleinwand umwickelt und mit Gummi bestrichen, was die Ägypter an Stelle von Leim zu verwenden pflegen. Nun holen die Angehörigen die Leiche ab, machen einen hölzernen Sarg in Menschengestalt und legen die Leiche hinein. So eingeschlossen wird sie in der Familiengrabkammer geborgen, aufrecht gegen die Wand gestellt. (II 86)

Das ist die Art, wie die Reichsten ihre Leichen behandeln. Wer die Kosten scheut und die *mittlere Einbalsamierungsart* vorzieht, verfährt folgendermaßen. Man füllt die Klystierspritze mit Zedernöl und führt das Öl in den Leib der Leiche ein, ohne ihn jedoch aufzuschneiden und die Eingeweide herauszunehmen. Man spritzt es vielmehr durch den After hinein und verhindert den Ausfluß. Dann wird die Leiche die vorgeschriebene Anzahl von Tagen eingelegt. Am letzten Tage läßt man das vorher eingeführte Zedernöl wieder heraus, das eine so große Kraft hat, daß Magen und Eingeweide aufgelöst und mit heraus gespült werden. Das Fleisch wird durch die Natronlauge aufgelöst, so daß von der Leiche nur Haut und Knochen übrigbleiben. Danach wird die Leiche zurückgegeben, und es geschieht nichts weiter mit ihr. (II 87)

Die *dritte, von den Ärmeren angewandte Art* der Einbalsamierung ist folgende. Der Leib wird mit Rettigöl ausgespült und die Leiche dann die 70 Tage eingelegt. Dann wird sie zurückgegeben. Die Frauen angesehener Männer werden nicht gleich nach dem Tode zur Einbalsamierung fortgegeben, auch schöne oder sonst hervorragende nicht. Man übergibt sie den Balsamierern erst drei oder vier Tage später; und zwar geschieht das deswegen, damit sich die Balsamierer nicht an den Frauen vergehen. Es sei einmal einer wegen der Schändung einer frischen Frauenleiche bestraft worden, den ein Berufsgenosse angezeigt hatte. (II 88-89)

Wie Aleida Assmann bemerkt, zeichnet sich Ägypten durch die besondere Gemeinschaft der Lebenden mit den Toten aus (2010, 32 ff.). Ganz anders als in Griechenland, wo die Leiche möglichst bald aus den Augen der Lebenden verschwinden muß, zielen die ägyptischen Mumifizierungsriten darauf, die Leiber der Toten in der kulturellen Gemeinschaft dauerhaft sichtbar zu halten. Der Verstorbene soll der

Gemeinschaft nicht verlorengehen.[35] Auf dieses Ziel hin sind auch die Totenkulte der indischen Padaier ausgerichtet:

> Wenn ein Mitglied des Stammes [der Padaier, ein Nomadenvolk im östlichen Indien] krank wird, eine Frau oder ein Mann, so wird es, wenn es ein Mann ist, von den nächsten männlichen Freunden getötet. Denn, sagen sie, die Krankheit zehrt sein Fleisch auf, *so daß es uns verlorengeht*. Auch wenn er seine Krankheit ableugnet, töten und verzehren sie ihn, ohne auf ihn zu hören. Ist es eine Frau, die krank wird, so tun die nächsten weiblichen Verwandten dasselbe mit ihr wie die Männer mit den Männern. (III 99)

Die Padaier stellen ihre Gemeinschaft insgesamt als sozialen Leib vor. Die von Herodot beschriebene Androphagie verfolgt den Zweck, daß dieser soziale Leib nicht verkleinert wird oder abnimmt. Der drohende oder reale Verlust eines Mitglieds der Gemeinschaft wird (übrigens auch bei dem indischen Volk der Kallatier)[36] dadurch kompensiert, daß die Lebenden sich die Kranken bzw. die Toten einverleiben. Bei genauerem Hinsehen zeigt sich, daß es sich bei dem sozialen Leib der Padaier eigentlich um zwei Leiber handelt: einen männlichen und einen weiblichen Leib. Die lebenden Männer und Frauen sorgen sich geschlechtsspezifisch um die je korrespondierenden Leiber. Während die Ägypter die Toten also für die Lebenden bewahren, indem sie jene sichtbar halten, leben die Toten der Padaier in ihren Leibern weiter fort. In beiden Fällen erhält sich der soziale Leib als ein zeitloses Ganzes, an dem sich das individuell vergängliche Leben orientiert. Eine ähnliche Zielrichtung verfolgen, obzwar mit reversen Methoden, die Skythen. Wenn ein skythischer König stirbt, wird sein Leichnam mit Wachs überzogen, der Bauch geöffnet und gereinigt, mit Gewürzen gefüllt und wieder zugenäht:

---

35 Vgl. Lévi-Strauss (1973), 45-48: Eine konstitutive Variante von Riten liegt demnach in dem im mythischen Denken stets virulenten Verhältnis der Lebenden zu den Toten: Stirbt ein Mitglied einer Population, so besteht damit die (neue) Situation eines Ungleichgewichts zwischen den Lebenden und den Toten.

36 Vgl. Hdt. III 38: „Einmal ließ Dareios während seiner Herrschaft [522-486 v. Chr.] diejenigen Griechen, die an seinem Hof weilten, herbeirufen und fragte sie, für welchen Preis sie bereit wären, ihre verstorbenen Väter zu verspeisen. Sie aber sagten, sie würden dies für keinen Preis tun. Daraufhin ließ Dareios von den *Indern die sogenannten Kallatier* rufen, die ihre Eltern verspeisen, und fragte sie (während die Griechen dabei standen und mit Hilfe eines Dolmetschers verstehen konnten, was gesprochen wurde), für welchen Preis sie bereit wären, ihre verstorbenen Väter im Feuer zu verbrennen. Diese schrieen laut auf und sprachen, er solle gottlose Äußerungen vermeiden." Herodot kommentiert dies wie folgt: „So verhält es sich also mit den Sitten und Pindar scheint mir recht zu haben, wenn er dichtet, der *Nomos* sei König über alle."

> Die Leiche wird nun von Stamm zu Stamm geführt. Jeder Stamm, zu dem sie gelangt, tut dasselbe, was die königlichen Skythen zuerst tun: jeder schneidet ein Stück von seinen Ohren ab, schert seine Haare, macht einen Schnitt rum um den Arm, ritzt Stirn und Nase und sticht einen Pfeil durch die linke Hand. Dann geht es zum nächsten Stamm, und der vorhergehende gibt der Leiche das Geleit. (IV 71)

In diesem Falle verleiben sich nicht die Lebenden den Leib der Toten ein. Vielmehr wird dem Toten ein Stück Leib der Lebenden mit-geteilt. Aber auch bei den Skythen sorgt der Ritus dafür, daß die Gemeinschaft der Lebenden mit den Toten nicht durch den Verlust eines führenden Mitglieds der Gemeinschaft vermindert und verkleinert wird. Im toten Leib des Königs leben die Untertanen anteilig weiter. Dieser Gedanke findet Anwendung auch im Falle des Todes eines einfachen Gesellschaftsmitglieds der Skythen. Jeder Verwandte des Toten bewirtet den Leichnam und setzt ihm Essen vor (IV 73). Dann erfolgt die Bestattung nach folgendem Reglement:

> Haben sie einen [Toten] begraben, so reinigen sich die Skythen auf folgende Art: [...] Sie stellen drei Stangen auf, gegeneinander gelehnt, um die spannen sie Filzdecken aus Wolle, und nachdem sie die sorgfältig abgedichtet haben, werfen sie im Feuer erhitzte Steine in eine Wanne, die unter den Stangen und Filzdecken in der Mitte steht. Es wächst bei ihnen im Lande nun auch Hanf, der ist dem Flachs sehr ähnlich, bis auf die Dicke und Größe; darin übertrifft ihn der Hanf bei weitem. Der wächst wild oder wird angebaut, und aus ihm machen die Thraker auch Stoff für Kleider, der ganz ähnlich ist wie Leinen. [...] Von diesem Hanf also nehmen die Skythen den Samen, schlüpfen unter den Filz und werfen dann den Samen auf die glühendheißen Steine. Der Samen qualmt auf in den Steinen und verbreitet einen solchen Dampf, daß kein griechisches Schwitzbad dagegen aufkommt. Den Skythen aber wird's richtig wohlig in dem Dampfbad, und sie heulen. (IV 74-75)

IV   Fazit

Meine Überlegungen sollten verdeutlichen, daß die Darstellung von leiblichen Aspekten und Körperriten bei Herodot in der Regel auf soziale und kulturelle Funktionen verweist. Diese Funktionen lassen sich als die epistemische Substruktur des Textes dekodieren. Im Kontext der Berichte über die Sexualriten wurde deutlich, daß der sexuelle Gebrauch des Leibes bei Herodot ein Indikator für seine Verortung der Ethnien in dem Ranking ,wild' – ,zivilisiert' ist. Der Passus zur Beschneidung der Ägypter ergab, daß Herodot diese Praktiken unter das übergeordnete Ensemble von hygienischen Vorschriften subsumiert, wobei dieser Hygeia-Komplex dann wieder normative Wertungen erlaubt. Die Analyse der Textstellen über die Toten und Trauernden zeigte, daß die frühen Kulturen ihre Gemeinschaft als einen sozialen Leib ansehen, der das einzelne Individuum überdauert; als einen sozialen Leib, der durch den Verlust eines Mitglieds der Gemeinschaft nicht vermindert oder beschädigt werden darf. Das Set ließe sich noch erweitern. So stehen unter den von Herodot beschriebenen somatischen Aspekten die *Ernährungsgewohnheiten* an erster Stelle. Eine Untersuchung dieses Aspekts würde allerdings unseren Rahmen sprengen; obwohl in Herodots Berichten über exotische Ernährungssitten seine normativen Wertungen (wild – zivilisiert) noch stärker durchschlagen. Ich darf abschließend nochmals betonen, daß es diesem Beitrag weniger um die *historische Wahrheit* von Herodots Berichten geht als vielmehr um seinen besonderen *Blick auf das Fremde*; oder (weniger metaphorisch gesagt) um die geistigen und begrifflichen Bedingungen, die seine Beschreibung des Fremden bestimmen. Die moderne Forschung hat in vielen Details gezeigt, wie weit Herodot von der tatsächlichen Realität entfernt war. Als historisches Zeugnis für fremde Kulturen ist Herodot eine insgesamt eher unzuverlässige Quelle; es war ihm letztlich nicht möglich, sich wirklich tief in die Religion und Riten fremder Kulturen einzufinden. Nur stellvertretend sei hier die Auffassung des britischen Ägyptologen Alan B. Lloyd über Herodots ägyptischen Logos wiedergegeben:

> Herodotus' account of the Egyptian culture spreads far and wide but is fundamental superficial in that it shows no understanding of its ideological underpinning and concentrates entirely on external observable phenomena, which he describes with variable accuracy. [...] Not surprisingly, Herodotus shows no ability to get inside the minds of ancient Egyptians, and he shows no understanding of the conceptual basis of their culture, though in this respect he is typical for all Greek and Roman writers on alien peoples. (Lloyd 2007, 738 bzw. 742)

Die *Historien* sind daher eher ein Zeugnis für die Art und Weise einer vorwissenschaftlichen Ethnologie; sie verraten viel von einer unreflektierten und unkritischen Methodik und gehören mithin selbst zu den Objekten der kulturwissenschaftlichen Forschung, insofern sie die Vorgeschichte des Fachs erhellen. Andererseits aber war Herodot über Jahrhunderte wenn nicht gar die einzige, so doch oft die maßgebliche Quelle für Sitten und Bräuche der von ihm beschriebenen Kulturen. Sein ethnographischer Blick war bis ins 19. Jahrhundert prägend für das Verständnis vieler außereuropäischer Kulturen und dann (als man sich, von neuen insb. archäologisch gestützten Erkenntnissen von dieser naiven Sicht abwenden mußte) auch für die Konstitution der späteren Wissenschaft.

Literatur

(i) Deutsche Textausgaben der Historien

Marg (1991): Walter Marg, Herodot. Historien. Übersetzt von Walter Marg. Mit einer Einführung von Detlev Fehling und Erläuterungen von Bernhard Zimmermann [Zwei Bände]. München.

Haussig (1971): H.W. Haussig (Hg.), Herodot. Historien. Deutsche Gesamtausgabe. Übersetzt von A. Horneffer. Neu herausgegeben und erläutert von H. W. Haussig. Mit einer Einleitung von W. F. Otto. Mit 4 Tafeln und 2 Karten. (4. Auflage) Stuttgart.

Brodersen (2005): Kai Brodersen (Hg.), Herodot. Historien. Zweites Buch. Griechisch/ Deutsch. übersetzt und herausgegeben von Kai Brodersen, Stuttgart.

Brodersen (2007a): Kai Brodersen (Hg.), Herodot. Historien. Erstes Buch. Griechisch/Deutsch. Übersetzt von Christine Ley-Hutton, (Ergänzte Auflage), Stuttgart.

Brodersen (2007b): Kai Brodersen (Hg.), Herodot. Historien. Drittes Buch. Griechisch/ Deutsch. Übersetzt von Christine Ley-Hutton, Stuttgart.

Brodersen (2013): Kai Brodersen (Hg.), Herodot. Historien. Viertes Buch. Griechisch/Deutsch. Übersetzt und herausgegeben von Kai Brodersen, Stuttgart.

(ii) Forschungsliteratur

Althoff (1993): Jochen Althoff, „Herodot und die griechische Medizin". In: K. Döring / G. Wöhrle (Hg.), *Antike Naturwissenschaft und ihre Rezeption*. Bamberg, 1-16.

Assmann (2010): Aleida Assmann, *Erinnerungsräume. Formen und Wandlungen des kulturellen Gedächtnisses*. Fünfte durchgesehenen Auflage [zuerst 1999], München.

Bichler (2001): Reinhold Bichler, *Herodots Welt. Der Aufbau der Historie am Bild der fremden Länder und Völker, ihrer Zivilisation und ihrer Geschichte*. Berlin.

Brandenburg (1976): Dietrich Brandenburg, *Medizinisches bei Herodot*. Berlin.

Brodersen (2007a): Kai Brodersen, „Einleitung [Zum Lebend Herodots]". In: K. Brodersen (Hg.), *Herodot. Historien. Erstes Buch. Griechisch/Deutsch*. Übersetzt von Christine Ley-Hutton, (Ergänzte Auflage), Stuttgart, 5-8.

Cobet (1977): Justus Cobet, „Wann wurde Herodots Darstellung der Perserkriege publiziert?" In: *Hermes 105*, 2-27.

Dewald (1981): Carolyn Dewald, "Women and Culture in Herodotus' Histories". In: H. Foley (Hg.), *Reflections of Women in Antiquity*. New York, 120-125.

Dewald (2007): Carolyn Dewald, "On Women and Marriage in Herodotus". In: R. B. Strassler (Hg.), *The Landmark Herodotus. The Histories*. A New Translation by Andrea L. Purvis with Maps, Annotations, Appendices, and Encyclopedic Index. Wirth an Introduction by Rosalind Thomas, New York, [Appendix U] 838-842.

Erbse (1956): Hartmut Erbse, „Der erste Satz im Werke Herodots". In: *Festschrift für Bruno Snell*, 209-222.

Evans-Pritchard (1954): Edward Evans-Pritchard, *The Institutions of Primitive Society*. Oxford.

Foley (1981): Helene Foley (Hg.), *Reflections of Women in Antiquity*. New York.

Foucart (1910): George Foucart, "Circumcision (Egyptian)". In: *ERE 3*, 670-677.

Ghalioungui (1963): Paul Ghalioungui, *Magic and medical science in ancient Egypt*. London.

Grapow (1954): Hermann Grapow, „Grundriß der Medizin der alten Ägypter". In: Ders., *Anatomie und Physiologie*. Berlin: Akademie-Verlag.

Heilen (2000): Stepahn Heilen, „Die Anfänge der wissenschaftlichen Geographie: Anaximander und Hekataios". In: Wolfgang Hübner (Hg.), *Geographie und verwandte Wissenschaften*. Stuttgart, 33-51.

Hommel (1981): Hildebrecht Hommel, „Herodots Einleitungssatz: Ein Schlüssel zur Analyse des Gesamtwerks". In: Gebhard Kurz / Dietram Müller / Walter Nicolai (Hg.), *Gnomosyne. Menschliches Denken und Handeln in der frühgriechischen Literatur. Festschrift für Walter Marg zum 70. Geburtstag*. München, 271 – 287.

How / Wells (1928): Walter W. How / Joseph Wells, *A Commentary on Herodotus with Introduction and Appendixes in two Volumes. Volume I (Books I – IV)*. Oxford (first printed 1912).

Jacoby (1912): Felix Jacoby, *Hekataios, RE*. 267-276.

Jacoby (1913): Felix Jacoby, *Herodotos, RE, Suppl. II*. 205-520.

Leven (2005): Karl-Heinz Leven, *Antike Medizin: Ein Lexikon*. München.

Lévi-Strauss (1971): Claude Lévi-Strauss, *Mythologica I. Das Rohe und das Gekochte* [zuerst frz. Paris 1964]. Frankfurt am Main.

Lévi-Strauss (1973): Claude Lévi-Strauss, *Das wilde Denken* [zuerst frz. Paris 1962]. Frankfurt am Main.

Lloyd (1976): Alan B. Lloyd, *Herodotus Book II, Commentary 61-98*. Leiden.

Lloyd (1988): Alan B. Lloyd, *Herodotus Book II, Commentary 99-182*. Leiden.

Lloyd (2007): Alan B. Lloyd, "The Account of Egypt: Herodotus Right and Wrong". In: R. B. Strassler (Hg.), *The Landmark Herodotus. The Histories*. A New Translation by Andrea L. Purvis with Maps, Annotations, Appendices, and Encyclopedic Index. Wirth an Introduction by Rosalind Thomas, New York, [Appendix C] 737-743.

Müller (1997): Klaus E. Müller, *Geschichte der Antiken Ethnologie*. Reinbek bei Hamburg.

Regenbogen (1965): Otto Regenbogen, „Herodot und sein Werk. Ein Versuch". In W. Marg (Hg.) *Herodot. Wege der Forschung*. Darmstadt, 57-108.

Schadewaldt (1970): Wolfgang Schadewaldt, „Lebenszeit und Greisenalter im frühen Griechentum". In: Ders., *Hellas und Hesperien, Bd. 1*. Zürich/Stuttgart, 41-59.

Sigerist (1963): Henry E. Sigerist: *Anfänge der Medizin: Von der primitiven und archaischen Medizin bis zum goldenen Zeitalter in Griechenland, Band I*. Zürich.

Strassler (2007): Robert B. Strassler (Hg.), *The Landmark Herodotus. The Histories*. A New Translation by Andrea L. Purvis with Maps, Annotations, Appendices, and Encyclopedic Index. Wirth an Introduction by Rosalind Thomas, New York.

Zoepffel (2006): Renate Zoepffel (Hg.), *Aristoteles, Oikonomika. Schriften zu Hauswirtschaft und Finanzwesen*. Übersetzt und erläutert von ders. (Aristoteles. Werke in deutscher Übersetzung Band 10/Teil 2), Berlin.

Ekstase und Selbstlazeration im Kontext von Mantik

Rüdiger Schmitt

I   Einführung

Riten der Selbstverletzung, Automutilation oder Selbstmarter[1], also Riten, bei denen sichtbare Alterierungen des Körpers durch dessen voluntäre Verletzung vorgenommen werden, sind aus vielen Religionen bekannt: Als prototypisch gelten vielfach die indischen Fakire (arab. Armer, Bettler), im Hinduismus selbst als Sadhus oder Yogins bezeichnet, die durch entsprechende Riten Erkenntnis oder Erlösung zu erlangen suchen.[2] Den Westernfreunden dürfte aus dem Film „Der Mann, den sie Pferd nannten" vielleicht das Sonnentanzritual der Lakota bekannt sein, wobei Piercings mit Haken an der Haut angebracht und schließlich herausgerissen werden.[3] Im schiitischen Islam werden zum Gedenken an den Tod Husain ibn Alis in der Schlacht von Kerbela 680, dem Aschura-Fest, rituelle Umzüge abgehalten, bei denen sich die Gläubigen der Flaggelation bzw. der Selbslazeration mit Rasierklingen, Messern und Schwertern unterziehen. Auch das Christentum kennt, sowohl historisch wie rezent zahlreiche Praktiken der Automutilation, angefangen von den z.T. exzentrischen Kasteiungen der spätantiken Wüstenasketen, über die mittelalterlichen Geißlerzüge bis hin zu Selbstgeißelungen von Laien und Priestern in ultra-konservativen Strömungen der katholischen Kirche der Gegenwart sowie im Volkskatholizismus. Eine besonders radikale Spielart stellt die rituelle Selbstentmannung in Ekstase im Kult der Dea Syria dar, wie von Lukian von Samosata geschildert (siehe dazu unten). Ein auch im Christentum

---

1   Vgl. Müller (1999).
2   Vgl. Golzio (1999).
3   Vgl. Bolz (1999).

nicht ganz unbekanntes Phänomen, wie die Selbstkastration des Origenes[4] oder Praktiken der Genitalverstümmelung in der russisch-orthodoxen Sekte der Skopzen[5] im vorrevolutionären Russland zeigt. Auf den ersten Blick mögen deratige Riten der Selbstmarter phänomenologisch ähnlich scheinen, so unterscheiden sie sich jedoch z.T. deutlich in ihrer sozio-religiösen Funktion in den jeweiligen religiösen Symbolsystemen. Unterscheiden lassen sich hier u. a folgende Funktionen, die sich z.T. auch überschneiden können:[6]

- Opfergabe, insbesondere als Blutopfer
- Selbstminderung als Zeichen der *devotio*
- Erfüllung eines Gelübdes
- Buße bzw. Sühnehandlung zur Tilgung individueller oder kollektiver Schuld
- Kommemorative Funktion
- Fürbitte
- Induzierung von Trance zur Visionssuche
- Induzierung von Trance zum Erreichen einer *unio mystica*
- Erzeugung ritueller *communitas* bei gemeinschaftlicher Selbstlazeration
- Strategie zur Körper- und Weltüberwindung
- Anerkennung von Tapferkeit bzw. Standfestigkeit, u. a. bei Initiationsriten
- Als Bestandteil von Trauerriten zur Induzierung von Schmerz bzw. zur äußerlichen Angleichung an die Toten.

An die Riten der Selbstverletzung sind in der Forschung weiter solche der Fremd- bzw. gegenseitigen Verletzung angeschlossen worden, vor allem im Kontext von Initions- und Passageriten, aber auch bei Fruchtbarkeits- bzw. bei Eliminationsritualen.[7]

Alterationen des Körpers durch Selbstlazeration, also durch Ritzen des Körpers, erscheinen im Alten Testament primär im Kontext von Trauerriten,[8] wie Lev 19,28, Dtn 14,1, Jer 16,6 und 41,5, ebenso das Schlagen auf die Brust in Jes 32,12 oder die Genitalregion in Jer 31,19. Die Induzierung von Schmerz unterstützt dabei insbesondere das ekstatische Wehklagen. Im Zusammenwirken mit anderen Alterationen

---

4   Vgl. Eusebius (1914), VI.8.1.

5   Vgl. Grass (1913).

6   Vgl. Müller (1999).

7   So z. B. Frazers Deutung des griechischen Festes der Thargelien, bei denen die Fruchtbarkeit von Männern und Frauen durch die Schläge auf die Genitalien eines menschlichen Sündenbockes mit frischen Zweigen gefördert werden solle: Frazer (1989 [1928]), 842ff.

8   Dazu ausführlich Olyan (2004); Albertz und Schmitt (2013), 433ff.

des Körpers wie dem Zerreißen der Kleidung (Lev 10,6; 21,10; 2 Sam 1,11; 3,31; Hi 1,20), dem Tragen des *śaq* (Gen 37,34; 2 Sam 3,31; Ez 27,31; Jdt 8,5), dem Lösen (Lev 10,6; 21,10; Jdt 10,3) und Abschneiden der Haare (Lev 21,5; Hi 1,20; Jer 16,6; Ez 7,18) bzw. dem Rasieren einer Stirnglatze (Lev 19,27; Dtn 14,1), dem Unterlassen von Körperpflege (2 Sam 14, 2; Jdt 10,3), dem sich Bestreuen mit Asche und im Dreck wälzen (Jos 7,6; 1 Sam 4,12; Jer 6,26), dienen diese Riten dazu, sich temporal mit den Toten gleichzumachen. Die Trauernden suspendieren so temporal ihren sozialen Status und existieren in einer Art liminalem Zustand zwischen Tod und Leben. Im Rahmen der exilisch-nachexilischen Identitätssicherung hat die deuteronomistische und priesterliche Legislation derartige ekstatische Trauerriten dann als unjahwistisch gebrandmarkt.

Neben den Trauerriten erscheinen Selbst- bzw. Fremdlazerationsriten im AT bei Propheten, insbesondere bei Angehörigen von Prophetengruppen in 1 Kön 18,28ff.; 1 Kön 20, 35-37 und Sach 13,2-6. Im Folgenden möchte ich daher die Funktion der Selbstlazeration im Kontext prophetischer Phänomene und ihre sozio-religiöse Funktion und Einbettung in das religiöse Symbolsystem untersuchen, und beginne daher mit einem Durchgang durch die atl. Befunde.[9]

II   Prophetische Selbst- und Fremdlazeration im Alten Testament

Das bekannteste Beispiel von Lazerationsriten wird nicht von Jahwe-Propheten berichtet, sondern von den 400 Propheten des Baal in 1 Kön 18,28f.:

26. (…) Sie tanzten hüpfend um den Altar, den sie gebaut hatten. (…)
28. Da schrien sie mit lauter Stimme und sie ritzten sich nach ihrer Sitte auf mit Schwertern und Lanzen, bis das Blut an ihnen herabfloß.
29. Als die Mittagszeit vorüber war, fielen sie in Ekstase (*nb' hitp*) und das dauerte bis zu der Zeit, zu der man das Speiseopfer darbringt.

In der Forschung ist der Text gerne im Lichte der Schilderungen des ekstatischen Kultes der Dea Syria durch Lukian und Apuleius als Musterbeispiel für die kanaanäische Praxis der Selbstlazeration angeführt worden, doch ist im Hinblick auf die religionsgeschichtliche Auswertung Vorsicht geboten. Es handelt sich hier um einen post-deuteronomistischen Text,[10] der sich an dem deuteronomistischen Verdikt

---

9   Vgl. hierzu Schmitt (2014), 53ff.; 161.
10   Vgl. Würthwein (1985), 207; Otto (2004) 248.

gegen die Selbstlazeration in Dtn 14,1–2 orientiert und daher keine unmittelbare Aussage über eine entsprechende Praxis von Fremdpropheten im Nordreich der EZ II B darstellt. Vielmehr dürfte sich die Konzeptualisierung der rituellen Ekstase auf eine zeitgenössische Praxis beziehen, wobei freilich offen bleiben muss, ob die Verfasser hier fremdreligiöse oder indigene Ekstaseriten im Blick hatten.

Der nächste Text, den ich hier erwähnen möchte ist, 1 Kön 20,35–37:

> 35. Einer von den *běnê hanĕbî'îm* sprach im Auftrag Jahwes zu seinem Genossen: Schlag mich doch! Und als dieser sich weigerte, ihn zu schlagen,
> 36. sprach er zu ihm: Weil du der Stimme Jahwes nicht gehorcht hast, wird dich ein Löwe töten, sobald du von mir weggegangen bist. Und als er sich von ihm entfernt hatte, da fiel ihn ein Löwe an und tötete ihn.
> 37. Hierauf traf er einen anderen Mann und befahl ihm: Schlag mich doch! Und der Mann schlug und verwundete ihn.
> 38. Und der Prophet ging hin und stellte sich dem König in den Weg und durch eine Binde über den Augen machte er sich unkenntlich.

Ein Mitglied einer Prophetengruppe bittet hier einen anderen, ihn zu schlagen. Als Grund hierfür ist die Induzierung einer Vision bzw. eines Orakels anzunehmen, da der Prophetenschüler, nachdem ihm die gewünschte Verletzung zugefügt wurde, sich dem König in den Weg stellt und ihm das Geschaute zum Zwecke der Rüge mitteilt. Die Binde über den Augen dient wohl dazu, die Verwundungen durch Schläge unkenntlich zu machen, durch die sich der Prophetenschüler sogleich als ein solcher entlarvt hätte. Dass dieser Text nicht orthodox im deuteronomistischen Sinne ist, ist deutlich. Er gehört vielmehr zu den zahlreichen post-deuteronomistischen Ergänzungen von Prophetenlegenden, wie die Erzählungskränze um Ahia von Schilo, sowie über die Wundertaten der Gottesmänner Elia und Elischa. Die post-dtr Ergänzer konnten hier offenbar unpolemisch die Erzählung von dem Selbst- bzw. Fremdlazerationsritus der *běnê hanĕbî'îm* in 1 Kön 20,35ff. übernehmen,[11] was auf eine Vertrautheit – auf jeden Fall aber Akzeptanz – derartiger Praktiken schließen lässt.

Im Zyklus der Elischa-Wundergeschichten wird über ekstatische Riten der *běnê hanĕbî'îm* nichts berichtet, wohingegen in 1 Sam 10 und 1 Sam 19,20ff. die Gruppen durch prophetische Verzückung gekennzeichnet sind. Dies lässt jedoch keinesfalls den Schluss zu, die mit Elischa verbunde Gruppe habe keine ekstatischen Riten praktiziert, vielmehr steht in den Wundergeschichten die Figur des Gottesmannes als wunderbarer Mittler im Vordergrund, während den Prophetenschülern nur

---

11 Vgl. Würthwein, ATD 11.2, 236; Otto, Jehu, 248ff.

eine Funktion als Statisten zukommt.[12] In 1 Kön 20,35ff., wo die *běnê haněbî'îm* als Protagonisten fungieren, wird die Visionssuche durch Selbst- und Fremdlazeration jedoch unpolemisch vorausgesetzt.

Der nächste Text, der hier hier in Augenschein genommen werden soll, ist Sach 13,2–6, eine sehr grundsätzliche Polemik gegen Propheten in der Deutero-Sacharja-Sammlung aus der hellenistischen Zeit:

> 2. Und an jenem Tag wird geschehen – Spruch Jahwes der Heerscharen – dass ich die Namen der Götzenbilder im Land ausrotten werde, so dass man sich nicht mehr an sie erinnert. Und auch die Propheten und den Geist der Unreinheit werde ich aus dem Land vertreiben. 3. Wenn dann noch einer als Prophet auftritt, werden sein Vater und seine Mutter, die ihn gezeugt haben, zu ihm sagen: Nicht wirst du am Leben bleiben, denn Trug hast du geredet im Namen Jahwes. Dann werden sein Vater und seine Mutter, die ihn gezeugt haben, ihn durchbohren, weil er prophezeit hat. 4. Und an jenem Tag wird geschehen, dass sich jeder Prophet schämen wird über die Schauungen, die er prophezeit hat. Und er wird nicht seinen härenen Mantel anziehen, um sich zu verleugnen. 5. Und er wird sagen: Ich bin kein Prophet, sondern ein Ackermann bin ich, denn der Ackerboden ist mein Besitz von Jugend an. 6. Und wenn man ihn dann fragt: Was sind das für Wunden zwischen deinen Schultern (wörtl. Händen)? Dann wird er antworten: Ich bin im Haus meiner Freunde verletzt worden.

Die Wunden (*makkôt*) des Propheten werden in der Forschung im Hinblick auf die in 1 Kön 18,28 beschriebenen Praktiken der Baalspropheten mit ekstatischen (fremdreligiösen) Praktiken der Selbstverletzung in Beziehung gesetzt, die dieser jetzt verbergen muss und auf privaten Zwist zurückführt, auch wenn sie tatsächlich von Selbstverletzung im Kreis prophetischer Konventikel, den ‚Freunden‘, herrühren sollten.[13] Da die Praxis der Selbstverletzung (vor allem im Kontext der Trauerriten aber auch zu anderen Anlässen)[14] bzw. gegenseitiger Verletzung (vgl. 1 Kön 20,35) unter israelitischen Propheten scheinbar gängige Praxis war, besteht kein Anlass, diese Automutilationen als fremdreligiöses Element zu bestimmen. Es handelt sich vielmehr um genuin israelitische Praktiken, die jedoch bereits in den Sach 13 vorliegenden Traditionen negativ als unjahwistisch gewertet werden. Interessant ist hier noch die Wendung *bet me'ahabay* ‚Haus meiner Freunde‘ in V. 6, die möglicherweise auf ein bestimmtes Haus für die Zusammenkünfte derartiger Konventikel schließen lässt. Inwieweit hier das Gemeinschaftserlebnis, mystische Erfahrung oder die Visionssuche im Vordergrund steht, ist zwar letztendlich nicht

---

12 Siehe dazu Schmitt (2014), 53.
13 Siehe dazu Schmitt (2014), 53f. mit Diskussion.
14 Vgl. Olyan (2004), 25ff.; Albertz und Schmitt (2012), 434.

zu sagen, im Kontext von Sach 13 dürfte es jedoch primär um die Induzierung von Offenbarungen durch Schmerz gehen.

Anzuschließen an die genannten Texte ist noch Sauls Verzückung im Prophetenhaus zu Rama, 1 Sam 19,23–24:

> 23. Als er von dort zur Wohnstätte [der Propheten] in Rama weiterging, kam auch über ihn der Geist Elohims und er ging in Ekstase weiter bis er zur Wohnstätte nach Rama kam.
> 24. Und er zog sogar seine Kleider aus und blieb auch vor Samuel in Ekstase. Und er lag nackt da den ganzen Tag und die ganze Nacht. Darum sagt man: Ist denn auch Saul unter den Propheten?

Hier geht es zwar nicht um Selbst- bzw. Fremdlazeration oder der Visionssuche, sondern eine andere Alteration des Körpers in Ekstase, nämlich die Nacktheit: Saul entkleidet sich in seiner Ekstase und bleibt auch noch in Trance eine längere Zeit nackt liegen. Inwiefern rituelle Nacktheit eine Rolle bei prophetischen Konventikeln spielt, ist schwer zu sagen. 1 Sam 19 ist freilich nicht an der Ekstase selbst interessiert, sondern daran, Saul despektierlich zu schildern: Könige liegen nicht nackt herum, und schon gar nicht im Angesicht des Propheten Samuel. Trotz dieser Tendenz ist es freilich nicht auszuschließen, dass die Erzählungskompostion über die frühe Königszeit hier auf eine tatsächliche Praxis der vorexilischen Gruppenprophetie rekurriert, zudem auch hier von einer Versammlungsstätte der Gruppenpropheten berichtet wird.

Noch in römischer Zeit berichtet Philo (*De vita contemplativa* 12) über ekstatische Praktiken im Kontext der Visionssuche bei der im Umkreis von Alexandria ansässigen asketischen jüdischen Sekte der Therapeuten:

> [...] because they are carried away by a certain heavenly love, give way to enthusiasm, behaving like so many revellers in bacchanalian or corybantian mysteries, until they see the object which they have been earnestly desiring.[15]

Philo erwähnt hier zwar nichts von Selbstlazeration, doch mag dies darin begründet liegen, diese jüdische Gruppierung mit Phänomenen der griechisch-römischen Religion in Verbindung zu bringen, nicht aber mit den syrischen Ekstatikern, die ja bei Lukian und Apuleius in äußerst ungünstigem Licht erscheinen.[16] Die Therapeuten stehen jedoch durchaus in ihrer Tätigkeit als rituelle Heiler in der Tradition

---

15   Philo (1995): 12.
16   Lucianus (2003): de Dea Syria 15; Apuleius (1998): Metamorphoses VIII 27f.

der atl. Gottesmänner und den ihnen angeschlossenen Prophetengruppen sowie Kultpropheten wie Jesaja in 1 Kön 20,1–11 bzw. Jes 38,1–8.

Interessant im Hinblick auf die Ekstase als Modus prophetischer Botschaftsinduzierung ist die Konzeptualisierung derselben im späteren Judentum: So ist die Weise der Offenbarung an Ezekiel in der Vision Ez 37,1 offenbar ebenfalls als Trancezustand einhergehend mit körperlichen Phänomen gedeutet worden: So zeigt das Fresko mit der Auferstehungsvision aus Ez 37 aus der Synagoge von Dura-Europos (Abb. 1) den Propheten in einer Art Trance-Tanz von der Hand Gottes berührt.[17]

Abb. 1 Fresko aus der Synagoge in Dura-Europos, 2. Jh. n. Chr.

IV   Ekstase und Automutilation in der Umwelt Israels

In der Keilschriftliteratur des 2. und 1. Jt. vor Chr. erscheint der *muḫḫūm* als ekstatischer Offenbarungsempfänger: In Mari belegen die Briefe ARM X 7,7 (*imma-ḫu*) und ARM X 8,7 (*im-ma-ḫi*) die Ekstase als Offenbarungsmittel, wiewohl der Zusammenhang zwischen Botschaftsempfang und Botschaftsübermittlung in Mari unklar bleibt.[18] Für Ugarit im 2. Jt. gibt es keine Hinweise auf Tranceprophetie und damit verbundene Selbstlazeration.[19]

---

17 Goodenough (1964), Fig. 348.

18   Noort (1977) 24f.

19 Gegen Huffmon (2000), 64, Anm. 42: Ug 5 162.11 mit der Erwähnung von *muḫḫūs* ist ein babylonischer Text und reflektiert nicht ugaritische Verhältnisse.

In assyrischen Prophetien des 1. Jt. erscheint anstatt des *muḫḫūm* zumeist der *raggimu* („Rufer")[20] bzw. die *raggintu* als Offenbarungsmittler, die von Parpola als Ekstasepropheten analog dem älteren *muḫḫūm* verstanden werden.[21] Dennoch ist der Stand des *muḫḫūm* auch neuassyrisch belegt,[22] freilich ohne dass etwas Genaues über Botschaftsinduzierung, Botschaftsempfang und Mitteilung ausgesagt wird. Die Texte selbst geben keine Hinweise auf die Art der Induzierung von Ekstase, wie z. B. auf Selbstlazeration. Das Phänomen gruppenbezogener Tranceprophetie ist im Mesopotamien des 1. Jt. nicht belegt.

Für den westsemitischen Raum in der Eisenzeit ist Ekstaseprophetie nur indirekt in der ägyptischen Wen-Amun-Erzählung bezeugt: Während eines Opferrituals des Königs von Byblos wird ein Seher von Amun ergriffen und in eine Ekstase versetzt, wobei Botschaftsempfang und Botschaftsübermittlung unmittelbar zusammenfallen:

> Als er [Tjekerbaal, der König von Byblos] nun einmal seinen Götter[n] opferte, [d]
> a ergriff (*ṯȝj*) der Gott einen seiner Ekstatiker ('*ḏd'ȝ*)[23] und versetzte ihn in Raserei
> (*ḫȝwt*), und er sagte zu ihm „Bring [den] Gott herauf, und bring den Boten, der ihn
> bei sich hat! Amun ist es, der ihn gesandt hat, er ist es, der ihn hat kommen lassen."
> Nun war der Ekstatiker gerade in der Nacht in Ekstase, in der ich ein Schiff mit der
> Bestimmung nach Ägypten gefunden und meine ganze Habe hineingeladen hatte und
> (nun) die Zeit verbrachte bis zur Dunkelheit mit den Worten: „Kommt sie herab, so
> werde ich (auch) den Gott verladen, damit kein anderes Auge ihn sieht."[24]

Automutilation im Kontext prophetischer Ekstase ist für den westsemitischen Raum im 1. Jt. nicht belegt. Spätere Belege für kultische Gruppen-Trancephänomene bieten Lukian von Samosata und Apuleius von Madaura mit ihren Schilderungen des ekstatischen Kultes der Dea Syria in römischer Zeit mit Elementen des Tanzes, der Selbst- und Fremdlazeration sowie der Selbstentmannung:

> Lukian, De Dea Syria 15:[25]
> An bestimmten Tagen sammelt sich die Menge im heiligen Bezirk, eine große Anzahl
> von Gallen und heiligen Männern feiern das Tanzfest, zerschneiden sich die Arme und
> schlagen sich gegenseitig auf den Rücken. Eine Menge von Musikanten steht dabei, die
> Flöte blasen, Pauke schlagen und begeistert heilige Lieder singen. Diese Zeremonie
> findet außerhalb des Tempels statt, und die Teilnehmer betreten ihn nicht. Bei dieser

---

20 Vgl. von Soden (1965-1981), 942.

21 Vgl. Parpola (1997), XLVf. Mit Anm. 141 u. 230.

22 Vgl. Borger (1967), 2: *šipir maḫḫê* : „(trance-)prophetische Orakel".

23 Zu '*ḏd'ȝ* als westsemitisches Lehnwort: Schipper (2005), 183ff.

24 Schipper (2005), 56ff.; 105 (Z. 1.38-1.42).

25 Übersetzung: Latte (1927), 43f. Gr. Text: Lucianus (2003).

Gelegenheit werden sie auch Gallen. Denn während die anderen Flöte spielen und tanzen, erfaßt viele Raserei, und mancher, der zum Zuschauen kam, hat nachher folgendes getan – doch ich will erzählen, wie sie verfahren. Der Jüngling, über den das verhängt ist, wirft seine Kleider ab, tritt in die Mitte und nimmt laut schreiend ein Schwert; die befinden sich dort zu diesem Zwecke, glaube ich, seit langen Jahren. Dann entmannt er sich damit und läuft durch die Stadt, das abgeschnittene Glied in der Hand. Er wirft es in ein beliebiges Haus und erhält aus ihm Weiberbekleidung und Weiberschmuck. So verfahren sie bei der Entmannung.

Apuleieus, Metamorphoses VIII 27 –28[26]
Anderntags legte die gesamte Priesterbande Gewänder von allerlei Farben an, putzte sich aufs lächerlichste heraus, schminkte sich zierlich das Gesicht, malte die Augenbrauen und zog in Prozession aus. Einige von ihnen mit safrangelben, leinenen, auch seidenen Binden, weiß und purpurn gestreiften Kleidern, von einer Schärpe umgürtet, und mit braunen Schuhen, setzten die Göttin, in einen seidenen Mantel gehüllt, auf meinen Rücken. Die Arme entblößt bis an die Schultern, große Schwerter und Äxte schwingend, hüpften sie jauchzend einher; von dem Geflüster der Flöten mehr und mehr zu ihrem ekstatischen Tanze ermuntert. Nachdem sie vor vielen schlechten Hütten vorübergezogen, kamen sie zum stolzen Landsitze eines vornehmen Reichen. Mit dem ersten Fuß, den sie hineinsetzten, gerieten sie in fanatische Wut, erhoben ein mißtönendes Geheul und machten das seltsamste Getümmel. Sie wirbelten lange mit gesenktem Haupte, den Hals aufs sonderbarste biegend und wendend und das lose Haar schüttelnd, im Kreise herum. Zuweilen bissen sie sich in die aufgeschwollenen Muskeln und zuletzt zerritzten sie sich gar die Arme mit ihren zweischneidigen Schwertern. Einer unter ihnen raste noch toller als die übrigen; er rollte fürchterlich die Augen, schnaufte, brauste, schäumte, stellte sich wahnsinnig: und das zum Beweis, dass er ganz göttlichen Geistes voll sei, gerade als müsse die Gegenwart der Götter den Menschen nicht stärken und erheben, sondern schwächen und niederdrücken. Aber hört, was die göttliche Vorsehung ihm dafür zum Lohne gab! Sie ließ ihn überlaut weissagen (*infit vaticinatione*); gleisnerisch sich selbst anklagen, als habe er sich gegen die heilige Religion einer Sünde schuldig gemacht, und zur Buße für sein schweres Verbrechen eigenhändige Kasteiung von sich selbst fordern. Nun nahm er eine Geißel, welche dergleichen dies Halbmannsgesindel immer führt, mit einer wollenen Schnur, die unten weit aufgefasert und mit vielen spitzen Schafknöchelchen versehen ist, und zerpeitschte sich damit gottserbärmlich, ohne jedoch die allergeringste Empfindlichkeit gegen den Schmerz erblicken zu lassen.

Apuleius und Lukian belegen – trotz des satirischen Charakters beider Texte – die Funktion von Auto- und Fremdmutilation zur Erzeugung ritueller communitas, im Rahmen der *devotio*, den Opfercharakter der Selbstmutilation bei der Selbstkastration sowie die Induzierung von Trance zur Visionssuche, wobei eine ältere westsemitische Tradition anzunehmen ist. Ein unmittelbarer Zusammenhang mit

---

26 Apuleius (1998): Metamorphoses VIII 27f.

Fruchtbarkeitsriten – wie in der älteren Forschung oftmals angenommen[27] – ist nicht erkennbar.

V   Zusammenfassung und Synthese

In der alttestamentlichen Forschung sind prophetische Ekstase und die damit verbundenen Riten der Automutilation zumeist als kanaanäisches *survival* angesehen worden, dass der „ethische Monotheismus" überwunden habe: Kennzeichnend für den *nabi* als religiösem Phänotyp gilt ein besonderes Charisma, das sich nach H. Gunkel insbesondere im Phänomen der Ekstase manifestiere.[28] W. Wundt hat in seiner einflussreichen „Völkerpsychologie" zwischen der exaltierten und der apathischen Ekstase unterschieden.[29] Letztere ist nach Hölscher kennzeichnend für die alttestamentliche Prophetie, während die ekstatische Prophetie ein Element der kanaanäischen Religion sei.[30] S. Mowinckel hat einen fundamentalen Gegensatz zwischen den aus der kanaanäischen Tradition herkommenden Kultpropheten und den „freien" Propheten als moralische Institution konstatiert.[31] Der Typus des Propheten ist auch nach Max Weber daher zu trennen vom Typus der Nebijim-Berufsekstatiker und von den Sehern der Frühzeit, die nicht viel mehr als Erwerbsmagier gewesen seien.[32] Fohrer hat die beiden Typen der Prophetie mit ihrem unterschiedlichen Herkommen begründet: Der urtümliche israelitische ‚Seher' gehöre zum nomadischen Bereich, wohingegen der ekstatische Prophet

---

27 Siehe Schmitt (2013a).

28 Vgl. u. a. Gunkel (1903); Hölscher (1914), 13; 46; Koch (1978), 36f.; van der Leeuw (1977), 24.

29 Wundt (1906), 97.

30 Vgl. Hölscher (1914) 129ff.

31 Vgl. Mowinckel (1961), 5ff. Mowinckel definiert den Kultpropheten folgendermaßen: „Ich verstehe unter einem ‚Propheten' hier einen, der im Auftrage sowohl der Gesellschaft als auch der Gottheit der Gemeinde auf Anfrage die nötige Auskunft in religiösen Dingen direkt aus göttlicher Quelle kraft einer übergewöhnlichen Machtausrüstung erteilt, einen, der in göttlichen Dingen Bescheid weiss, sei es, dass er inspi-riert ist oder Offenbarungen empfangen kann, sei es, dass ihm technische Mittel zur Verfügung stehen, durch die er den Willen und die Weisungen der Gottheit ermitteln und der Gemeinde als Antwort auf eine Frage oder auf ein Gebet derselben erteilen kann" (Zitat ebd., 9). Später hat Mowinckel diese Unterscheidung wesentlich revidiert, s. u. Anm. 60.

32 Weber (2005) 346f.; ders., (1971), 111ff.; 281ff.

mit den Fruchtbarkeits- und Vegationskulten im Kulturland sowie mit Tempel und Palast verknüpft sei.[33]

Der älteren Forschung ist – besonders im Hinblick auf die Wen-Amun-Erzählung – insofern zuzustimmen, dass Tranceprophetie und wohl auch die damit verbundenen Riten der Automutilation ein Phänomen der westsemitischen Religion darstellt. Die polemischen Vorbehalte der älteren Forschung gegenüber dem vermeintlich „sexuellen" oder „orgiastischen" Charakter der kanaanäischen Religion sind freilich apologetisch und obsolet.[34] Diese Phänomene können jedoch nicht einfach nur als ein ‚survival' bezeichnet werden, sondern hängen wohl mit einer generellen Tendenz in den westsemitischen Religionen des 1. Jt. v. Chr. (inklusive der israelitisch-judäischen Religion) mit der Bevorzugung bestimmter intuitiver mantischer Praktiken zusammen, die sich von der komplexen mesopotamischen Vorzeichenwissenschaft, der *bārutû*, unterscheiden, deren Kenntnis im Rahmen der Umbrüche der SBZ verloren gegangen ist.[35]

Die besprochenen Zeugnisse zur Auto- und Fremdmutilation im Alten Testament zeigen, dass entsprechende Praktiken noch in der nachexilischen Zeit (1 Kön 20,35-37) und der frühhellenistischen Zeit (Sach 13,2-6) in Yehud bekannt waren und praktiziert wurden. Es ist daher hier mit einer weitaus größeren Kontinuität der Gruppenprophetie und der hier praktizierten Riten der Auto- und Fremdmutilation zu rechnen. Die entsprechenden Verdikte in der deuteronomistischen Legislation (Dtn 14,1-2) und im Heiligkeitsgesetz (Lev 19,21) gegen Selbstlazeration können daher nicht pauschal als Zeugnisse der „Überwindung" kanaanäischer Praktiken bzw. ihren *survivals* gewertet werden, zumal gerade 1 Kön 20,35-37 keinerlei Kritik an diesen Praktiken erkennen lässt. Der vielfach in der Forschung hergestellte Zusammenhang zwischen Automutilation und Fruchtbarkeitskulten (u. a. Frazer, Fohrer) ist nicht evident zu machen. Automutilation ist in den genannten biblischen und außerbiblischen Texten eine Technik zur Induzierung von Trance und Ekstase durch Schmerz,[36] die primär – aber nicht ausschließlich – der Kommunikation mit der Gottheit dient: In der gruppenbezogenen Religion der *běnê haněbî'îm* der EZ II und der Perserzeit, sowie den prophetischen Konventikeln der hellenistischen Zeit (Sach 13,6) und sektenartigen Gruppen wie den alexandrinischen Therapeuten dienen Ekstasetechniken, wie u. a. Selbstlazeration und musikinduzierte Trance, nicht nur der Divination im engeren Sinn, sondern auch der *devotio* und der Erzeugung von *communitas* unter den Adepten (und zwar mit Turner sowohl im

---

33 Vgl. Fohrer (1969), 222ff.; ders., (1974), 8f.
34 Dazu: Schmitt (2013a; 2013b).
35 Vgl. Schmitt (2014), 163f.
36 Siehe dazu Vaitl (2012), 227f.

Hinblick auf die Erzeugung von existentieller bzw. spontaner *communitas* in der Ekstase bzw. Trance, auf die normative *communitas* zu Etablierung sozialer Kontrolle innerhalb der Gruppe, als auch auf die Erzeugung ideologischer *communitas* als gemeinschaftsbildender Lehre).[37] Ekstase und Selbst- bzw. Fremdmutilation in der Gruppenprophetie sind somit als Techniken der Kommunikation mit der Gottheit bzw. der Herstellung einer *unio mystica* anzusprechen, was als mantische Praxis im weiteren Sinne verstanden werden kann. Die Körperalteration durch das Zufügen von sichtbaren Wunden, dient, dies zeigen sowohl 1 Kön 20,35-37 und Sach 13,2-6, natürlich auch der Kommunikation an die Außenstehenden, durch die die Adepten ihre *devotio* innerhalb der Gesellschaft sichtbar demonstrieren.

Literatur

Albertz und Schmitt (2012): Rainer Albertz und Rüdiger Schmitt, *Family and Household Religion in Ancient Israel and the Levant*. Winona Lake, IN.

Apuleius (1998): Apuleios von Madaura, *Der Goldene Esel*. Düsseldorf u. a.

Bolz (1999): Peter Bolz, „Sonnentanz". In: Wolfgang Müller (Hg.), *Wörterbuch der Völkerkunde*, Berlin, S. 345.

Borger (1967): Rykle Borger, *Die Inschriften Asarhaddons König von Assyrien*. AfO Beiheft 9. 2. Auflage, Osnabrück.

Eusebius (1914): Eusebius von Cäsaräa, *Kirchengeschichte*, hg. von Eduard Schwarz, Leipzig.

Fohrer (1969): Georg Fohrer, *Geschichte der israelitischen Religion*. Berlin.

Fohrer (1974): Georg Fohrer, *Die Propheten des 8. Jahrhunderts: Die Propheten des Alten Testaments Bd. 1*. Gütersloh.

Frazer (1989): James George Frazer, *Der Goldene Zweig: Das Geheimnis von Glauben und Sitten der Völker*. RE 483. Nachdruck der Ausgabe von 1928, Reinbek.

Goodenough (1964): Erwin R. Goodenough, *Jewish Symbols in the Graeco-Roman Period Vol. 11: Symbolism in the Dura Synagogue*. Bollingen Series XXXVII. New York.

Golzio (1999): Karl Heinz Golzio, „Fakir". In: Wolfgang Müller (Hg.), *Wörterbuch der Völkerkunde*. Berlin, 119.

Grass (1913): Karl Konrad Grass, „Russische Sekten. Teil 4: Skopzen". In: *Religion in Geschichte und Gegenwart Bd. 5*, 1. Aufl., S. 74–90.

Hölscher (1914): Georg Hölscher, *Die Propheten: Untersuchungen zur Religionsgeschichte Israels*. Leipzig.

Huffmon (2000): Herbert B. Huffmon, „A Company of Prophets: Mari, Assyria, Israel". In: Marti Nissinen (Hg.), *Prophecy in its Ancient Near Eastern Context: Mesopotamian, Biblical and Arabian Perspectives*. SBL Symposium Series 13. Atlanta, GA, 47–70.

Koch (1978): Klaus Koch, *Die Profeten I*. UTB 280. Stuttgart u. a.

---

37   Vgl. Turner (1969): 132.

Latte (1927): Kurt Latte, *Die Religion der Römer und der Synkretismus der Kaiserzeit.* Religionsgeschichtliches Lesebuch Bd. 5. Tübingen.

Lucianus (2003): Lucianus of Samosata, *On the Syrian Goddess*, hg. von J. L. Lightfoot. Oxford.

Mowinckel (1961): Sigmund Mowinckel, *Kultprophetie und prophetische Psalmen. Psalmenstudien III.* Nachdruck der Ausgabe von 1923, Amsterdam.

Müller (1999): Wolfgang Müller, „Selbstmarter". In: Ders. (Hg.), *Wörterbuch der Völkerkunde.* Berlin, 339.

Noort (1977): Ed Noort, *Untersuchungen zum Gottesbescheid in Mari: Die „Mariprophetie" in der alttestamentlichen Forschung.* AOAT 20. Kevelaer und Neukirchen-Vluyn.

Olyan (2004): Saul M. Olyan, *Biblical Mourning: Ritual and Social Dimensions.* Oxford.

Otto (2001): Susanne Otto, *Jehu. Elia und Elisa: Die Erzählung von der Jehu-Revolution und die Komposition der Elia-Elisa-Erzählungen.* BWANT 152. Stuttgart u. a.

Parpola (1997): Simo Parpola, *Assyrian Prophecies.* SAA IX. Helsinki.

Philo (1995): Philo of Alexandria, *Works Vol. IX.* LCL 363. Cambridge, Mass/London.

Schmitt (2013a): Rüdger Schmitt, „Die ‚dunklen Seiten' Gottes in der Religionsgeschichte Israels". In: *MARG* 21 (2013), 3–15.

Schmitt (2013b): Rüdiger Schmitt, „Das Monotheismus/Polytheismus-Paradigma in der religionswissenschaftlichen Forschung des 19. und 20. Jh. und sein Einfluß auf die Theoriebildung der Gegenwart". In: MARG 21 (2013), 323–335.

Schmitt (2014): Rüdiger Schmitt, *Mantik im Alten Testament.* AOAT 411, Münster.

Stökl (2012): Jonathan Stökl, *Prophecy in the Ancient Near East: A Philological and Sociological Comparison.* CHANE 56. Leiden und Boston.

Turner (1969): Victor W., *The Ritual Process: Structure and Anti-Structure.* London.

van der Leeuw (1977): Gerardus van der Leeuw, *Phänomenologie der Religion.* 4. Auflage, Tübingen.

Vaitl (2012): Dieter Vaitl, *Veränderte Bewusstseinszustände: Grundlagen-Techniken-Phänomenologie.* Stuttgart.

von Soden (1965–1981): Wolfram von Soden, *Akkadisches Handwörterbuch.* Wiesbaden.

Weber (1971): Max Weber, *Gesammelte Aufsätze zur Religionssoziologie III.* 5. Auflage, Tübingen.

Weber (2005): Max Weber, *Wirtschaft und Gesellschaft.* Frankfurt am Main.

Wundt (1906): Wilhelm Wundt, *Völkerpsychologie: Eine Untersuchung der Entwicklungsgesetze von Sprache, Mythos und Sitte, Bd. II.2: Mythos und Religion.* Leipzig.

Würthwein (1985): Ernst Würtweihn, *Die Bücher der Könige.* ATD 11/2. Göttingen.

Add, Subtract or Do Neither:
The Role of Somatic Manipulations
in Biblical Rites of Reclassification

Saul M. Olyan

Representations of ritual reclassification are commonplace in biblical texts. Polluted persons become clean through purification rites such as bathing and washing clothes, as in Lev 15,13; non-mourners are transformed into mourners through ritual actions such as tearing the garment and shaving the head, as Job 1,20 illustrates; a person who has undertaken the Nazirite vow returns to non-Nazirite status by means of a series of ritual acts which include shaving the head and burning the consecrated hair on the altar of burnt offerings (Num 6,18); the alien war captive becomes wife of her Israelite captor after she removes her captive's garment, cuts her nails and shaves her head, among other ritual actions (Deut 21,12-13); and the uncircumcised resident alien or slave becomes qualified to eat the Passover sacrifice by undergoing circumcision according to Exod 12,44. 48. In each of these examples and in many others, ritual reclassification is accomplished at least in part by means of one or more manipulations of the body. Somatic manipulations may be undertaken by an actor undergoing a reclassification him- or herself, as with the Nazirite who cuts and burns his or her own hair, or by another ritual agent who is present, as in the case of Moses, who dresses and anoints Aaron for priestly service in Exod 29,5–7. Persons whose ritual status is changing may bathe, shave themselves or be shaven, or be circumcised. They may strip off their clothes and remove their shoes, thereby exposing parts of the body normally concealed. They may lacerate or tattoo themselves, cut their nails, be sprinkled with oil, daubed with the blood of sacrificial offerings, or immerse themselves in cleansing waters. They may strew ashes or dirt on their heads, allow their hair to grow long, sit on the ground or move the body back and forth in a prescribed manner. Somatic manipulations such as these typically play a central role in rites of reclassification, including those associated with tripartite transitions (*rites de passage*).[1] Along with other rites that do

---

1   I view rites of transition (*rites de passage*) as one type of ritual reclassification, given that many reclassifications lack tripartite rites of separation, liminality and aggregation. On

not focus on manipulating the body—rites such as laundering clothes and offering sacrifices to the deity—somatic manipulations function both to make real and to mark status changes.[2] It is my purpose in this paper to explore the role played by bodily manipulation in several examples of ritual reclassification as represented in biblical texts. I am especially interested in the ways in which a body might be manipulated through the removal or subtraction of characteristic features (e. g., hair), body parts (e. g., the male foreskin) and coverings (e. g., clothes, shoes) as well as through the introduction or addition of ritual substances such as oil, sacrificial blood, and cleansing water; markings such as tattoos; and coverings such as new clothes. I will also consider somatic manipulations that neither add nor subtract (e. g., the physical movement of the body to a different locus) and non-somatic rites such as laundering and offering sacrifices.[3] Three texts from the Priestly Writing (= P) will be my focus: Exod 29,1-37, which concerns the sanctification of Aaron and his sons to serve as priests; Lev 14,1-20, which describes the aggregation rites of the person whose skin disease has been healed; and Num 8,5-22, which speaks of the separation of the Levites for cultic service.[4]

---

the *rites de passage*, see the classic works of van Gennep (1909) and Turner (1995); some of the model's limitations, particularly with respect to gender, are brought into relief by Lincoln (1991) and Bynum (1984). A biblical example of a ritual reclassification that lacks a tripartite structure is Jehoiakim's animal-like burial in Jer 22,18-19. In this text, the dead king's corpse is given an ass's burial, "dragged and tossed beyond the gates of Jerusalem." There is no liminal phase nor is there evidence of any kind of aggregation. The dead king is ritually reclassified as an animal through the casting of his corpse as if it were an ass's carcass. On this text, see Olyan (2014).

2 And they do so not only for living persons. Corpses are subject to somatic manipulation in the context of rites of reclassification in any number of texts, as are inscribed stelae and statuary. On the ritual manipulation of corpses, see, e. g., Olyan (2014); for such treatment of stelae and statuary, see, e. g., Levtow (2013).

3 Though such adding and subtracting are common ways to ritually manipulate bodies, they are not the only ways. Other somatic manipulations such as moving bodies to another place or moving them in a particular manner in the same locus involve neither the removal of characteristic features, body parts or coverings nor the introduction of ritual substances, markings, new clothes or other novel items.

4 On the hypothetical source called the Priestly Writing (or P) by biblical scholars, see, e. g., Gertz (2006).

I

My first passage of interest is Exod 29,1-37, a text of the Priestly Writing.[5] In it, Moses is ordered by Yhwh to undertake a series of rites intended to sanctify Aaron and his descendants so that they might serve as priests for Yhwh (V. 1). First, Aaron and his sons are brought near to the entry of the deity's shrine and washed with water; Aaron is clothed by Moses in the garments of the high priest and anointing oil is poured on his head; Aaron's sons are clothed in priestly garments (VV. 4-9). Then Aaron and his sons are ordained (literally, their hands are filled [*mallē 'yādām*]): Three sacrificial animals are slaughtered and their blood is manipulated at the altar of burnt offerings. Some of the blood from the third animal, a ram, is applied by Moses to the lobe of Aaron's right ear and those of his sons as well as to their right thumbs and right big toes. (The reason for this favoring of the right hand remains wholly obscure, but occurs in other biblical ritual texts, as we shall see.) Blood and anointing oil are then sprinkled upon Aaron, his sons and their vestments (VV. 9b-21). The text goes on to state that Aaron, his sons and their garments are holy (V. 21). The next stage of the ritual focuses on the processing of the remaining parts of the second ram: The fat portions, some of the organs, the right thigh and several unleavened baked items, some mixed with oil, are put in the hands of Aaron and his sons (VV. 23-24). These are then burned on the altar before Yhwh by Moses (V. 25). The text ends with the notation that the rites of ordination are to last seven days (V. 35).[6]

According to this text, the sanctification and ordination of Aaron and his sons for the priesthood are the goals of the elaborate series of rites narrated (VV. 1. 35). In the Priestly Writing's ideology, only priests among persons are holy, and their holiness sets them apart from other cultic servants and the rest of Israel (Exod 16,1-17,5), securing for them various privileges (e.g., eating holy items [V. 33]). Thus, establishing the holiness of the priests is a prime objective for the Priestly Writing. Moses is the primary ritual actor in this narrative, performing various somatic manipulations as instructed by Yhwh. For their part, Aaron and his sons are portrayed primarily as the recipients of Moses's ritual actions and thus patients rather than agents, though they do lay their hands upon the head of each sacrificial

---

5   Some elements of this narrative are presented again in Exod 40,12-15 though with differences, e.g., 40,15, where the sons of Aaron are anointed, in contrast to 29,8, which speaks only of Aaron's anointing. The narrative in 29,1-37 is anticipated briefly in 28,41, also with the anointing of Aaron's sons, as in 40,15.

6   Verses 27-28 state that in the future, the Aaronid priests will receive the breast and right thigh as their due from offerings of well being; in addition, VV. 31-34 note that the priests receive unleavened baked items as well as meat.

victim, suggesting agency or at least its potential (VV. 10. 15. 19). Although not all of the somatic manipulations narrated add or subtract in the manner I have described, they almost all do so. (An exception is V. 4, in which Aaron and his sons are brought near to Yhwh, a case of physical movement rather than adding or subtracting.)

Rites that both add and subtract function to realize the status change of the priests, though in the case of this particular narrative, the vast majority of the rites introduce something new, whether it be a ritual substance such as anointing oil, water, or sacrificial blood, or new clothes. Water is used to wash the bodies of Aaron and his sons and they are dressed in priestly attire; anointing oil is poured on Aaron's head and also sprinkled on Aaron and his priestly garments as well as on his sons and their priestly vestments. The blood of a slaughtered ram is daubed on the right ear, thumb and big toe of Aaron and his sons as well as sprinkled on them and on their priestly garb. Liquids play the primary role in this narrative, applied to bodies in various ways: oil through pouring and sprinkling, blood through daubing and sprinkling, water presumably through immersion. Clothes are also recipients of sprinkled blood and anointing oil. Water presumably purifies the bodies of the priests, while the oil and blood contribute to their transformation into holy persons.[7] In addition to the ritual application of liquids, the changing of garments plays an important role in the sanctification and ordination of the priests, and the new priestly vestments are tied to the oil and blood directly through the act of sprinkling. The only subtraction I notice in the narrative is the unmentioned removal of the previous, non-priestly garments of Aaron and his sons (V. 5). This is striking, given the importance of subtractions such as hair removal or the exposure of the body or its constituent parts in other narratives of ritual reclassification.

II

Lev 14,1-20, another text of the Priestly Writing, presents the "teaching" concerning the cleansing rites of the person formerly afflicted with highly polluting skin disease.[8] When his disease is healed, an elaborate ritual process for the person "purifying

---

7   Verse 4 states that Moses "shall wash them in water." Though nothing is said explicitly about purification, this procedure is not infrequently associated with such a goal, as in Lev 14,8; 15,5; etc.

8   The identity of this skin affliction is unknown and it does not present in the manner of leprosy (Hansen's Disease). Thus, I refer to it simply as "skin disease." On this, see further Milgrom (1991), 816-17.

himself" (*hammiṭṭahēr*) begins. It is overseen by a priest, who also plays a central role in the rites. Its goal: To allow the formerly afflicted and polluted person to fully reenter the community from which he was expelled (see Lev 13,45-46). Two clean living birds, cedar wood, scarlet stuff and hyssop are brought to the priest. One bird is slaughtered over a clay vessel containing "living water" and the other, living bird, the hyssop, the cedar wood and the scarlet stuff are dipped together by the priest into the blood of the slaughtered bird (VV. 5-6).[9] The blood is then sprinkled seven times on the one "purifying himself" and the living bird is freed to fly away (V. 7). The person undergoing the cleansing washes his garments, shaves his entire body, bathes in water and is said to be clean. He may now enter the camp but must remain outside his dwelling seven days (V. 8). On the seventh day, the one "purifying himself" once again shaves off all his hair, including that of the head, beard and eyebrows, launders his clothes, and bathes. Once again, he is said to be clean (V. 9). Prescribed sacrifices and oil and grain offerings follow on the eighth day: The priest stations the person being cleansed before Yhwh at the entry to the Tent of Meeting and sacrifices and offerings are made by the priest, who also daubs some of the blood of the reparation offering (*'āšām*) on the right ear lobe, right thumb and right big toe of the one who is "purifying himself" (V. 14). Then some of the oil brought as an offering is sprinkled seven times "before Yhwh" by the priest; some is applied on the right ear lobe, right thumb and right toe of the person being cleansed, over the blood of the sacrificial victim; and the remainder is put on the head of the one "purifying himself" (VV. 15-18). The ritual ends with additional sacrifices and offerings to Yhwh and the notation that the man is now clean (VV. 19-20).

Lev 14,1-20 is a challenging text, raising a number of questions which are difficult to answer (e. g., Why is the man said to be clean at several points in the cleansing process?[10]) Nonetheless, the text lends itself to an analysis that focuses on the role played by bodily manipulations in the process of reclassifying the man who is "purifying himself." As was the case with Exod 29,1-37, Lev 14,1-20 is a representation of a ritual which includes somatic manipulations that add, those that subtract, and those that do neither, as well as non-somatic rites. Oil, sacrificial blood and water are introduced to the body of the person undergoing cleansing: The slaughtered bird's blood is sprinkled seven times on him by the priest (V. 7); he bathes his body in water (VV. 8. 9); the blood of the reparation sacrifice is daubed

---

9   "Living water" may refer to flowing water as some have maintained (e. g., Milgrom [1991], 836-37).

10   The text seems to be speaking of stages of purification. On this, see the discussion in Olyan (2000), 46-47, 53-54.

on his right ear lobe, thumb and big toe (V. 14); oil is applied over that blood (V. 17); and oil is placed on the man's head (V. 18). Besides these acts that introduce ritual substances such as water, oil and blood, there are rites that remove a characteristic somatic feature: The man shaves off the hair of his entire body twice (VV. 8. 9). Finally, there are somatic rites that neither add nor subtract (e. g., the man being cleansed is stationed before Yhwh by the priest) and a number of rites that do not involve bodily manipulation at all. The latter include washing garments (VV. 8. 9), the release of the living bird (V. 7), the sprinkling of oil seven times before Yhwh (V. 16), and the offering of sacrifices.

In contrast to the Aaronids of Exod 29,1-37, the person being cleansed in Lev 14,1-20 is an active ritual participant. Though the officiating priest performs a number of the rites on the man "purifying himself," he shaves his own body twice, bathes in water twice and launders his clothes twice (VV. 8. 9). His active role in his cleansing is also signaled by the reflexive Hebrew participle used several times by the text to describe him: "the man purifying himself" (*hammiṭṭahēr*). Nonetheless, there are close similarities between the two texts as well. Both passages include the ritual manipulation of oil and sacrificial blood and the use of purifying water; both describe the acts of sprinkling and daubing, and include sacrifices. The parallel between the daubing of blood on the right ear lobe, right thumb and right big toe is quite striking, especially given its contrasting purpose in each text: In Exod 29,20, sacrificial blood is daubed in these three places in order to sanctify and ordain Aaron and his sons; in Lev 14,14, it is applied in order to purify the man whose skin disease has disappeared.[11] In each text, all of the rites in combination function to reclassify the individual(s) in question: Aaron and his sons become priests and the man "purifying himself" becomes clean. The rites of Lev 14,1-20 represent a classic example of aggregation following rites of separation and a period of liminality (see Lev 13,45-46).

---

11 In Lev 14,17, some of the oil is also daubed over the blood on the man's right earlobe, right thumb and right toe, as already noted. This additional ritual act creates a contrast with Exod 29,20, where only blood is daubed in these three places.

III

Num 8,5-22, a third text of the Priestly Writing, describes the separation of the Levites for cultic service. According to the P, the Levites serve the Aaronid priests, undertaking secondary, non-priestly cultic responsibilities (Num 3,5-10; 8,19).[12] In contrast to the priestly sons of Aaron, the Levites are not holy, but share common status with the rest of Israel (Num 16,5. 7). Thus, the Levites of Num 8,5-22 are not sanctified as are the sons of Aaron in Exod 29,1-37; they are only separated from Israel for cultic service, as stated explicitly in V. 14. As in Exod 29,1-37 and Lev 14,1-20, a series of rites accomplish a larger goal, in this case the separation and purification of the Levites for cultic service. And as with the preceding two examples I have discussed, these rites include somatic manipulations that add, that subtract, and that do neither, as well as rites that do not involve manipulation of the body. The Levites are sprinkled with purifying waters by Moses (V. 7), who also brings them before Yhwh, at the entry of the Tent of Meeting (VV. 9. 10) and then stations them before Aaron and his sons (V. 13). The Levites shave themselves as part of their purification process and wash their clothes (VV. 7. 21). The Israelites lay their hands on the Levites (V. 10), and Aaron "offers" them as an elevation offering (VV. 11. 13. 21). For their part, the Levites lay their hands on the heads of two sacrificial bulls which are then sacrificed to Yhwh in order to make expiation for the Levites (V. 12).

Unlike the other two narratives I have discussed previously, Num 8,5-22 has fewer rites which introduce ritual substances to the bodies of those undergoing reclassification. In this case, it is only purifying water that is sprinkled on the Levites; there are no manipulations of sacrificial blood or of oil. In contrast, the movement of the Levites from one locus to another, a somatic manipulation that neither adds nor subtracts, plays a prominent role in this narrative, as I have noted. (Its role in the other two narratives is far less significant.) Also, the laying of hands, both by the Israelites on the Levites and by the Levites on sacrificial bulls, neither adds nor subtracts, though it is a somatic manipulation in both instances. And as with the man "purifying himself" in Lev 14,1-20, the Levites of Num 8,5-22 shave themselves, though unlike that man, the Levites do not bathe. The Levites also wash their clothes, a non-somatic manipulation, as the man "purifying himself" from skin disease does in Lev 14,8. 9. Agency in Num 8,5-22 is shared by Moses, Aaron and the Aaronids, the people and the Levites themselves: Moses sprinkles the Levites

---

12 According to a number of non-P texts, all males of the Levi tribe are potential priests and there exists no secondary class of cultic servants called "Levites" (e. g., Num 26,58a; Deut 18,1; 33,8-11; Jer 33,18. 21-22). On this, see Cross (1973), 195-215.

with purifying water and stations them first before Yhwh and then before Aaron and his sons; the children of Israel lay their hands on the Levites; Aaron "offers" the Levites as an elevation offering to Yhwh; the Levites shave themselves, launder their clothes and lay their hands on the sacrificial bulls. This contrasts with Exod 29,1-37, where agency is almost entirely concentrated in Moses, but resembles Lev 14,1-20, where agency is more widely spread among the various ritual participants.

IV

What might be concluded from this investigation of the role of somatic manipulations, particularly those that add or subtract, in three biblical representations of rites of reclassification, all of which come from the Priestly Writing (P)? For one thing, it seems clear that different combinations of somatic rites are possible: In some settings, the introduction of ritual substances (e. g., liquids) and other forms of addition (e. g., new clothes) might play the primary role, as in Exod 29,1-37; in other contexts, rites that add and those that subtract are more balanced, as in Lev 14,1-20; in still others, the introduction of ritual substances plays a lesser role, while rites that remove characteristic features such as hair are prominent, as in Num 8,5-22. Furthermore, the texts under consideration suggest that both somatic rites that neither add nor subtract and non-somatic rites also play a part—often significant— in rituals of reclassification. In Num 8,5-22, the physical movement of the Levites by Moses, a somatic rite that neither adds nor subtracts, plays a central part in their reclassification; in all three examples, non-somatic rites such as laundering and sacrifice are significant components of the ritual process. Agency also varies among the texts, with those undergoing a ritual reclassification virtually lacking agency in a text such as Exod 29,1-37, in contrast to texts such as Lev 14,1-20 and Num 8,5-22, where the man "purifying himself" and the Levites perform significant ritual actions themselves. Nonetheless, all three texts suggest that somatic manipulations, whether they add, subtract or do neither, play an important role in rites of reclassification as they are represented in biblical texts. In fact, it would be impossible to imagine the rites described in these three texts were one to remove all forms of somatic manipulation from them. Little besides sacrifices and washing clothing would be left! The body plays a central part—arguably, the central part— in the rites of the three texts discussed here and other rites of reclassification mentioned at the beginning of this essay. Its potential to be manipulated in a variety of ways which might result in different ritual configurations allows for both the realization and communication of a range of status changes. In a word, in the body's very plasticity resides its potential as a site of social inscription.

Bibliography

Bynum (1984): Caroline W. Bynum, "Women's Stories, Women's Symbols: A Critique of Victor Turner's Theory of Liminality". In: Moore, R. L. and Reynolds, F. E., (hg.), Anthropology and the Study of Religion. Chicago, 105–25.
Cross (1973): Frank Moore Cross, Canaanite Myth and Hebrew Epic: Essays in the History of the Religion of Israel. Cambridge, Mass.
Van Gennep (1909): Arnold van Gennep, Les rites de passage. Paris.
Gertz (2006): Jan Christian Gertz, "Tora und Vordere Propheten". In: Grundinformation Altes Testament. Göttingen, 230-40.
Levtow (2013): Nathaniel B. Levtow, "Artifact Burial in the Ancient Near East". In: Johnson Hodge, C. et al., (hg.), "The One Who Sows Bountifully": Essays in Honor of Stanley K. Stowers. Providence, 141-51.
Lincoln (1991): Bruce Lincoln, Emerging from the Chrysalis: Rituals of Women's Initiation. New York.
Milgrom (1991): Jacob Milgrom, Leviticus 1-16. (Anchor Bible Bd. 3). New York.
Olyan (2000): Saul M. Olyan, Rites and Rank: Hierarchy in Biblical Representations of Cult. Princeton.
Olyan (2014): ders., "Jehoiakim's Dehumanizing Interment as a Ritual Act of Reclassification". In: Journal of Biblical Literature 133:271-79.
Turner (1995): Victor Turner, "Liminality and Communitas". In: ders., The Ritual Process: Structure and Anti-Structure. New York, 94-130.

Weibliche Askese und christliche Identität im 2. Jh. n. Chr.

Judith Hartenstein

I   Einführung[1]

Zu den vielen Möglichkeiten, wie religiöse oder philosophische Überzeugungen sich auf den Körper auswirken können, gehört auch Askese. Ich verstehe unter Askese die Kontrolle von oder den bewussten Verzicht auf bestimmte körperliche Bedürfnisse, besonders häufig in Bezug auf Nahrungsaufnahme und Sexualität. Das ist ein modernes Verständnis des Begriffs, in seiner ursprünglichen griechischen Verwendung steckt der Gedanke der Übung, des Trainings. Askese hat so eine bestimmte Funktion. Sie ist ein kultur- und religionsübergreifendes Phänomen, es ist nicht spezifisch christlich und auch das entstehende Christentum hebt sich mit seiner teilweise asketischen Praxis nicht grundsätzlich von seiner spätantiken Umgebung ab.[2]

Ich will mich im Folgenden mit asketischen Ansätzen und ihrer Begründung im 2. Jh befassen. Dabei ist vor allem sexuelle Askese ein Thema. Ab dem 4. Jh. gibt es eine Vielzahl von Texten von Kirchenvätern und aus dem entstehenden Mönchtum (Wüstenväter und mütter) zum Thema und die Askese nimmt feste und klare Formen an.[3] Mir geht es allerdings um die Frühzeit, in der sich die Praxis und ihre Begründung im christlichen Kontext erst entwickelt, vieles noch offen ist und eine Vielzahl von verschiedenen Ansätzen auch nebeneinander existiert. Ein Problem ist dabei die Quellenlage. Mein Zugang sind vor allem apokryphe Schriften zum Neuen Testament, die zum Teil explizit und offensiv (apokryphe Apostelakten) Askese zum Thema machen, sich zum Teil aber nur sehr indirekt

1   Ich danke Angelica Dinger für Ihre Mitwirkung an der Ausgestaltung dieses Beitrags.
2   Vgl. Horn (1997), 86f.; Bergman/Gribomont (1979), 195-197, 207-212.
3   Vgl. die Monographien von Elm (1994) und Shaw (1998) mit ihrer Schwerpunktsetzung. Brown (1988) geht dagegen ausführlich auf die frühere Zeit ein.

und umstritten in der Deutung darauf beziehen (apokryphe Evangelien). In jedem
Fall sind Rückschlüsse auf die konkrete Praxis der Zeit nicht einfach. Auffällig ist,
dass in vielen der möglicherweise oder eindeutig asketischen Schriften der Zeit
Frauen eine große und teilweise den Männern gleichberechtigte Rolle spielen und
dabei konventionelle Frauenrollen hinter sich lassen. Diesem Phänomen und den
theologischen Hintergründen will ich gerne nachgehen.

Zunächst ist aber das kulturelle Umfeld in den Blick zu nehmen: In verschiedenen
philosophischen Strömungen der Zeit hat Askese im Sinne einer Kontrolle über
Begierden einen hohen Wert. Da die wichtigsten Werte geistig sind und Unabhän-
gigkeit von äußeren (materiellen) Umständen erstrebenswert ist, gehört Askese zu
einem philosophischen Leben dazu. Askese ist ein Training für den Körper (und
mitunter auch für die Seele), das die grundlegenden Überzeugungen in die Praxis
umsetzt.[4] Dabei scheint gerade das 2. Jh. die Zeit zu sein, in der sich eine gemäßigte
Form von Askese, die der gesellschaftlichen Ordnung zuträglich ist, durchsetzt,
auch in Abgrenzung von radikaleren Asketen, die diese Ordnung gefährden und
durch ihre persönliche Autorität suspekt sind.[5] Wichtige dieser gemäßigt asketischen Philosophen sind z. B. Musonius Rufus oder Epiktet. Sie fordern in Bezug auf
Sexualität keine radikale Enthaltsamkeit, sondern halten eine Ehe, in der Sexualität
nicht getrieben von Begierde, sondern als Akt des Willens mit dem Ziel der Kinder-
zeugung gelebt wird, als für die männlichen Philosophen angemessen. Eine ganz
ähnliche Position vertritt z. B. auch der Kirchenvater Clemens von Alexandrien
in Auseinandersetzung mit radikaleren asketischen Gruppen. Dies scheint eine
typische Position von gebildeten, gesellschaftlich etablierten Männern zu sein.[6]

Das Thema sexuelle Askese hat einen doppelten Genderbezug: Bei Aufforderungen zur Askese, die an Männer gerichtet sind, steht die Warnung vor konkreten
Frauen oder dem Verkehr mit der Weiblichkeit im Zentrum. Gleichzeitig werden
die Begriffe „männlich" und „weiblich" auch metaphorisch verwendet und stehen
für „geistig" versus „materiell" bzw. „körperlich" oder andere Beschreibungen der
beiden hierarchisch aufeinander bezogenen Pole.[7] Die Zuwendung zu geistigen
Werten kann deshalb grundsätzlich als Abkehr vom Weiblichen beschrieben
werden. „Flieh die Weiblichkeit!" ist doppeldeutig.[8]

---

4   Vgl. Dillon (1995), 80.

5   Vgl. Francis (1995), 181-183.

6   Vgl. Francis (1995), 11-19.

7   Dabei ist weiblich-männlich nicht komplementär gedacht wie heute meistens, sondern
    als eine Abstufung mit klar unterschiedlicher Wertigkeit.

8   Vgl. Wisse (1988), 300f. am Beispiel des Zostrianus.

Diese Ambiguität macht Askese in der Theorie für reale Frauen schwieriger, aber auch sie können im antiken Kontext ihre Weiblichkeit hinter sich lassen und zum männlich-geistigen Sein aufsteigen. Etliche apokryphe Schriften thematisieren diese Frage auch ausdrücklich, was dafür spricht, dass Frauen nennenswert beteiligt waren. Gravierender sind außerdem die sozialen Konsequenzen von Askese, da Familie und Kinder einen großen Anteil am üblichen weiblichen Leben ausmachen, während Männer viele weitere Rollen ausfüllen.[9]

Askese bezieht sich auf den Körper, kann ihn aber sehr unterschiedlich werten: Eine Kontrolle und Einschränkung von körperlichen Bedürfnissen kann in einer grundsätzlichen Abwertung und Verachtung des Körpers begründet sein, sie kann aber auch auf einer positiveren Sicht beruhen, nach der der Körper durch die entsprechende Übung gerade zu einem würdigen Gefäß der Seele wird. Beide Sichtweisen gehen schon auf Plato (Phaed. 67A) zurück.[10] Dieser Unterschied ist wichtig für die christlichen Schriften, deren Sicht des Körpers klar variiert. Eine Abwertung des Körpers, wie sie vor allem in sog. gnostischen Schriften zu finden ist, ist aber weder die Voraussetzung für Askese noch führt sie zwingend zu ihr, sie ist aber für die jeweilige Begründungsstruktur wichtig.

II   Frühchristliche Apostelgeschichten – Theklaakten

Ein erstes Beispiel sind die Theklaakten, die einen in sich geschlossenen, selbstständig (und vielfach) überlieferten Teil der Paulusakten bilden. Entstehungszeit ist wohl die 2. Hälfte des 2. Jh. Thekla ist eine junge Frau, die am Fenster des Hauses ihrer Mutter eine Predigt des Paulus hört, deren zentraler Inhalt die Aufforderung zur Enthaltsamkeit ist.

> Und als Paulus im Hause des Onesiphorus eingekehrt war, herrschte (dort) große Freude; die Knie wurden gebeugt und das Brot gebrochen und das Wort Gottes von der Enthaltsamkeit und der Auferstehung (verkündet), indem Paulus sprach:
>
> „Selig sind, die reines Herzens sind, denn sie werden Gott schauen.
>
> Selig sind, die ihr Fleisch rein bewahrt haben, denn sie werden ein Tempel Gottes werden.
>
> Selig sind die Enthaltsamen, denn Gott wird zu ihnen reden.
>
> Selig sind, die dieser Welt entsagt haben, denn sie werden Gott wohlgefallen.

---

9   Vgl. Wire (1988), 309.
10 Vgl. Dillon (1995), 80-82.

Selig sind, die Frauen haben als hätten sie nicht, denn sie werden Gott beerben.

Selig sind, die Gottesfurcht haben, denn sie werden Engel Gottes werden.

Selig sind, die vor den Worten Gottes zittern, denn sie werden getröstet werden.

Selig sind, die die Weisheit Jesu Christi ergriffen haben, denn sie werden Söhne des Höchsten heißen.

Selig sind, die die Taufe bewahrt haben, denn sie werden bei dem Vater und dem Sohn ausruhen.

Selig sind, die das Verständnis Jesu Christi erfasst haben, denn sie werden im Lichte sein.

Selig sind, die um der Liebe Gottes willen das weltliche Wesen verlassen haben, denn sie werden Engel richten und zur Rechten des Vaters gesegnet werden.

Selig sind die Barmherzigen, denn sie werden Barmherzigkeit erlangen, und den bitteren Tag des Gerichts werden sie nicht sehen.

Selig sind die Leiber der Jungfrauen, denn sie werden Gott wohlgefallen, und sie werden den Lohn ihrer Keuschheit nicht verlieren. Denn das Wort des Vaters wird ihnen zum Werk der Rettung auf den Tag des Sohnes werden, und sie werden Ruhe finden in alle Ewigkeit.“[11]

Diese Predigt entspricht nicht dem, was wir von Paulus aus seinen Briefen wissen, knüpft aber durchaus an seine Lehre an, sowohl in einzelnen Formulierungen als auch in der Grundtendenz. Paulus hält z. B. in 1Kor 7 ein sexuell enthaltsames Leben für erstrebenswert und lebt selbst so, sieht aber eheliche Sexualität als einen Schutz gegen unbedingt zu vermeidende Unzucht an. Beides ist also möglich. Die kanonische Paulustradition – die Pastoralbriefe – ziehen dann die Ehe der strengen Askese vor (1Tim 4,1-5 u.ö.).

In den Theklaakten wird die Heldin sofort zu einem enthaltsamen Leben bekehrt, sehr zum Ärger von ihrem Verlobten und ihrer Mutter. Sie wird von diesen sogar vor Gericht gebracht und zum Tierkampf verurteilt, den sie wunderbar überlebt – später nochmals in einer anderen Stadt. Sie setzt sich aber durch und wirkt schließlich in Männerkleidern als wandernde Missionarin. Paulus spielt nur am Rande eine Rolle.

Sexuelle Askese ist also ein zentrales Thema der Schrift. Sie ist als eine Grundbedingung christlichen Lebens dargestellt, ohne Enthaltsamkeit keine Auferstehung, also kein Heil.

11　ActThecla, 5f. Übersetzung: Schneemelcher (1999), 216f.

(…) Er [Paulus] macht aber Jünglingen die Frauen und Jungfrauen die Männer abspenstig, indem er sagt: ‚Auf andere Weise gibt es für euch keine Auferstehung, es sei denn, dass ihr rein bleibt und das Fleisch nicht befleckt, sondern es keusch bewahrt.'[12]

Der Körper ist dabei durchaus positiv gesehen, jedenfalls so lange er rein bewahrt wird. Gerade dies scheint die Voraussetzung für eine vermutlich körperlich gedachte Auferstehung zu sein. Als Konsequenz dieser Botschaft gibt Thekla die übliche weibliche Rolle (Heirat) auf und wird eine selbstständige Apostelin.

Auch in anderen Apostelakten steht die Aufforderung zur sexuellen Askese im Zentrum der Botschaft der jeweiligen Apostel (Andreas, Thomas, Petrus). Häufig werden davon gerade Frauen angesprochen, die sich, wenn sie verheiratet sind, dann ihren Ehemännern verweigern, mit erheblichen Komplikationen und bedrohlichen Folgen für den Apostel. Frauen verlassen so ihre üblichen Rollen, eine so deutliche Überschreitung der weiblichen Sphäre wie bei Thekla findet sich in den anderen Akten allerdings nicht. Askese scheint gerade Frauen anzusprechen und sie kann eine neue Rollenperspektive eröffnen,[13] aber sie führt nicht zwingend dazu. Der Hauptzweck bleibt die Enthaltsamkeit als Vorbedingung für das ewige Leben (im weitesten Sinne). Ein asketischer Lebensstil scheint in dem Teil des frühen Christentums, der durch die Apostelakten repräsentiert wird, die einzig akzeptable Möglichkeit zu sein, wird aber nicht immer mit theologischen Argumenten untermauert. Es gibt auch eher banale, praktische Begründungen für Askese, etwa die Mühe und der Ärger, die das Großziehen von Kindern mit sich bringt (ActThom 12).[14] Askese steht so sehr im Zentrum, dass sie ihren Zweck in sich zu haben scheint.

---

12 ActThecla 12. Übersetzung: Schneemelcher (1999), 218. Allerdings fasst an dieser Stelle ein Gegner des Paulus seine Lehre zusammen, es ist nicht eindeutig, ob dies wirklich der Lehre der Schrift entspricht.

13 Vgl. Petersen (1999), 325-328.

14 Gedenket, meine Kinder, an das, was mein Bruder mit euch geredet und wem er euch befohlen hat, und erkennet, daß ihr, wenn ihr euch von diesem schmutzigen Verkehr befreit, heilige Tempel, rein und solche werdet, die von Leiden und Schmerzen, offenbaren und nicht offenbaren, befreit sind; und ihr werdet euch nicht Sorgen für Leben und Kinder auflegen, *deren Ende Verderben* (Phil 3,19) ist. Wenn ihr euch aber viele Kinder anschafft, so werdet ihr um ihretwillen Räuber und Habsüchtige, die Waisen schinden und Witwen übervorteilen, und indem ihr dies tut, unterwerft ihr euch sehr schlimmen Strafen. Denn die meisten Kinder werden unnütz, von bösen Geistern besessen, die einen offenbar, die anderen auch unsichtbar. … Denn sie werden entweder in Ehebruch oder Mord oder Diebstahl oder Unkeuschheit gefunden, und durch dies alles werdet ihr in Betrübnis versetzte werden. Wenn ihr aber gehorcht und eure Seelen Gott rein bewahrt, werden euch lebendige Kinder werden, die von diesen Schäden unberührt bleiben, und werdet ohne Sorgen sein, indem ihr ein unbeschwertes Leben ohne Schmerz und

III   Apokryphe Evangelien –
Ägypterevangelium, Thomasevangelium

Eine ganz andere Art von Text sind das Ägypterevangelium und das Thomasevangelium, es handelt sich um Evangelien, die aus einzelnen Sprüchen Jesu und kurzen Dialogen bestehen. Beim EvThom, das vollständig im Fund von Nag Hammadi in koptischer Übersetzung vorliegt,[15] gibt es praktisch keine Erzählungen, weder Wunder noch Passionsgeschichte. Beim EvÄg ist nicht die ganze Schrift, sondern es sind nur kurze Stücke erhalten. Der Kirchenvater Clemens von Alexandrien zitiert sie und setzt sich dabei gleichzeitig mit ihrer Deutung in einer streng asketisch lebenden Gruppe auseinander – die verschiedenen Ebenen sind nicht immer leicht zu trennen.[16] Diese Zitate bilden ein Gespräch zwischen Jesus und Salome, in dem es um das Zerstören der „Werke des Weiblichen" geht, die Jesus beabsichtigt. Zwei dieser Stücke bieten einen Eindruck von den Schwierigkeiten der Deutung:

> Diejenigen, die sich der Schöpfung Gottes mittels der wohlklingenden Enthaltsamkeit (ἐγκράτεια) widersetzen, zitieren auch die an Salome gesprochenen Worte, die wir früher erwähnt haben. Sie stehen aber, wie ich denke, im Evangelium nach den Ägyptern. Sie sagen nämlich, dass der Heiland selbst sprach: ‚Ich bin gekommen, die Werke des Weiblichen zu zerstören'. ‚Des Weiblichen', das heißt: der Begierde; ‚die Werke', das heißt aber: Werden und Vergehen.[17]

Später heißt es bei Clemens:

> Deshalb sagt nun Cassianus: ‚Als Salome fragte, wann man das, was sie erfragt hatte, erkennen werde, sprach der Herr: ‚Wenn ihr das Gewand der Scham mit Füßen treten werdet und wenn die zwei eins werden und das Männliche mit dem Weiblichen und weder männlich noch weiblich (sein wird).' Erstens nun haben wir den Ausspruch nicht in den uns überlieferten vier Evangelien, sondern in dem (Evangelium) nach den Ägyptern. Weiter aber scheint er (Cassianus) mir nicht zu wissen, dass mit dem männlichen Trieb der Zorn und mit dem weiblichen die Begierde angedeutet ist; wenn aber diese wirksam geworden sind, dann folgt Reue und Scham.[18]

Sorge verlebt und jene unvergängliche und wahrhaftige Hochzeit (als euch gebührend) erwartet zu empfangen, und werdet bei ihr als Brautführer mit hineingehen in jenes Brautgemach, <das voll von> Unsterblichkeit und Licht ist. (ActThom 12. Übersetzung: Drijvers [1999], 308)

15   Layton (1989).

16   Vgl. Markschies (2012), 667.

17   Clem., *Strom.* III 63,1f. (Markschies [2012], 667).

18   Clem., *Strom.* III 92,2-93,1 (Markschies [2012], 671).

Das „Zerstören der Werke des Weiblichen" scheint das Durchbrechen der Abfolge von Werden und Vergehen zu meinen. Der Tod hat Macht, solange Frauen gebären. Überwunden wird dieser Zustand durch eine Verbindung der geschlechtlichen Zweiheit zu einer Einheit ohne Geschlechterdifferenz. Diese Worte werden durch eine enkratitische Gruppe bzw. Cassianus als Aufforderung zur völligen Enthaltsamkeit gelesen, während Clemens ihnen eine weitere Deutung gibt. Er bezieht die „Werke des Weiblichen" nicht ausschließlich auf Sexualität und Geburt, sondern auf weitere Untugenden wie Eitelkeit und Völlerei. Zwar ist der Text des EvÄg nicht insgesamt erhalten, durch das Zeugnis des Clemens wird aber deutlich, dass diese Schrift zur Begründung einer asketischen Position genutzt wurde (Cassianus, die enkratitische Gruppe). Gleichzeitig zeigt Clemens auch, dass eine andere Interpretation möglich war: Er selbst versteht das Zerstören der Werke der Weiblichkeit nicht als Aufforderung zur Enthaltsamkeit und zum Verzicht auf Kinder, sondern sieht durch die Termini „männlich" und „weiblich" umfassend Begierden und Untugenden angesprochen, die im christlichen Leben aufgehoben sind. Er kommt dadurch zu einer gemäßigten Askese mit Kontrolle von Begierden, aber ohne Verzicht auf Fortpflanzung (diese ist vielmehr eine Rechtfertigung für eheliche Sexualität.) Die Schöpfung ist für Clemens gut! Er nennt an anderer Stelle aber auch eine Interpretation durch die christlich-gnostische Gruppe der Valentinianer, die die Werke der Weiblichkeit mythologisch, nicht auf menschliches Verhalten, deutet.[19]

Vorstellungen über die Werke des Weiblichen, in der Gebären als Ursache für Vergehen und Tod erscheint, und auch die Aufhebung der Geschlechterdifferenz zu einer umfassenden Einheit begegnen in etlichen weiteren Schriften. Im Hintergrund steht vermutlich eine Schöpfungsvorstellung in Etappen, wie sie z. B. Philo von Alexandrien entwickelt hat. Der ursprüngliche Mensch ist ein geistiges, nicht geschlechtlich differenziertes, unsterbliches Wesen. Erst in einem weiteren Schritt wird es körperlich und geschlechtlich und sterblich, der Zustand, in dem wir bis heute leben.[20]

Für das EvÄg und auch für das EvThom ist durch Jesu Wirken eine Rückkehr in den Ursprungszustand möglich und vielleicht sogar schon erreicht. In einer solchen geistigen, nicht differenzierten Existenz hat sich sexuelle Anziehungskraft einfach erübrigt und Askese ist die logische Folge. Anders als in den Theklaakten scheint hier die Veränderung des menschlichen Seins der Ausgangspunkt zu sein, daraus folgt dann Askese, sie ist nicht der Weg zum Ziel Auferstehung. Deshalb fehlen in diesen Schriften die ausdrücklichen Aufforderungen. Es ist auch deutlich, dass der Körper eine geringere Bedeutung hat, das neue Sein ist eine geistige Existenz,

---

19 Clem., *exc. Thdot.* 67,2-4.
20 Vgl. Philo, *op. mund.* 134f.151 und dazu die Erklärung bei Heininger (2010), 386-388.

eine besondere Würdigkeit des Körpers ist kein Thema. Ob das angestrebte Ziel aber als Auferstehung oder Rückkehr in den ursprünglichen Schöpfungszustand beschrieben wird, ist dagegen nur eine geringe Differenz, beides kann sehr ähnlich verstanden werden.

Im EvThom finden sich viele dieser Gedanken: Auch hier geht es um die Rückkehr zu einer ursprünglichen Einheit mit Aufhebung der Geschlechterdifferenz (22),[21] um Schöpfung und um eine Ablehnung von Reproduktion (79). Körperlichkeit kommt eher abwertend vor, die eigentlichen Werte sind eindeutig geistig, die materielle Welt ist für die, die sie erkennen, ein Leichnam. Ausdrückliche Aufforderungen zum asketischen Leben finden sich jedoch nicht (in Bezug auf Nahrung wird Fasten sogar ausdrücklich abgelehnt). Es ist deshalb durchaus strittig, ob EvThom Askese fordert oder nicht.[22] Verschiedene Indizien sprechen allerdings dafür, hinzu kommt noch, dass sich zwei etwas spätere und klar asketische Schriften ausdrücklich auf das EvThom beziehen (LibThom und ActTho). Zumindest ist das EvThom wie das EvÄg in asketischen Kreisen gelesen worden, auch wenn dies nicht die einzige mögliche Interpretation sein muss. Askese ist aber jedenfalls nicht ein Hauptanliegen, sondern eher eine Auswirkung der vertretenden Theologie.

Im EvThom lässt sich aber relativ gut der soziale Kontext erheben. Seine Position lässt sich als Wanderradikalismus beschreiben – mit diesem Begriff wird eigentlich der Lebensstil Jesu und der Gruppe um ihn bezeichnet, der in der Darstellung der kanonischen Evangelien aber schon deutlich abgeschwächt ist. Anders als diese richtet sich das EvThom nicht an sesshafte Gemeinden, sondern an Einzelne, die sich um der Botschaft willen von allen gesellschaftlichen (und auch religiösen) Bindungen gelöst haben und wohl missionierend umherziehen. Familie, Besitz, Berufstätigkeit, auch religiöse Bräuche werden im EvThom abgelehnt. Es ist ein Evangelium für gesellschaftliche Außenseiter, die keine Kompromisse mit der vorfindlichen Welt eingehen (auch wenn es denkbar ist, dass das EvThom in einem spirituellen Sinne verstanden wurde, ohne die sozial radikalen Konsequenzen).

---

21 EvThom 22:
    (4) Jesus sprach zu ihnen: „Wenn ihr die zwei zu einem macht und wenn ihr das Innere wie das Äußere macht und das Äußere wie das Innere und das Obere wie das Untere, (5) und zwar damit ihr das Männliche und das Weibliche zu einem Einzigen macht, auf dass das Männliche nicht männlich und das Weibliche nicht weiblich sein wird (6) wenn ihr Augen macht anstelle eines Auges und eine Hand anstelle einer Hand und einen Fuß anstelle eines Fußes, eine Gestalt anstelle einer Gestalt, (7) dann werdet ihr eingehen in [das Königreich]."
    Übersetzung: Plisch (2007), 90.
22 Vgl. zur Diskussion Uro (1998).

Der Bruch mit gesellschaftlichen Konventionen schlägt sich auch in der Rolle von Frauen im EvThom nieder. Von fünf namentlich genannten JüngerInnen sind drei Männer und zwei Frauen – wobei zwei der Männer als wenig vorbildliche Figuren dargestellt sind. Eine Jüngerin – sie nennt sich ausdrücklich so – ist wie im EvÄg Salome. Die andere ist Maria (Magdalena), an ihrer Person wird am Ende des EvThom ausdrücklich diskutiert, ob Frauen eine gleichberechtigte Teilhabe gestattet ist. Und an dieser Stelle kommt indirekt auch wieder Askese ins Spiel:

> (1) Simon Petrus sprach zu ihnen: „Maria soll von uns weggehen, denn die Frauen sind des Lebens nicht wert."(2) Jesus sprach: „Siehe, ich werde sie ziehen, auf dass ich sie männlich mache, damit auch sie ein lebendiger, euch gleichender, männlicher Geist werde." (3) (Ich sage euch aber): „Jede Frau, wenn sie sich männlich macht, wird eingehen in das Königreich der Himmel."[23]

Hier wird anders als in EvThom 22 nicht von einer Aufhebung von beiden Geschlechtern gesprochen, sondern es geht am Beispiel von Maria um die Aufhebung von Weiblichkeit. Maria wird zu einem männlich-geistigen Wesen. Die Formulierung ist aus heutiger Sicht gewöhnungsbedürftig, im antiken Kontext aber gut verständlich. Männlich und weiblich sind eben nicht nur biologische Geschlechter, sondern auch metaphorische Beschreibungen, die in einem hierarchischen Verhältnis zueinander stehen. Männlich wird mit geistig assoziiert wie weiblich mit materiell. Maria erreicht einen neuen Zustand, den einer geschlechtslosen (deshalb männlichen) Geistigkeit. Das ist eine Form von Leben, die nicht mehr vom Tod bedroht ist. Und dies ist auch für andere Frauen möglich, deshalb können sie nicht als Geschlecht ausgeschlossen werden.[24]

In diesem neuen Sein erübrigt sich sexuelle Aktivität, in der Gruppe von männlich-geistigen Wesen findet sie nicht statt, auch wenn dies nicht ausdrücklich gefordert ist. Ich denke nicht, dass Frauen sich auf diesen Status wie Maria berufen und gleichzeitig Familie und Kinder haben konnten. Dagegen spricht auch der nicht sesshafte Lebenswandel. Für Maria eröffnen sich wie für Thekla neue Handlungsräume jenseits des normalen weiblichen Lebens. Das Männlich-Machen lässt sich vielleicht auch ganz konkret und wie bei Thekla als Wanderleben in Männerkleidung verstehen.

Ich habe versucht, aus diesen – und noch einigen anderen – Schriften ein theologisches Konzept zu entnehmen, in dem die Rückkehr in einen ursprünglichen Schöpfungszustand im Zentrum steht, durch die ein Ausstieg aus dem Werden und Vergehen sowie dem Tod und der Körperlichkeit samt Geschlechtsdifferenz

---

23 EvThom 114. Übersetzung: Plisch (2007), 260f.
24 Vgl. Petersen (1999), 314f.; dagegen Heininger (2010), 361.

möglich ist. Dies setzt eine mehr oder weniger präsentische Eschatologie – die Angesprochenen sind schon in diesem neuen Zustand – und eine Geringschätzung des Körpers und der Welt bzw. der Schöpfung voraus. Zu den Konsequenzen gehören ein asketischer Lebensstil (der nicht ausdrücklich gefordert wird, aber sich logisch ergibt), ein radikaler Bruch mit gesellschaftlichen Konventionen und eine starke und gleichberechtigte Beteiligung von Frauen. Dieses Muster ist vermutlich deutlich älter als das 2. Jh: Der 1Kor zeigt m.E. klar, dass Paulus sich mit ähnlichen Vorstellungen in der Gemeinde in Korinth auseinandersetzt. Am 1Kor zeigt sich aber auch, dass eine spezifisch gnostische Theologie dabei nicht nötig ist. Auch in diesem Fall gehört Askese aber in ein Modell christlicher Identität, das wesentlich durch Abgrenzung von der Welt und der Gesellschaft bestimmt ist.[25]

Die Apostelakten zeigen dagegen eine andere Sicht des Körpers, so dass Askese vor allem zu seiner Reinhaltung dient. Askese ist in ihnen eine primäre Forderung, der Weg zu einem Ziel (ewiges Leben), nicht die Konsequenz aus einem theologischen Konzept (in dem neues Sein schon erlangt ist). Auch hier zeigen sich sozial radikale Auswirkungen, besonders allerdings bei den Theklaakten. In Bezug auf die Frauenrolle gibt es hier mehr Parallelen zum EvThom und vergleichbaren Schriften als in den anderen Akten.

IV Nichtasketische Schriften – Pastoralbriefe, Philippusevangelium

Es gibt also Askese-affine frühchristliche Schriften, die sich in der Bedeutung der Askese und in der grundsätzlichen Sicht des Körpers unterscheiden, aber sozial und besonders in Bezug auf Frauen radikal sind. Diese Vielfalt findet sich auch auf der anderen Seite:

Die Pastoralbriefe (1/2Tim, Tit) sind vermutlich im 2. Jh im Namen des Paulus auch mit dem Ziel geschrieben worden, eine bestimmte Deutung seiner Briefe durchzusetzen. Sie lehnen Askese ab (1Tim 4,3-5) und fordern insbesondere Frauen ausdrücklich zum Heiraten und Kinderkriegen auf (1Tim 5,9-14). Gleichzeitig wird Frauen das Lehren verboten (1Tim 2,11-15, mit Bezug auf die Schöpfungsordnung!) – und es werden Vorstellungen abgelehnt, nach denen die Auferstehung schon geschehen sei (2Tim 2,18). Es finden sich hier also die gleichen Elemente, nur mit anderem Vorzeichen. In den Pastoralbriefen steht dies im Zusammenhang mit dem

---

25 Vgl. Merklein (1997), 235, 243f.

Versuch der gesellschaftlichen Integration.[26] Auch in den christlichen Gemeinden sollen die Tugenden der Mehrheitsgesellschaft gelten, sie grenzen sich nicht ab, sondern bilden einen positiven und verlässlichen Teil des römischen Staates. Das ist ein Hauptanliegen dieser Briefe. Der Körper wird als Teil der guten Schöpfung positiv gewertet.

Die Pastoralbriefe stehen klar in der Entwicklungslinie, die zur Mehrheitsmeinung in der entstehenden Kirche wird. Es gibt aber auch in ganz anderen Teilen des frühen Christentums eine durchaus ähnliche Tendenz: Die gnostische Gruppe der Valentinianer weicht insbesondere in der Gotteslehre und in der Sicht der Schöpfung, die nicht auf den obersten Gott zurückgeht, deutlich von der späteren Mehrheitsmeinung ab. Innerhalb des Christentums stehen sie wohl eher am Rand. Gesellschaftlich sind sie aber vermutlich besser integriert als viele andere Teile, sie schätzen z. B. philosophische Bildung. Und auch in Bezug auf Askese sind sie, trotz einer negativen Sicht des Körpers als Werk von feindlichen Archonten, jedenfalls nicht extrem. Sexualität ist möglich und Fortpflanzung möglicherweise sogar ein Ziel, auch wenn eine Beherrschung von Begierden wichtig ist und die zentralen Werte geistig sind.[27] Und bei ihnen finden sich weitaus weniger Frauen in starken, leitenden Rollen – das gilt zumindest für das EvPhil.[28]

Eine ähnliche Sicht bietet das ebenfalls gnostische (aber eher sethianische) Apokryphon des Johannes, das im Wesentlichen aus einer Auslegung der Schöpfungsgeschichten in Gen 1-3 besteht. Die Schöpfung findet hier in drei Etappen statt seelisch – geistig – körperlich, letzteres ist ein Versuch den (geistigen) Menschen an die Materie zu fesseln, auch Sexualität dient der Versklavung. Begierde entfernt Adam von seinen himmlischen Ursprüngen. Allerdings ist die Schöpfung der Frau und die Differenzierung der Geschlechter hier positiv.[29] Trotzdem hat diese Schrift keine Indizien für eine radikal asketische Haltung, sie vertritt wohl eher die moderate Kontrolle von Begierden. Dazu gibt weder Hinweise auf neue Frauenrollen und ihre gleichwertige Beteiligung noch Aussagen dagegen.

Es gibt also in frühchristlichen Gruppen Körperabwertung mit radikaler und gemäßigter Askese sowie eine positive Sicht von Körper und Schöpfung mit ra-

---

26 Vgl. Merklein (1997), 255-257.

27 Vgl. als Beispiel das Motiv des „Brautgemachs" und die Brautmetaphorik im EvPhil (EvPhil 67f.70.81f. u. a.). Vgl. dazu Zimmermann (2001), 586; Buckley (1986), 120f; Heimola (2011), 283f.

28 Die Frage der Rolle von Frauen in gnostischen Gruppen ist problematisch, weil Attraktivität für Frauen ein beliebter Topos der Polemik ist. Deshalb hier der Verweis auf das EvPhil, nicht auf Kirchenväter.

29 Vgl. BG 55 (Waldstein/Wisse [1995], 120-122).

dikaler, gemäßigter oder auch ganz ohne Askese. Verschieden sind sie in ihrem Interesse an gesellschaftlicher Integration. Christliche Gruppen, die sich als radikale Außenseiter verstehen, vertreten eher eine strenge Askese (oder führt umgekehrt möglicherweise radikale Askese zu einer Außenseiterposition?). Wo sich christliche Gruppen in die bestehende Gesellschaft integrieren wollen, werden Askese und die Rollen von Frauen reduziert, wie in den Pastoralbriefen, bei Clemens und möglicherweise auch bei manchen gnostischen Gruppen zu sehen war.[30]

V Zusammenfassung

Askese ist eine Praxis, die sich am Körper äußert, aber damit gerade nicht den Körper meint, sondern auf ein geistiges Ziel ausgerichtet ist. Ob es um die Reinheit des Körpers als Vorbedingung der Auferstehung geht oder ob ein neuer Seinszustand den Körper schon transzendiert, überflüssig macht oder Sexualität erübrigt, die körperliche Seite ist ein Nebenschauplatz, der nur dann großen Raum einnimmt, wenn er Weg zum Ziel ist. Askese kann auch einfach die Konsequenz sein, die sich aus einem neuen Sein so zwingend ergibt, dass sie kaum thematisiert werden muss.

Steht diese aus den theologischen Vorstellungen folgende Askese am Anfang der Entwicklung? Zumindest ist eine solche Position schon früh nachweisbar, Mitte des 1. Jh in Korinth. Aber eine einfache Ableitung wird dem komplexen Befund in den Schriften nicht gerecht, ein schon erreichtes neues Sein und der Weg dorthin lassen sich nicht immer klar trennen. Zudem sprechen die Beziehungen zur Praxis der Umwelt und den dort verbreiteten Begründungen gegen den Versuch, eine nur innerchristliche Entwicklung nachzuzeichnen.

Askese in ihrer radikalen Form und mit theologischer Begründung hat ein gesellschaftskritisches Potential. Dies wird besonders an neuen Frauenrollen deutlich, stellt aber auch sonst die „natürliche" Ordnung in Frage. Sie findet sich entsprechend vor allem in christlichen Strömungen, die nicht an der Integration in die Mehrheitsgesellschaft interessiert sind. Das heißt nicht, dass sie theologisch besonders wenig rechtgläubig wären, eine nur gemäßigte Askese und die Neigung zur gesellschaftlichen Integration finden sich auch in häretischen Gruppen wie den Valentianern. Umgekehrt ist radikale Askese nicht zwingend (nicht einmal besonders häufig) mit gnostischen Vorstellungen verbunden. Die Hauptströmung sucht in der Beziehung zur umgebenden Gesellschaft einen Mittelweg mit Eigenständigkeit, aber ohne zu starke Konfrontation. Askese wird in gemäßigter Form übernommen

---

30 Vgl. Merklein (1997), 257f.; Francis (1995), 167-177.

(für die Mehrheit) und in radikaler Form an den Rand gedrängt (als Sonderweg für Einzelne). Das befreiende Potential für Frauen geht dabei weitgehend verloren.

M.E. gibt es eine deutliche Korrelation zwischen radikaler Askese und einer gleichberechtigten Beteiligung von Frauen. Anscheinend brauchen Frauen im antiken Kontext diese Praxis und/oder die theologischen Vorstellungen von der Aufhebung der Geschlechterdifferenz, um neue Rollen zu übernehmen. Vielleicht war auch innerhalb von Gruppen für Frauen die Askese nötiger als für die Männer.

Es bleibt zu fragen, was aus der Askese in späterer Zeit wird. Ich habe den Eindruck, dass sich aus der Vielfalt von Gruppen, die teilweise ja Askese für alle fordern, eine gemäßigte Mehrheitsströmung herausbildet, die radikale Askese an ihren Rändern integrieren kann. Für die dort beteiligten Frauen und Männer ist das gesellschaftskritische Potential, konkret die Trennung und Abgrenzung von der Gesellschaft (z. B. durch Rückzug in die Wüste oder in ein Kloster) weiter aktuell.

Literatur

Bergman / Gribomont (1979): Jan Bergman, Jean Gribomont, Manfred Seitz u. a., Art. „Askese". Theologische Realenzyklopädie. Berlin, New York. Bd. 4:195-259.

Brown (1988): Peter Brown, The Body and Society: Men, Women and Sexual Renunciation in Early Christianity. New York.

Buckley (1986): Jorunn J. Buckley, Female Fault and Fulfillment in Gnosticism. Studies in Religion. Chapel Hill.

Dillon (1995): John M. Dillon, „Rejecting the Body, Refining the Body. Some Remarks on the Development of Platonist Asceticism." In: Vincent L. Wimbush, Richard Valantasis (ed.), Asceticism, Oxford, 80-87.

Drijvers (1999): Han J.W. Drijvers, „Thomasakten." In: Wilhelm Schneemelcher (Hg.), Neutestamentliche Apokryphen in deutscher Übersetzung. 2. Bd.: Apostolisches, Apokalypsen und Verwandtes. Studienausgabe. 6. Aufl. Tübingen, 289-367.

Elm (1994): Susanna Elm, ‚Virgins of God'. The Making of Asceticism in Late Antiquity. Oxford.

Francis (1995): James A. Francis, Subversive Virtue. Asceticism and Authority in the Second-Century Pagan World. University Park.

Heimola (2011): Minna Heimola, Christian Identity in the Gospel of Philip. Publications of the Finnish Exegetical Society 102. Helsinki.

Heininger (2010): Bernhard Heininger, „Jenseits von männlich und weiblich. Das Thomasevangelium im frühchristlichen Diskurs der Geschlechter. Zugleich ein Beitrag zur Geschichte der Jesustradition." In: ders., Die Inkulturation des Christentums. Aufsätze und Studien zum Neuen Testament und seiner Umwelt. Wissenschaftliche Untersuchungen zum Neuen Testament 255. Tübingen, 357-395.

Horn (1997): Christoph Horn, Art. „Askese". Der Neue Pauly. Stuttgart, Weimar, Bd. 2:86f.

Layton (1989): Bentley Layton, Thomas O. Lambdin, Helmut Koester u. a., „The Gospel According to Thomas." In: Bentley Layton (ed.), Nag Hammadi Codex II,2-7. Nag Hammadi Studies 20. Leiden, Vol. 1:38-128.

Markschies (2012): Christoph Markschies, „Das Evangelium nach den Ägyptern." In: ders. und Jens Schröter (hg.), Antike christliche Apokryphen in deutscher Übersetzung. I. Band: Evangelien und Verwandtes. Tübingen, Bd. 1:661-682.

Merklein (1997): Helmut Merklein, „Im Spannungsfeld von Protologie und Eschatologie. Zur kurzen Geschichte der aktiven Beteiligung von Frauen in paulinischen Gemeinden." In: ders., Martin Evang, Michael Wolter (Hg.), Eschatologie und Schöpfung. FS für Erich Gräßer zum siebzigste Geburtstag. Beihefte zur Zeitschrift für die neutestamentliche Wissenschaft 89. Berlin, New York, 231-259.

Petersen (1999): Silke Petersen, „Zerstört die Werke der Weiblichkeit!" Maria Magdalena, Salome und andere Jüngerinnen Jesu in christlich-gnostischen Schriften. Nag Hammadi and Manichaean Studies 48. Leiden, Bosten u. a.

Plisch (2007): Uwe-Karsten Plisch, Das Thomasevangelium. Originaltext mit Kommentar. Stuttgart.

Schneemelcher (1999): Wilhelm Schneemelcher, Rodolphe Kasser, „Paulusakten." In: Wilhelm Schneemelcher (Hg.), Neutestamentliche Apokryphen in deutscher Übersetzung. 2. Bd.: Apostolisches, Apokalypsen und Verwandtes. Studienausgabe. 6. Aufl. Tübingen, 193-243.

Shaw (1998):Teresa M. Shaw, The Burden of the Flesh. Fasting and Sexuality in Early Christianity. Minneapolis.

Uro (1998): Risto Uro, „Is Thomas an Encratite Gospel?" In: ders. (ed.), Thomas at the Crossroads. Essays on the Gospel of Thomas. Studies of the New Testament and Its World. Edinburgh, 140-162.

Waldstein / Wisse (1995): Michael Waldstein, Frederik Wisse (ed.), The Apocryphon of John. Synopsis of Nag Hammadi Codices II,1; III,1; and IV,1; with BG 8502,2. Nag Hammadi and Manichean Studies 33. Leiden.

Wire (1988): Antoinette C. Wire, „The Social Function of Women's Asceticism in the Roman East." In: Karen L. King (ed.), Images of the Feminine in Gnosticism. Studies in Antiquity and Christianity. Philadelphia, 308-323.

Wisse (1988): Frederik Wisse, „Flee Femininity. Antifemininity in Gnostic Texts and the Question of Social Milieu." In: Karen L. King (ed.), Images of the Feminine in Gnosticism. Studies in Antiquity and Christianity. Philadelphia, 297-307.

Zimmermann (2001): Ruben Zimmermann, Geschlechtermetaphorik und Gottesverhältnis. Traditionsgeschichte und Theologie eines Bildfelds in Urchristentum und antiker Umwelt. Wissenschaftliche Untersuchungen zum Neuen Testament, 2. Reihe 122. Tübingen.

Beschneidung in der Hebräischen Bibel und ihre literarische Begründung in Genesis 17

Thomas Römer

I Einführung

Die Beschneidung junger Männer bzw. männlicher Neugeborener ist nach van Gennep ein „rite de passage"[1], der weltweit aber nicht überall praktiziert wird. Die Erklärungen dieses Brauchs, der einen Eingriff in die körperliche Integrität darstellt und eine Transformation des Geschlechtsteils zur Folge hat, sind vielfältig und bedienen sich soziologischer, psychoanalytischer, medizinischer, ritueller und religiöser Argumentationen, wobei die Schwierigkeit oft darin liegt, dass bei den Völkern und Gruppen, die diesen Brauch praktizieren, eine Erklärung desselben nicht immer leicht zu finden ist.

Für die Hebräische Bibel, die insbesondere in Bezug auf den Pentateuch die Grundlage des sich ab der Mitte der Perserzeit entstehenden Judentums bildet und allein die männliche Beschneidung kennt, auf welche ich mich deswegen im Folgenden beschränken werde, findet sich in der Tat ein Text, der eine theologische Erklärung der Beschneidung liefert und diese zum Zeichen des Bundes zwischen Jhwh und Abraham erklärt, nämlich Genesis 17. Allerdings stellt nach Zeugnissen anderer Texte die Beschneidung keineswegs ein Spezifikum Israels dar. Deswegen soll vor der Besprechung von Genesis 17 eine kurze Übersicht über die Verbreitung der Beschneidung im Alten Orient nach Zeugnis der Bibel und anderer Quellen stehen.

---

1 Van Gennep (1909).

II   Die Beschneidung im Umfeld Israels

Die Autoren der biblischen Texte wussten darum, dass die Beschneidung ein verbreiteter Brauch war.[2] So findet sich am Ende des Kapitels 9 des Buches Jeremia eine Androhung des Gerichts Jhwhs gegenüber den beschnittenen Völkern: „Siehe, es kommen Tage, Spruch Jhwhs, da suche ich alle heim, die an der Vorhaut beschnitten sind (מול בערלה): Ägypten und Juda und Edom und die Ammoniter und Moab und alle mit gestutztem Haarrand, die in der Wüste wohnen, denn alle Nationen sind unbeschnitten, und das ganze Haus Israel hat ein unbeschnittenes Herz (V. 24-25)". Jhwh droht hier durch den Propheten ein Gericht über alle beschnittenen Völker an, zu denen neben den Judäern die Ägypter, Edomiter, Ammoniter, Moabiter und die arabischen Stämme, die hier durch ihre spezielle Haartracht charakterisiert werden, gehören.[3] Alle diese Völker werden als „beschnitten an der Vorhaut" bezeichnet, die Kombination ist redundant und einmalig, benutzt aber die beiden *termini technici* für den betreffenden Ritus. Festzuhalten ist hier zunächst, dass der Autor die Beschneidung als einen in der Levante und in Ägypten geübten Brauch betrachtet, wobei interessanterweise Ägypten in der Aufzählung vor Juda zu stehen kommt. Steht hinter dieser Aufreihung die Auffassung Herodots II, 104,1-3, nach dem die Beschneidung zunächst von den Ägyptern praktiziert wurde,[4] und die Phönizier und die in Palästina lebenden Syrer sie von diesen gelernt haben? Für Ägypten ist die Beschneidung in der Tat bereits im 3. Jt. v. u. Z. auf einem Grabrelief aus Saqqara bezeugt,[5] auf welchem man Priester, die junge Männer beschneiden, sieht. Allerdings ist in Ägypten die Beschneidung wohl nie generell praktiziert worden;[6] für den Autor von Jer 9 gehören die Ägypter jedoch eindeutig zu den „Beschnittenen". Den in Jer 9,25 aufgezählten Völkern wird vorgeworfen, dass sie zwar in Bezug auf die Vorhaut beschnitten sind, aber im Herzen unbeschnitten. Hier bereitet sich eine Kritik an dem Ritus der Beschneidung vor, auf den später noch kurz einzugehen ist. Dass die Beschneidung in der Levante weit verbreitet war, bezeugen auch aus dem 3. Jt. v. u. Z. stammende nackte männliche Figuren aus Tell Judaidah in Ana-

---

2   Vgl. zum folgenden auch Blaschke (1998), *passim*.

3   Der Ausdruck „gestutzt an der Schläfe" findet sich in der hebräischen Bibel nur in Jer, noch in 25,23 und 49,32.

4   Herodot spricht von Kolchiern, Ägyptern und Äthiopiern.

5   Das Relief kann konsultiert werden unter http://www.circlist.de/beschneidung.jpg. S. Art. Bauks, Abb. 1 in diesem Band.

6   Quack (2012), 561-65.

tolien, deren Penis eindeutig beschnitten ist.[7] Das bedeutet, dass zumindest bis zur Zeit des babylonischen Exils oder zu Anfang der Perserzeit die Beschneidung ein selbstverständliches Ritual war, das man in Israel und Juda mit anderen Völkern der Levante teilte. Allerdings gab es bereits zu dieser Zeit eine Unterscheidung von Beschnittenen und Unbeschnittenen,[8] wobei die letzteren, die „Vorhäutigen" (ערלים), in Israel und Juda mit negativen Konnotationen besetzt waren. Dies zeigt sich zunächst an zwei Texten aus dem Buch Ezechiel. In Kap. 28 wird dem Fürsten von Tyrus angedroht, dass er den Tod von Unbeschnittenen sterben müsse (V. 10), woraus hervorgeht, dass auch die Phönizier die Beschneidung praktizierten. Eine ähnliche Strategie wird in Ez 32,17-32 auf den Pharao angewandt. In dieser Parodie einer Totenklage wird refrainartig berichtet, dass dieser sich im Totenreich mit Unbeschnittenen niederlegen müsse: „Denn ich habe Schrecken vor mir verbreitet im Land der Lebenden. Und inmitten von Unbeschnittenen wird er niedergelegt, mit den vom Schwert Erschlagenen, der Pharao und seine gesamte Menschenmenge!" (V. 32). Diese Texte legen nahe, dass deren Verfasser die Nicht-Beschnittenen als verachtenswert betrachteten, von denen sich die „zivilisierten" Kulturen, zu denen Judas und Israels Nachbarn, sowie die Ägypter und Phönizier gehörten, abzugrenzen hatten. Das Heilsorakel in Jesaja 52,1 setzt sogar „unbeschnitten" und „unrein" parallel: „Kleide dich mit den Kleidern deiner Herrlichkeit, Jerusalem, heilige Stadt! Denn kein Unbeschnittener und kein Unreiner (ערל וטמא) wird dich jemals mehr betreten" und auch Ez 44,7-9 versteht die Präsenz von Unbeschnittenen im Heiligtum als dessen Entweihung.

Die exemplarischen Unbeschnittenen in der Hebräischen Bibel, die ebenfalls als exemplarische Feinde im Kontext der Entstehung des israelitischen Königtums erscheinen, sind die Philister, für welche die Autoren der Samsongeschichte und des Samuelbuches das Lexem „Vorhäutige" als Identitätsmarkierer benutzen.[9] Die wohl aus der Ägäis stammenden Philister haben sich in vieler Hinsicht schnell in Kanaan assimiliert, jedoch den Brauch der Beschneidung nicht übernommen. So muss der junge David, der Sauls Tochter ehelichen will, seinem künftigen Schwiegervater als Brautpreis hundert philistäische Vorhäute überbringen. David führt diesen Auftrag prompt aus, dem masoretischen Text zufolge verdoppelt er sogar Sauls

---

7 Die Statuen sind unter http://www-news.uchicago.edu/releases/05/050112.oi-objects. shtml abgebildet. Deshalb schlägt Sasson (1966), 473-476, vor, dass die Beschneidung in Syrien entstand und dann nach Ägypten kam.

8 Vgl. dazu auch Olyan (2000), 64-68.

9 Ri 14,3; 15,8; 1 Sam 14,6; 17,26.36; 31,4 (= 1 Chr 10,4); 2 Sam 1,20.

Forderung und überbringt ihm 200 Vorhäute der Philister[10] (1 Sam 18,20-27). Ob hinter dieser Episode nur ein gewisser „ethnic humour" (R. Klein)[11] steht, oder ob sich dahinter reale Praktiken widerspiegeln, ist schwer auszumachen. Immerhin sei auf die Reliefs des Totentempels Ramses III (1186-1155) in Medinet Habu verwiesen, von denen eines die Zählung abgetrennter Penisse getöteter Feinde darstellt.[12] Eine ähnliche Praxis könnte auch hinter 1 Sam 18 stehen, da David die Vorhäute als Beweis bringen soll, dass er hundert Philistern das Leben genommen hat; um eine „Zwangsbeschneidung" handelt es sich hier sicher (noch) nicht.[13]

Abschließend ist zu diesem Teil der Untersuchung festzustellen, dass die Unterscheidung von Beschnittenen und Nichtbeschnitten zunächst keineswegs als eine Spezifizität Israels und Judas verstanden wurde, sondern dass diese Unterscheidung die Völker der Levante und Ägyptens von den aus dem griechischen Bereich stammenden Philistern unterschied, wobei Männer mit Vorhaut als minderwertig bzw. unrein galten. Dass dieselbe Unterscheidung auch in gegensätzlicher Richtung funktioniert, zeigt eine griechische Vase aus dem 6./5. Jh. v .u. Z., welche Herakles als Sieger des mythischen ägyptischen Königs Busiris zeigt.[14] Dabei werden in dieser Darstellung der unbeschnittene Penis Herakles' und der beschnittene Penis des Ägypters kontrastiert, was im Gegensatz zu den biblischen Texten die Superiorität der griechischen Kultur der Nichtbeschneidung darstellt.

III   Die Theologisierung der Beschneidung in der „Priesterschrift"

Aus den obigen Ausführungen ergibt sich, dass die Beschneidung in Israel und Juda in der ersten Hälfte des 1. Jt. v. u. Z. praktiziert wurde, ohne dass sich dafür theologische oder andere Begründungen finden. Im Grunde galt der Ritus wohl als selbstverständlich und wurde nicht weiter kommentiert. So finden sich in biblischen

---

10 Wie oft in Samuel ist dem griechischen Text (LXX[B]), der von hundert Vorhäuten spricht, der Vorzug zu geben.

11 Klein (1983), 190.

12 Das Candeias Sales (2012), 85-116.

13 Solche Praktiken werden von verschiedenen Quellen den Hasmonäern zugeschrieben, vgl. Josephus, *Antiquitates Judaicae* 13:257. Theodotos zitiert von Eusebius, *Praeparatio evangelica*, 9.22.

14 http://fr.wikipedia.org/wiki/Busiris_%28mythologie%29#mediaviewer/File:NAMA_H %C3%A9racl%C3%A8s_%26_Busiris.jpg

Texten, die in die Zeit vor der Zerstörung Jerusalems gehören, weder Vorschriften zur Beschneidung noch der Versuch einer Begründung dieses Rituals. Bisweilen wird der in diesem Symposium von Michaela Bauks besprochene mysteriöse Text in Exodus 4,24-26, der von einer symbolischen Beschneidung Moses durch seine Frau handelt, als archaischer Rest aus vorstaatlicher Zeit interpretiert, aber es handelt sich in diesen Versen wohl doch um einen recht jungen, nicht vor dem 6. Jh. v. u. Z. entstandenen Text.[15]

So ist davon auszugehen, dass die theologische Begründung der Beschneidung erst spät einsetzt, und zwar nach der Zerstörung Jerusalems und seines Tempels im Jahr 587 und der Deportierung der judäischen Eliten nach Babylon. Die Konfrontation mit der mesopotamischen Zivilisation machte aus der Beschneidung ein Differenzierungskriterium für die judäischen Exilierten, da nach Ausweis der uns bekannten Quellen weder Assyrer noch Babylonier noch die Perser die Beschneidung praktizierten. Dies wird auch zumindest teilweise durch den bereits erwähnten Text von Ez 32 bestätigt, der in der Beschreibung der Unterwelt indirekt die Assyrer und direkt die Elamiter, und das mysteriöse Volk Meschech-Tubal, das vielleicht ebenfalls in Mesopotamien zu verorten ist, als Nichtbeschnittene bezeichnet. Der direkte Kontakt mit Babyloniern und Persern provozierte in priesterlichen judäischen Kreisen eine Reflexion über den Sinn der Beschneidung und möglicherweise auch eine Modifizierung der gängigen Praxis. Diese Erneuerungen spiegeln sich in Kapitel 17 des Buches Genesis wider, ein Text, der von Exegeten so gut wie einmütig der sogenannten Priesterschrift (P) zugeschrieben wird. Umstritten ist dabei, ob P zunächst als eine selbstständige Schriftrolle existierte oder von vornherein als Redaktion funktionierte, d. h., dass die Priester ihre Texte in eine Rolle mit älteren Überlieferungen einschrieben.[16] Für unsere Untersuchung ist jedoch die Entscheidung dieser Frage nicht zentral, wenn auch m. E. sich gewichtigere Argumente für die Existenz einer ursprünglich autonomen Priesterschrift finden.[17]

Die priesterlichen Autoren verankern die Beschneidung in Genesis 17 in der Patriarchenzeit, und erklären ihren Ursprung als Zeichen eines von Jhwh mit Abraham in Kraft gesetzten Bundes.

Der Text lautet in deutscher Übersetzung wie folgt:

---

15 Diebner (1984), 119-126; Blum und Blum (1990), 41-54; Römer 1994, 1-12.
16 Zur Diskussion vgl. Shectman und Baden (2009).
17 Römer (2014), 90-93.

1   Als Abram neunundneunzig Jahre alt war, erschien Jhwh dem Abram und sprach
     zu ihm: Ich bin El Schaddai.[18] Wandle vor mir und sei vollkommen.

2   Ich will meinen Bund stiften zwischen mir und dir und dich über alle Massen
     mehren.

3   Da fiel Abram nieder auf sein Angesicht. Und Gott redete mit ihm und sprach:

4   Was dich betrifft, das ist mein Bund mit dir: Du wirst zum Vater einer Vielzahl
     von Völkern werden.

5   Man wird dich nicht mehr Abram nennen, sondern Abraham wird dein Name
     sein, denn zum Vater einer Vielzahl (*'ab hamon*) von Völkern mache ich dich.

6   Ich mache dich über alle Massen fruchtbar und lasse dich zu Völkern werden,
     und Könige werden aus dir hervorgehen.

7   Ich richte meinen Bund auf zwischen mir und dir und deinen Nachkommen
     nach dir, in allen ihren Generationen, als einen dauernden Bund, dass ich dir
     und deinen Nachkommen Gott sei.

8   Und ich gebe dir und deinen Nachkommen das Land, in dem du als Fremder
     weilst, das ganze Land Kanaan, zu dauerndem Besitz, und ich will ihnen Gott
     sein.

9   Und Gott sprach zu Abraham: Was dich betrifft, halte meinen Bund, du und
     deine Nachkommen nach dir, in allen ihren Generationen.

10   Dies ist mein Bund zwischen mir und euch und deinen Nachkommen nach dir,
     den ihr halten sollt: Es soll bei euch beschnitten werden alles, was männlich ist.

11   Das Fleisch eurer Vorhaut sollt ihr beschneiden lassen. Das soll das Zeichen sein
     des Bundes zwischen mir und euch.

12   Im Alter von acht Tagen soll alles bei euch, was männlich ist, beschnitten werden,
     in allen Generationen, der im Haus geborene und der von irgendeinem Fremden
     um Silber gekaufte Sklave, der nicht zu deinen Nachkommen gehört.

13   Es soll auch der in deinem Haus geborene und der von dir um Silber gekaufte
     Sklave sich beschneiden lassen. So soll mein Bund in eurem Fleisch ein dauernder
     Bund sein.

14   Ein männlicher Unbeschnittener aber, der nicht am Fleisch seiner Vorhaut
     beschnitten ist, soll von seinem Volk abgeschnitten werden; meinen Bund hat
     er gebrochen.

15   Und Gott sprach zu Abraham: Was Sarai, deine Frau, betrifft, sollst du sie nicht
     mehr Sarai nennen, sondern Sara soll ihr Name sein.

16   Ich will sie segnen, und auch von ihr will ich dir einen Sohn geben. Ich will
     sie segnen, und sie soll zu Völkern werden. Könige von Völkern sollen von ihr
     werden.

17   Da fiel Abraham nieder auf sein Angesicht und lachte. Er sagte sich: Kann
     einem Hundertjährigen noch ein Sohn geboren werden, und kann Sara, eine
     Neunzigjährige, gebären?

---

18   Nach priesterschriftlicher Auffassung offenbart sich Jhwh unter seinem wahren Namen
     erst dem Mose (vgl. Ex 6,2-3). Für die Patriarchenzeit erscheint er als El Schaddai, „Gott
     der Felder", eine von arabischen Stämmen verehrte Gottheit; vgl. Knauf (1985), 97-105.

18 Und Abraham sprach zur Gottheit: möge Ismael vor dir leben.

19 Gott aber sprach: In der Tat wird Sara, deine Frau, dir einen Sohn gebären, und du sollst ihn Isaak nennen. Und ich werde meinen Bund mit ihm aufrichten als einen dauernden Bund für seine Nachkommen.

20 Was Ismael betrifft erhöre ich dich: Sieh, ich segne ihn und mache ihn fruchtbar und mehre ihn über alle Massen. Zwölf Fürsten wird er zeugen, und ich werde ihn zu einem großen Volk machen.

21 Meinen Bund aber richte ich auf mit Isaak, den Sara dir gebären wird um diese Zeit im nächsten Jahr.

22 Und als er aufgehört hatte, zu ihm zu reden, fuhr Gott auf, weg von Abraham.

23 Da nahm Abraham seinen Sohn Ismael und alle Sklaven, die in seinem Haus geboren und die von ihm um Silber gekauft waren, alles, was männlich war unter den Leuten vom Haus Abrahams, und beschnitt das Fleisch ihrer Vorhaut am selben Tag, wie Gott ihm gesagt hatte.

24 Abraham war neunundneunzig Jahre alt, als er sich am Fleisch seiner Vorhaut beschneiden ließ.

25 Sein Sohn Ismael aber war dreizehn Jahre alt, als er am Fleisch seiner Vorhaut beschnitten wurde.

26 Am selben Tag wurde Abraham beschnitten sowie sein Sohn Ismael.

27 Und alle Männer seines Hauses, die im Haus geborenen und die von Fremden um Silber gekauften Sklaven wurden mit ihm beschnitten.

Der doppelt gerahmte Text (in V. 1 und V. 24-25 durch chronologische Angaben zum Alter Abrahams –99 Jahre– und Ismaels –13 Jahre–; in V. 1 und V. 22 durch das Erscheinen und Hinauffahren Gottes), der hauptsächlich aus Reden besteht, stellt den Höhepunkt der priesterlichen Abrahamerzählung dar. Durch den göttlichen Bundesschluss erhält das Erzelternpaar eine neue Identität, die sich bereits im Namenswechsel (Abram wird Abraham, Sarai wird Sarah) manifestiert, wobei allein der neue Name Abrahams erklärt wird. Dabei ist wahrscheinlich ein Wortspiel mit der Bezeichnung „Vater einer großen Menge" intendiert. In der Tat wird in den folgenden priesterlichen Texten der Patriarchenerzählungen auf die zahlreichen Nachkommen Abrahams abgehoben, die nicht auf Israel beschränkt sind, sondern auch die Nachkommen Ismaels (die arabischen Stämme) und die Nachkommen Esaus (die Edomiter) miteinbeziehen.[19]

In Bezug auf die Entstehungsgeschichte des Textes wurde bisweilen die These aufgestellt, dass die Beschneidungsforderung in V. 9-14 einen sekundären Zusatz zu dem ursprünglichen Bericht darstellt. Grünwaldt[20] bemerkt, dass der Ausdruck „den Bund halten" in V. 9 und 10 nicht typisch für priesterlichen Sprachgebrauch

---

19 De Pury (2000), 163-181.
20 Grünwaldt (1992).

und Theologie ist, da für P der Bund von Gott gestiftet ist. Demzufolge passt auch die Idee eines Bundesbruches von Seiten der Menschen nicht in das priesterliche Welt- und Gottesbild. Dass Menschen den Bund halten müssen, ihn aber auch brechen können, sind Ansichten der sogenannten deuteronomistischen Theologie, die mit den priesterlichen Anschauungen in Konflikt steht. Demnach wäre nach Grünwaldt die gesamte Gottesrede in V. 9-14 aus dem ursprünglichen Text zu entfernen. Dann wird aber nicht verständlich wie Abraham auf die Idee kommt sich und alle männlichen Glieder seines Hauses zu beschneiden. Insofern ist der Vorschlag Wöhrles[21] konsequenter die Anordnung sowie die Ausführung der Beschneidung in Gen 17 insgesamt als sekundär zu erklären. Damit ergibt sich aber das Problem, dass dann im Gegensatz zum ersten priesterlichen Bundesschluss mit Noah in Gen 9 keinerlei Zeichen des mit Abraham gestifteten Bundes existieren würde. Deshalb scheint es logischer das Beschneidungsthema in der ursprünglichen Version von Gen 17 zu belassen.[22] Grünwaldt und Wöhrle bemerken jedoch zu Recht, dass das Thema des Haltens und des Brechens des Bundes sich nicht zu P fügt. Auch die Idee, dass jemand sich nicht beschneiden lassen will, passt nicht in das ausgehende 6. oder frühe 5. Jh., in welche man die Priesterschrift datiert, sondern setzt wohl die hellenistische Zeit voraus. In 1 Makk 1,15 wird in der Tat berichtet, dass nach der Hellenisierung Jerusalems die Beschneidung von Teilen der unter griechischem Einfluss stehenden judäischen Elite verpönt und als Makel angesehen wurde. 1 Makk 1,15 berichtet: „sie stellten künstlich ihre Vorhaut wieder her und fielen vom heiligen Bund ab, sie passten sich den andern Völkern an und gaben sich dazu her, allen Lastern zu frönen."

Weiter kann in Gen 17 auch die in V. 12, 13 und 23 begegnende Ausweitung der Beschneidungsformel auf „den im Haus geborenen und den von Fremden um Silber gekauften Sklaven" einer späteren Überarbeitung zugeschrieben werden. Hier ist wie in dem späten Zusatz zur Passahvorschrift in Exodus 12,43-44[23] („43 Jhwh sprach zu Mose und Aaron: Dies ist die Ordnung für das Passa: Kein Fremder darf davon essen. 44 Jeden mit Silber gekauften Sklaven[24] aber sollst du beschneiden. Dann darf er davon essen.") die Integration von Außenstehenden in die Kultgemeinschaft im Blick und vielleicht bereits die Existenz von Proselyten. Schließlich ist auch V. 21, der den Gottesbund auf den Abrahamssohn Isaak einschränken will, eine Dublette zu V. 19 und unterbricht den Erzählzusammenhang zwischen

---

21 Wöhrle (2011), 71-87.

22 Blenkinsopp (2009), 225-241.

23 Johnstone (2003), 99-114.

24 Dies ist der einzige andere Text der Hebräischen Bibel, der wie in Gen 17,12-13.23 den Ausdruck „mit Silber gekauft" verwendet.

V. 20 und 22ff.[25] Damit dürfte die ursprüngliche priesterliche Erzählung vom Gottesbund mit Abraham und der Ätiologie der Beschneidung *grosso modo* aus den Versen 17,1-8.11.12a.15-20.22.23a*.24-25.26(?) bestehen. Bevor wir uns näher mit der darin vorliegenden Begründung und der Innovation der Beschneidungspraxis beschäftigen, ist es sinnvoll kurz den Platz von Gen 17 im Rahmen der priesterlichen Offenbarungstheologie zu kommentieren.

IV Die Beschneidung zwischen Bluttabu und Offenbarung des Jhwh-Namens

In Gen 17 offenbart sich Jhwh Abraham nicht unter seinem Namen Jhwh,[26] sondern als „El Shaddai". Dies entspricht der im priesterlichen Text Exodus 6,2-3 präzisierten Offenbarungstheologie: „Gott redete mit Mose und sprach zu ihm: Ich bin Jhwh. Dem Abraham, Isaak und Jakob bin ich als El Schaddai erschienen, mit meinem Namen Jhwh aber habe ich mich ihnen nicht kundgetan". Mit dieser Rede und anderen Texten konstruieren die Verfasser der Priesterschrift eine Theorie der göttlichen Offenbarung in drei Stufen. In Bezug auf die gesamte Schöpfung offenbart sich Gott als „Elohim", an Abraham und alle seine Nachkommen als „El Schaddai" und erst an Mose und damit auch an Israel unter seinem wahren Namen Jhwh. Damit vertritt die Priesterschrift einen „inklusiven Monotheismus", demzufolge alle Völker denselben Gott verehren, auch wenn sie seinen wahren Namen nicht kennen und sie im Gegensatz zu Israel ihm nicht den adäquaten Kult darbringen können. Jede der Etappen der Offenbarung Gottes ist mit einem Bundeszeichen und gewissen Praktiken verbunden.

1. Nach der Sintflut richtet Gott einen Bund mit Noah und damit mit der nachsintflutlichen Menschheit auf, durch welchen er sich verpflichtet, nicht noch einmal seine Schöpfung durch eine Flut zu vernichten (Gen 9,1-17). Das Zeichen des Bundes ist der Regenbogen, welcher Gottes ausgestreckten Bogen repräsentiert, mit dem er die chaotischen Wassermassen bekämpft. Verbunden mit diesem

---

25 Dieser Ausdruck wurde eingefügt, als man Gen 17 mit der älteren Geburtsansage in Gen 18 verband, da der seltene „um diese Zeit im nächsten Jahr" das Geburtsorakel in Gen 18 vorbereitet.

26 Bisweilen wird dabei störend empfunden, dass in Gen 17,1a von Jhwh die Rede ist. Dabei handelt es sich jedoch nicht um eine Mitteilung an Abraham sondern um eine Information an die Adressaten, die sicherstellen will, dass der im folgenden erwähnte Gott mit Jhwh zu identifizieren ist.

Bund ist aber auch das Tabu des Blutverzehrs, welches nach P für die gesamte Menschheit gelten sollte: „ Alles, was sich regt und lebt, soll eure Nahrung sein. Wie das grüne Kraut übergebe ich euch alles. Nur das Fleisch, in dem noch Blut und Leben ist, dürft ihr nicht essen. Euer eigenes Blut aber will ich einfordern. Von allen Tieren will ich es einfordern, und von den Menschen untereinander will ich es einfordern" (Gen 9,3-5). Die Tabuisierung des Blutes wird begründet durch die Idee, dass sich im Blut das Leben eines Tieres oder eines Menschen befindet, weshalb sich an das Tabu die Errichtung der Todesstrafe für menschliches Blutvergießen anfügt.

2. Der in Gen 17 errichtete Bund mit Abraham und das mit ihm verbundene Zeichen der Beschneidung gilt für alle Nachkommen Abrahams, was die priesterlichen Autoren vor die Frage stellt, wie zwischen Ismael und Isaak, durch welchen die „Israel-Linie" weitergeführt wird, zu differenzieren sei. Sie beantworteten diese dadurch, dass sie in V. 2-6 von einem Bund mit Abraham sprechen, in V. 7 von einem Bund mit Abraham und seinen Nachkommen, und danach die dort jeweils gebrauchten Lexeme auf Isaak und Ismael anwenden wie folgender Überblick zeigt:

- *Abraham* in V. 2-6: geben (*natan*), vermehren (*rabah*), fruchtbar sein (*rabah*), Vater einer großen Menge (*'ab hamon*).
- *Abraham und seine Nachkommen in V. 7*: AUFRICHTEN, DAUERNDER BUND (BERÎT ʿOLAM).
- *Isaak in V. 19*: AUFRICHTEN, DAUERNDER BUND (BERÎT ʿOLAM).
- *Ismael in V. 20*: vermehren (*rabah*), fruchtbar sein (*parah*), geben (*natan*), ein großes Volk (*goy gadol*).

Mit dieser Wortwahl wird suggeriert, dass die Reihe der Nachkommen Abrahams, die zu Israel führt, mit Isaak weitergeführt wird, wohingegen Ismael mit Abraham parallelisiert wird und damit angedeutet wird, dass er wie sein Vater der Gründer eines großen Volkes wird, nämlich der gesamten arabischen Stämme. Allerdings bedeutet das, wie sich noch zeigen wird, keinesfalls den Ausschluss Ismaels aus dem Bundeszeichen der Beschneidung, allerdings wird auch in dieser Hinsicht zwischen ihm und Isaak differenziert.

3. Die dritte Stufe der Offenbarung, die sich auf die Israellinie beschränkt, wird durch die Offenbarung des Jhwh-Namens an Mose in Ex 6 eingeleitet. Durch Moses erhält Israel dann am Sinai Kenntnis aller Gebote und Rituale, welche die Spezifizität Israels ausmachen. Jedoch umfasst die Beschneidung, wiewohl sie ein Identitätsmarkierer ist, einen weiteren Kreis.

V   Die Beschneidung am achten Tag

Die Theologisierung der Beschneidung in Gen 17 ist, wie bereits ausgeführt, als eine Reaktion auf ein neues Umfeld verständlich, in welchem dieses Ritual nicht praktiziert wird. Soziologisch kann eine solche Reaktion leicht als Aufarbeitung einer neuen Minderheitssituation erklärt werden, wofür es auch heute verschiedene Parallelen gibt. Aber scheinbar haben sich die priesterlichen Autoren nicht damit begnügt der Beschneidungspraxis eine Ätiologie zu geben. Sie haben wohl auch aus einem Pubertätsritus einen Ritus für männliche Neugeborene gemacht. Leider besitzen wir so gut wie keine Informationen über das Alter, in welchem die Beschneidung in der Levante durchgeführt wurde. Die ägyptischen Quellen belegen, dass diese in der Regel in der Pubertät praktiziert wurde. Dass in Gen 17 ein Wechsel vorgenommen wird, zeigt auch die Bemerkung, dass Ismael 13 Jahre bei seiner Beschneidung war, was das normale Alter für einen Ritus, der den Eintritt in das Mannesalter markiert, darstellt. Die 99 Jahre Abrahams bei seiner Beschneidung können natürlich nicht in diesem Sinne erklärt werden, sondern sind durch das chronologische Gerüst der Priesterschrift zu verstehen, nach welchem Abraham bei der Geburt Isaaks 100 Jahre alt ist. Die priesterlichen Autoren wollten suggerieren, dass Abraham bei der Zeugung Isaaks bereits beschnitten war.[27]

Die Verlegung des Beschneidungsalter auf den achten Tag verändert auch die Funktion des Rituals; bei einem Neugeborenen kann die Beschneidung natürlich nicht mehr den Eintritt in das Mannesalter symbolisieren, allerdings kann auch hier die Beschneidung apotropäischen Charakter besitzen (was auch in Ex 4,24-26 suggeriert wird). Vielleicht reflektiert aber Gen 17 auch einen Zusammenhang zwischen Beschneidung und Fruchtbarkeit, da die erste Gottesrede an Abraham die Zusage von Mehrung und Fruchtbarkeit enthält. Oder soll das Ritual, wie die Taufe, bereits in Gen 17 die Aufnahme eines Neugeborenen in eine religiöse Gemeinschaft symbolisieren, wie es das Judentum heute versteht?

Die Festsetzung des achten Tags für die Beschneidung kann durch den mesopotamischen Kontext bedingt sein, da sie dort als eine Glückszahl galt. Im biblischen Kontext kann die Zahl „8" auch mit der Schöpfungsgeschichte der Priesterschrift in Gen 1,1-2,3 in Zusammenhang gebracht werden. Vielleicht hat man aufgrund der Schöpfung in sieben Tagen die erste Woche des Neugeborenen als tabu oder

---

27 Jedoch haben sie sich wohl kaum um die Frage gekümmert wie Abraham beschnitten wurde. Nach der griechischen Übersetzung und den Targumim beschnitt Abraham sich selbst. Der *Midrasch Tanchuma* A (5. Jh) interpretiert die Nif'al-Form als Passiv und berichtet, dass ein Skorpion Abrahams Vorhaut abgeschnitten habe, wohingegen *Tanchuma B* Gott eingreifen lässt, der Abraham bei der Beschneidung unterstützt.

als notwendige Wartefrist angesehen, aufgrund der hohen Sterblichkeitsrate von neugeborenen Säuglingen. Schließlich ist die Zahl 8 auch in der Sintflutgeschichte positiv belegt, da die geretteten Menschen, Noahs Familie, acht an der Zahl sind.

Vielleicht ist die Zahl 8 aber auch im Zusammenhang von Quarantäne zu verstehen, da in Lev 15 für Menstruation, verschiedene Ausflüsse und sexuelle Unreinheiten eine siebentägige Isolierung vorgeschrieben wird. Dieses Verständnis findet sich ebenfalls in dem priesterlichen Text Lev 12,1-3[28], der die Beschneidung in die Sinaioffenbarung eingliedert: „2b Wenn eine Frau Mutter wird und einen Knaben gebärt, ist sie sieben Tage lang unrein. Wie in den Tagen ihrer monatlichen Blutung ist sie unrein. 3 Und am achten Tag soll seine Vorhaut beschnitten werden".

Mit der Verlegung der Beschneidung auf den achten Tag hat die Priesterschrift vielleicht die ursprüngliche Bedeutung der Beschneidung verdeckt, indem sie diese nun zu einem exkludierenden Unterscheidungsmerkmal gegenüber der mesopotamischen Umwelt stilisiert und sich ebenso von den in der Levante und in Ägypten die Beschneidung praktizierenden Völkern durch eine Verschiebung des Alters singularisiert.

VI   Die Beschneidung des Herzens

Verschiedene Texte der hebräischen Bibel sprechen von Beschneidung in einem metaphorischen Sinn.[29] In der priesterlichen Berufungserzählung des Mose verweigert sich dieser zunächst dem göttlichen Auftrag mit dem Einwand, dass die Israeliten nicht auf ihn hören würden, da er „unbeschnittene Lippen" (ערל שפתים) habe (6,12 vgl. auch 6,30). Im Kontext dieser Szene zielt die Metapher vielleicht darauf ab, dass Mose sich unwürdig oder unfähig fühlt seinen Auftrag durchzuführen. Wenn, wie bisweilen postuliert wird, der Ausdruck besagen will, dass Mose ungeschickt im Reden sei,[30] würde das bedeuten, dass für priesterliches Denken die Vorhaut ein Hindernis darstellt, das beseitigt werden muss, sei es im konkreten wie im übertragenen Sinn.

---

28 Wenn unsere Annahme richtig ist, dass der achte Tag eine von P eingebrachte Neuerung ist, lässt sich die von Nihan (2007), 281f. u. a. vertretene Idee, wonach 12,1-7 ein älteres P vorliegendes Ritual darstellt nicht halten; es sei denn man vertritt die Lösung in V. 3 einen Zusatz von P zu sehen.

29 Vgl. zum folgenden auch Bernat (2009), 83ff.

30 So die meisten Kommentare zum Buch Exodus.

In Lev 19 wird vorgeschrieben neu gepflanzte Obstbäume als Vorhäute zu behandeln: „Und wenn ihr in das Land kommt und Bäume mit essbaren Früchten pflanzt, sollt ihr ihre Früchte wie eine Vorhaut behandeln (וערלתם ערלתו את־פריו). Drei Jahre sollen sie euch als unbeschnitten gelten, sie dürfen nicht gegessen werden. Und im vierten Jahr sollen alle ihre Früchte als Festgabe Jhwh geweiht werden. Im fünften Jahr aber dürft ihr ihre Früchte essen, so werden sie euch weiter ihren Ertrag geben. Ich bin Jhwh, euer Gott." (19,23-25). In diesem Bild werden die Früchte mit einer Vorhaut verglichen, der Baum mit einem Penis.[31] Die Beschneidung des Baums bedeutet, dass er zunächst für Jhwh und dann für die Israeliten brauchbar wird. Wie ein unbeschnittener Neugeborener noch nicht in die Jhwhgemeinde integriert ist, so werden auch die Früchte eines neu gepflanzten Baumes in den ersten drei Jahren als unbenutzbar angesehen, bevor sie Jhwh geweiht werden können, und dann den Israeliten zum Gebrauch frei stehen. Die in Ex 6 und Lev 19 verwendeten Metaphern lassen sich gut mit priesterlichem Denken vereinbaren. Wie steht es aber mit Texten, die von der „Beschneidung des Herzens" sprechen?

In Dtn 10,16 werden die Adressaten der Moserede aufgefordert ihre Herzen zu beschneiden: „So beschneidet eure Herzen, und seid fortan nicht mehr widerspenstig"; was in diesem Kontext bedeutet sich „frei" zu machen, um Jhwhs Gebote halten zu können. Ähnlich sagt Dtn 30,6 den Hörern zu, dass Jhwh selbst ihre Herzen beschneiden wird, damit sie sich wieder zu ihm kehren und er sie aus dem Exil in ihr Land zurückbringen kann: „Jhwh, dein Gott, wird dein Herz und das Herz deiner Nachkommen beschneiden, so dass du Jhwh, deinen Gott, liebst von ganzem Herzen und von ganzer Seele, um deines Lebens willen". Beide Passagen gehören zu den jüngsten Texten des Buches Deuteronomium[32], die bereits die Priesterschrift kennen und auf diese reagieren. Von daher kann man sich fragen, ob die Rede von der Beschneidung des Herzens nicht eine leise Kritik an der rituellen Konzeption der priesterlichen Texte enthält, welche ohne die Beschneidung direkt in Frage zu stellen, gleichzeitig eine „spirituelle" Beschneidung fordert.[33]

---

31 Dieses Bild kann durch konkrete Assoziationen hervorgerufen sein, da die an einem Baum hängenden Früchte mit der am Penis hängenden Vorhaut assoziiert werden können.

32 Otto (2012), 762.

33 Eine vergleichbare Idee findet sich am Ende des sog. Heiligkeitsgesetzes in Lev 26, welches priesterliches und dtr Denken zu kombinieren sucht: in V. 41 geht es darum, dass sich das unbeschnittene Herz der Exilierten demütigen muss. Auch Jer 4,4, womöglich ein später Einschub in die Orakelsammlung Jer 2-6, spricht von der Beschneidung des Herzens: „Beschneidet euch für Jhwh und entfernt die Vorhaut eures Herzens, Mann aus Juda und ihr Bewohner Jerusalems, damit mein Zorn nicht ausbricht wie Feuer und brennt".

Diese wurde in der rabbinischen Tradition in die Beschneidungspraxis integriert, wobei jedoch dem Ritual immer und bis heute der Vorrang gegeben wurde.

VII   Kurze Zusammenfassung

In der ersten Hälfte des ersten Jahrtausends v. u. Z. haben Israel und Juda die Beschneidung wie ihre Nachbarn, wohl im Jünglingsalter praktiziert, ohne eine Theorie für diese Praxis schriftlich niederzulegen. Beschnittenheit war ein Distinktivkriterium gegenüber den Philistern, deren Nichtbeschnittenheit als ein Makel verstanden wurde. Erst im Kontext des babylonischen Exils wurde die Beschnittenheit der Judäer inmitten der neuen babylonischen Umwelt zu einem Unterschied, welchen man theologisch zu erklären und zu legitimieren versuchte. Die Beschneidung wurde nun zu einem Bundeszeichen, ohne dass ältere Konnotationen wie Fruchtbarkeit, Schutz, Initiation o. ä. völlig verschwanden. Durch die Verlegung der Beschneidung auf den achten Tag nach der Geburt markierte man nun auch einen Unterschied zu den Ägyptern, den arabischen Völkern und anderen Nachbarn der Levante. Dass man Abraham zum Begründer der Beschneidung machte, erlaubte auch andere Völker in die Beschneidung zu integrieren und doch gleichzeitig die Partikularität der Israellinie zu betonen.

Literatur

Bernat (2009): David A. Bernat, Sign of the Covenant: Circumcision in the Priestly Tradition (AIL 3). Atlanta.

Blaschke (1998): Andreas Blaschke, Beschneidung. Zeugnisse der Bibel und verwandter Texte (TANZ 28). Tübingen – Basel.

Blenkinsopp (2009): Joseph Blenkinsopp, "Abraham as Paradigm in the Priestly History in Genesis". In JBL 128, 225-241.

Blum und Blum (1990): Ruth Blum und Erhard Blum, „Zippora und ihr *ḥtn dmym*". In Erhard Blum et al. (Hg.), Die Hebräische Bibel und ihre zweifache Nachgeschichte (FS R.Rendtorff). Neukirchen-Vluyn, 41-54.

Das Candeias Sales (2012): José das Candeias Sales, "The Smiting of the Enemies Scenes in the Mortuary Temple of Ramses III at Medinet Habu". In Oriental Studies 1, 85-116.

De Pury (2000): Albert de Pury, "Abraham: The Priestly Writer's 'Ecumenical' Ancestor". In Steven L. McKenzie und Thomas Römer (Hg.), Rethinking the Foundations. Historiography in the Ancient World and in the Bible. Essays in Honour of John Van Seters (BZAW 294). Berlin – New York, 163-181.

Diebner (1984): Bernd J. Diebner, „Ein Blutsverwandter der Beschneidung. Überlegungen zu Ex 4,24-26". In DBAT 18, 119-126.

Grünwaldt (1992): Klaus Grünwaldt, Exil und Identität. Beschneidung, Passa und Sabbat in der Priesterschrift (BBB 85). Frankfurt/M.

Johnstone (2003): William Johnstone, "The Revision of Festivals in Exodus 1-24 in the Persian Period and the Preservation of Jewish Identity in the Diaspora". In Rainer Albertz und Bob Becking (Hg.), Yahwism after Exile. Perspectives on Israelite Religion in the Persian Era (Studies in Theology and Religion 5). Assen, 99-114.

Klein (1983): Ralph W. Klein, 1 Samuel (WBC). Waco/Texas.

Knauf (1985): Ernst A. Knauf, „El Šaddai – der Gott Abrahams ?". In BZ NF 29, 97-105.

Nihan (2007): Christophe Nihan, From Priestly Torah to Pentateuch: A Study in the Composition of the Book of Leviticus (FAT II/25). Tübingen.

Olyan (2000): Saul M. Olyan, Rites and Rank : Hierarchy in Biblical Representations of Cult. Princeton.

Otto (2012): Eckart Otto, Deuteronomium 1-11 (HK.AT), Freiburg i.B. et al.

Quack (2012): Joachim Friedrich Quack, „Zur Beschneidung im Alten Ägypten". In Angelika Berlejung, Jan Dietrich und Joachim Friedrich Quack (Hg.), Menschenbilder und Körperkonzepte im Alten Israel, in Ägypten und im Alten Orient (Orientalische Religionen in der Antike 9). Tübingen, 561-651.

Römer (1994): Thomas Römer, „De l'archaïque au subversif: le cas d'Exode 4/24-26". In ETR 69, 1-12.

Römer (2014): Thomas Römer, „Der Pentateuch". In Walter Dietrich, Hans-Peter Mathys, Thomas Römer und Rudolf Smend, Die Entstehung des Alten Testaments (Theologische Wissenschaft 1). Stuttgart, 52-166.

Sasson (1965): Sasson, Jack, "Circumcision in the Ancient Near East". In JBL 85, 473-476.

Shectman und Baden (2009): Sarah Shectman und Joel S. Baden (Hg.), The Strata of the Priestly Writings. Contemporary Debate and Future Directions (AThANT 95). Zürich.

Van Gennep (1909): Arnold van Gennep, Les rites de passage : étude systématique des rites (1909). Paris (Nachdruck 2011).

Wöhrle (2011): Jakob Wöhrle, "The Integrative Function of the Law of Circumcision". In Reinhard Achenbach et al. (Hg.), Legal Distinctions Regarding Foreigners in the Hebrew Bible and the Ancient Near East (BZAR 16). Wiesbaden, 71-87.

Beschneidung zwischen Identitätsmarkierung und substituierter Opferhandlung: Kulturelle Deutungen eines schwierigen Ritualtexts (Exodus 4,24-26)

Michaela Bauks

I   Ex 4,24-26 in der kulturwissenschaftlichen Diskussion

Die religiöse Beschneidung von Jungen ist wahrscheinlich eines der bekanntesten und zuletzt in der deutschen Öffentlichkeit auch wieder sehr diskutierten Körperrituale. Die religiösen Gemeinschaften verstehen das Ritual als in den Körper eingeschriebene Identität. In kulturwissenschaftlicher Perspektive ist dieses Körperritual indes nur selten in Bezug auf Identitätsstiftung hin reflektiert worden. Einschlägig wäre hier Mary Douglas[1] zu nennen mit ihrer Definition des Körpers als eines „natürlichen Symbols": Sie begreift den Körper einerseits als den aller Kultur vorgängigen Ausdruck des Menschseins, der andererseits – meist unbewusst und universell – immer auch soziale Strukturen und Situationen *symbolisch* zum Ausdruck bringt. Der physische Körper wird somit zu einer Einschreibfläche gesellschaftlicher Strukturen.[2]

> Der Körper als soziales Gebilde steuert die Art und Weise, wie der Körper als physisches Gebilde wahrgenommen wird; und andererseits wird in der durch soziale Kategorien modifizierten physischen Wahrnehmung des Körpers eine bestimmte Gesellschaftsauffassung manifest. Zwischen dem sozialen und dem physischen Körpererlebnis findet ein ständiger Austausch von Bedeutungsgehalten statt, bei dem sich die Kategorien beider wechselseitig stärken. In den ihm eigenen Formen der Ruhe und Bewegung kommt der soziale Druck auf mannigfaltige Weise zum Ausdruck. Die Sorgfalt, die auf seine Pflege verwendet wird, ... kurz all die kulturell geprägten Kategorien, die die Wahrnehmung des Körpers determinieren, müssen den Kategorien, in denen die

1   Douglas (1986), 104ff.

2   Vgl. Gugutzer (2004), 83: Denn im „symbolischen Ausdrucksverhalten des Körpers zeigt sich die Verflochtenheit von Natur und Kultur". S. auch Ackermann, in diesem Band.

Gesellschaft wahrgenommen wird, eng korrespondieren, weil und insofern auch diese sich aus den kulturell verbreiteten Körpervorstellungen ableiten.[3]

Douglas wählt als historisches Beispiel die Konflikte in der jüdischen Gesellschaft angesichts hellenistischer Maßnahmen im 2. Jh. v. Chr., welche die kulturellen Unterschiede der verschiedenen Bevölkerungsgruppen in Palästina unter anderem durch ein Beschneidungsverbot nivellieren wollten. Sie führt aus, wie verstärkt Angst vor körperlicher Verunreinigung aufkommt, als der seleukidische König Antiochus IV. Epiphanes (175-164 v. Chr.) neben dem Genuss von Schweinefleisch (vgl. Jos.Bell. I,34) auch ein Beschneidungsverbot (vgl. 2Makk 6,6.10) erzwingen will. Der Ungehorsam gegen die auferlegten Maßnahmen entwickelt sich zu einem entscheidenden Symbol der Gruppenloyalität und -identität.[4]

Bereits R. Hendel wies darauf hin, dass es für die Rekonstruktion historischer Anthropologie fruchtbar sei, M. Douglas' Voraussetzung des Zusammenspiels von implizitem Wissen (*implicite knowledge*), kultureller Befangenheit (*cultural bias*) und Denkstilen (*thought styles*) zu rezipieren und auszuwerten.[5] Es geht dabei einerseits um die Verdeutlichung, wie sehr Formen sozialen Lebens und moralischen Urteils bezüglich der Körper- und Personwahrnehmung verwoben sind. Andererseits ist die vorauszusetzende Spannung von universalen und partikularen Körpervorstellungen im Blick.

Einen etwas kritischeren Blick wirft der britische Soziologe Bryan S. Turner[6] auf das Verhältnis von Körper und Ritual. Er beschreibt, dass in traditionellen Gesellschaften die Geburt allein nicht ausreicht, um in das Kollektiv aufgenommen zu werden, und wie Beschneidung (neben Taufe oder Scarifizierung) zu einer Aufnahme*bedingung* werden kann; denn der Zutritt von Natur zu Kultur erfolgt durch Rituale, die vor allem der sozialen Inklusion dienen.[7] Der Transfer geschieht über das Waschen, Brennen oder Schneiden und geht häufig einher mit einer persönlichen Namengebung, die die soziale Zugehörigkeit schließlich definiert.

---

3   Douglas (1986), 99.

4   Douglas (1986), 60f. Letzten Forschungsergebnissen zufolge, die das Gewicht auf ein politisch-ethnographisches Interesse an Riten legten, hat die Beschneidung erst nach den Bar-Kochba-Aufständen 135 n. Chr. und dem Beginn der rabbinischen Zeit besondere religiöse Motivierung erfahren (vgl. Eckhardt [2012], 2-5 u.ö.).

5   Hendel (2008a), art. 8; Hendel (2008b), 3-15.

6   Turner (2008), 103.

7   Turner (2008), 173: „one has to be transferred from nature to culture by rituals of social inclusion".

In einer anderen Weise kritisch und den neurotischen Charakter von Körperriten betonend hat sich der Wiener Psychoanalytiker und Arzt Sigmund Freud dem Thema gewidmet. Er geht in seinen Schriften mehrfach auf den Ritus der Beschneidung ein:

So äußert er z. B., dass es trotz einer generellen Blutscheu nicht gelungen sei „Gebräuche wie die Beschneidung der Knaben und die noch grausamere der Mädchen … zu unterdrücken".[8] Beschneidung erzeuge bei Kindern Kastrationsangst in Form einer infantilen Neurose, die besonders unter Juden verbreitet sei.[9] Er situiert den Ritus historisch folgendermaßen:

> Die in der Urzeit und bei primitiven Völkern so häufige Beschneidung gehört dem Zeitpunkt der Männerweihe an, wo sie ihre Bedeutung finden muß, und ist erst sekundär in frühere Lebenszeiten zurückgeschoben worden. Es ist überaus interessant, daß die Beschneidung bei den Primitiven mit Haarabschneiden und Zahnausschlagen kombiniert oder durch sie ersetzt ist, und daß unsere Kinder, die von diesem Sachverhalt nichts wissen können, in ihren Angstreaktionen diese beiden Operationen wirklich wie Äquivalente der Kastration behandeln.[10]

Insbesondere in „Der Mann Moses und die monotheistische Religion" (1938) äußert er sich zu der ägyptischen Herkunft des Beschneidungsritus, den der Ägypter und Echnaton-Anhänger Mose nach seiner Flucht in den am ägyptischen Vorbild angelehnten JHWH-Kult importiert habe.[11] Freud geht davon aus, dass die biblischen Texte diese Herkunft verschleiern wollten, indem sie den Ritus narrativ in die Zeit Abrahams vordatierten (s. den Beitrag von T. Römer). Für Freud war Mose, der als Ägypter selbst beschnitten war, stolz darauf und maß dem Ritus ein hohes Ansehen bei.

> Die Juden, mit denen er [Mose] das Vaterland verließ, sollten ihm ein besserer Ersatz für die Ägypter sein, die er im Land zurückließ [und die vom Echnaton-Kult abtrünnig geworden waren]. … Ein ‚geheiligtes Volk' wollte er aus ihnen machen […], und als Zeichen solcher Weihe führte er auch bei ihnen die Sitte ein, die sie den Ägyptern mindestens gleichstellte. Auch konnte es ihm nur willkommen sein, wenn sie durch ein solches Zeichen isoliert und von der Vermischung mit den Fremdvölkern abgehalten wurden.[12]

Freuds falsche und historistisch anmutende Prämisse, dass der Aton-Jünger Mose Israel den ägyptischen Monotheismus gebracht habe, der Monotheismus also

---

8 Freud (1999a), 166.
9 Freud (1999b), 119f.
10 Freud (1999c), 184 mit Anm. 1.
11 Freud (1950), 124f.
12 Freud (1950) , 128f.

eine ägyptische „Erfindung" sei, ist dem Wissensstand seiner Zeit geschuldet und braucht uns nicht weiter zu interessieren.[13] Religionsgeschichtlich scheint mir aber die Einsicht wichtig, dass das Alte Israel zu Beginn des 20. Jahrhunderts nicht mehr wie ein hermetisch abgeriegelter Kulturraum, sondern in vielerlei Form (Recht, Verwaltung, Kult und Brauchtum) als von den umliegenden Großmächten beeinflusst verstanden wurde.

Dass die Beschneidung in Ägypten bereits seit dem 4. Jahrtausend[14] weit verbreitet war, lässt sich sowohl textlich als auch ikonographisch und archäologisch (vgl. insbesondere Mumienfunde) in einer Reihe von Dokumenten belegen.[15]

Altägyptische Beschneidungsszene. Relief aus dem Grab des Anchmahor, Sakkara; 6. Dynastie, 2350-2000 v.Chr. (© Deutsche Bibelgesellschaft, Stuttgart)

13 Er rekurriert auf Arbeiten der alttestamentlichen Kollegen E. Meyer (1906) und E. Sellin (1992), die in Anlehnung an die Forschungsperspektiven der Religionsgeschichtlichen Schule die Entstehung des israelitischen Monotheismus untersuchten. Vgl. dazu auch Van Ruiten/Vandermeersch (1994), 269-291, bes. 277ff.

14 J. Sasson weist darauf hin, dass in Kanaan im 'Amuq-Tal die hebräische Art der Beschneidung seit dem 3. Jt. anhand von Figurinen belegt werden kann in der Art, dass die Figur nicht nur einen Einschnitt in der Vorhaut, sondern die vollständige Entfernung erkennen lässt (Sasson (1966), 473-476); kritisch äußerst sich Quack (2012a), 624f. zu der Folgerung, dass die Beschneidung wegen dieser Funde kanaanäischen und nicht – wie seit Herodot angenommen – ägyptisch-äthiopischen Ursprungs sei; wenigstens reichen die ägyptischen Belege bis in das vierte Jahrtausend. V. Wagner hält die Beschneidung sogar für einen erst in nachexilischer Zeit gängigen Brauch in Palästina (Wagner [2010]). Dem widersprechen aber die bei Olyan (2000) 65ff. aufgeführten Belege wie Gen 34 und 1 Kön 18, die „Unbeschnittensein" geradezu als Emblem für Fremdheit erkennen lassen und auch Römer in diesem Band.

15 Quack (2012a), bes. 562-568 geht sehr genau die diversen Mumienfunde durch, um zu dem Schluss zu kommen, dass trotz einer Reihe von Unwägbarkeiten davon auszugehen ist, dass bis in die Spätzeit im alten Ägypten in der Regel Jungen zwischen 12 und 14 Jahren beschnitten worden sind. Antike Textquellen lassen erkennen, dass das Alter vorverlagert werden konnte (s. Anm. 20); vgl. auch die Kurzfassung Quack (2012b) 17-22.

J.F. Quack interpretiert den Vorgang in der linken Bildhälfte als Schamhaarrasur, die der Szene in der rechten Hälfte, der Beschneidung eines Totenpriesters, vorausgeht.[16] Im Grabkontext geht es um die Darstellung notwendiger kultischer Reinheitsvorschriften, um den Toten versorgen zu können.

Die ägyptischen Belege für die Beschneidung bezeugen einen Eingriff, der vornehmlich in den Bereich der Alltagserfahrung fällt. Ausführlichere Ritualangaben finden sich erst in griechisch-römischer Zeit.[17] Ein deutlicher Rückgang der Beschneidung vollzog sich in Ägypten wohl erst in der römischen Kaiserzeit unter Hadrian (76-138).[18] Als dieser die Beschneidung im Römischen Reich verbot,[19] wurden den ägyptischen Provinzen auf Antrag hin Ausnahmeregelungen für ihre Priester eingeräumt, damit diese die gebotenen kultischen Reinheitsvoraussetzungen erfüllen konnten. In diesem Kontext finden sich auch Angaben zum Alter der zu beschneidenden Jungen[20], die vom 1. bis zum 12. Lebensjahr reichen, was davon ausgehen lässt, dass der Ritus vorzugsweise im Kindes- bzw. Jugendalter vorgenommen worden ist, wobei über die Motive nur Mutmaßungen angestellt werden können, da sie nirgends expliziert werden.[21]

Für Freud stand fest, dass Israel die Beschneidung aus Ägypten mitgebracht hat:

> Mit den Rückkehrern hatte man es nicht so leicht, sie ließen sich den Auszug aus Ägypten, den Mann Moses und die Beschneidung nicht rauben. Den Mann Moses erledigte man [...] Die Beschneidung, das gravierendste Anzeichen der Abhängigkeit von Ägypten, mußte man beibehalten, aber man versäumte die Versuche nicht, diese Sitte aller Evidenz zum Trotz von Ägypten abzulösen. Nur als absichtlichen Widerspruch gegen den verräterischen Sachverhalt kann man die rätselhafte, unver-

---

16 Quack (2012a), 569-73 in kritischer Aufnahme von Grunert (2002), 137-151, die in der Darstellung keine Beschneidungsszene, sondern eine Reinigungsszene sieht. Eine zweite deutlich erkennbare Beschneidungsdarstellung liegt aus dem Mut-Bezirk in Karnak zum Thema der Geburt des Gott-Königs vor (dazu Quack [2012a], 573f.).

17 Jördens (2010), 326-330; der einzige ältere Ritualtext aus pEbner 732,88,10-12 (Hanning/Witthuhn (2010), 217-273, hier 265) handelt nicht von Beschneidung, sondern vom Herausoperieren eines Dorns (s. Westendorf [2010], 209 mit Anm. 34); vgl. dazu weiterhin Quack (2012a), 598f., der Jördens Beispiele für weibliche Beschneidung (UPZ I 2 und pOxy. LXVI 4542 und 4543) entkräftet. S. zur weiblichen Beschneidung auch Feucht (2003).

18 Vgl. Quack (2012a), 597-599.

19 Ob diese Maßnahme den Bar-Kochba-Aufstand 132 auslöste, ist umstritten; vgl Kuhlmann (2002), 133-136; Schafer (1981), 38-50 und Oppenheimer (2003), 55-69 und Abusch (2003), 71-91 und zuletzt Bazzana (2010), 85-109.

20 Vgl. dazu Quack (2012a), 598.

21 Vgl. dazu ausführlich Quack (2012a), 610-622 (Rite de passage, generelle Volkssitte, Reinheitsgebot). S. auch den Beitrag von Meyer in diesem Band.

ständlich stilisierte Stelle in Exodus auffassen, daß Jahve einst dem Moses gezürnt, weil er die Beschneidung vernachlässigt hatte, und daß sein midianitisches Weib durch schleunige Ausführung der Operation sein Leben gerettet! […] Und hier findet sich der Anlaß zu einem entscheidenden Streich gegen die ägyptische Herkunft der Beschneidungssitte. Jahve hat sie bereits von Abraham verlangt, hat sie als Zeichen des Bundes zwischen sich und Abrahams Nachkommen eingesetzt.[22]

Nach Freud übernahmen die Israeliten mit dem fremden Zeichen aber auch die ihm innewohnende Ambivalenz: Denn einerseits erinnere Beschneidung an Ägypten als das Land der Knechtschaft, andererseits markiere es die (vermeintliche) Kontinuität mit dem ägyptischen Monotheismus und somit die Übernahme eines ägyptischen Gottesbildes.

Freud folgert zudem, dass die Beschneidung als Bundeszeichen insofern „eine besonders ungeschickte Erfindung" gewesen sei, als durchaus bekannt war (vgl. Jos 5,2), dass viele andere Völker im Süden Israels die Beschneidung kannten (144f.). Letztlich suggeriert Freud in seinen Ausführungen die Zwanghaftigkeit der Übernahme eines ursprünglich gar nicht exkludierend wirkenden Ritus. Die Zwanghaftigkeit erklärt Freud zudem damit, dass Beschneidung die archaische Unterwerfung unter den Willen des Vaters bzw. Vatergottes darstelle:

Die Beschneidung ist der symbolische Ersatz der Kastration, die der Urvater einst aus der Fülle seiner Machtvollkommenheit über die Söhne verhängt hatte, und wer dies Symbol annahm, zeigte damit, daß er bereit war, sich dem Willen des Vaters zu unterwerfen, auch wenn er ihm das schmerzliche Opfer auferlegte. (230)

Freud schließt mit dem Fazit, dass „die Moses-Religion ihre Wirkung auf das jüdische Volk erst als Tradition durchgesetzt hat" (256), und schließt auf eine längere Latenzzeit, die das religiöse Konzept brauchte, um Strahlkraft zu erlangen.[23]

Es ist nicht verwunderlich, dass Freud in seinem Mose-Traktat wiederholt auf eine kleine Notiz rekurriert, die allerdings zu den rätselhaftesten Texten der Hebräischen Bibel zählt und deshalb nur bedingt verwertbar ist.

---

22 Freud (1950), 144f.
23 Diesem Prozess ging zuletzt Assmann (2001) ausführlich nach.

II   Ex 4,24-26 im Kontext der Mosegeschichte

18 Es ging *Mose* und kehrte zu seinem Schwiegervater Jeter zurück und sprach zu ihm: Ich will nun hingehen und zu meinen Brüdern, die in Ägypten sind, zurückkehren und (nach)sehen, ob sie noch am Leben sind. Jitro sprach zu Mose: „Geh in Frieden." 19 Da sprach JHWH zu *Mose* in Midian: „Geh, kehr um (nach) Ägypten, denn alle Leute, die nach deinem Leben getrachtet haben, sind gestorben." 20 Da nahm *Mose* seine Frau, seine Söhne, setzte sie auf einen Esel und kehrte in das Land Ägypten zurück.

Und *Mose* nahm den Stab Gottes in seine Hand (vgl. Ex 4,17; 14,16 *maṭeh*). 21 Da sprach JHWH zu *Mose*: „Sobald du gehst um nach Ägypten zurückzukehren, sieh alle Wunderzeichen (*mophetîm*), die ich in deine Hand gelegt habe, und praktiziere sie vor Pharao. Aber ich werde sein Herz schwer machen (verstocken), so dass er das Volk nicht fortschickt. 22 Dann sprich zu Pharao: „So spricht JHWH: Mein erstgeborener Sohn ist Israel. 23 Ich aber sprach zu dir: Schicke meinen Sohn [LXX: mein Volk] weg, damit er mir diene." (Ex 3,18) Du aber weigertest dich (*m'n*) ihn fortzuschicken – Siehe, ich töte deinen erstgeborenen Sohn."

24 Es geschah auf dem Weg im Nachtlager (*malôn*[24]), da begegnete[25] *ihm* JHWH und trachtete danach[26] *ihn* sterben zu lassen (*môt* hif; vgl. 2,15; 4,19 + *hrg*). 25 Da nahm Zippora einen Feuerstein (bzw. scharfkantigen Stein; akk. ṣurru; vgl. Ägypten HAL 985; ÄHw *sf*), schnitt die Vorhaut ihres Sohnes ab (*krt*), berührte *seine* [=Moses oder des Sohnes?] Füße (= sein Geschlecht) und sprach: „Ja, *du* bist mir ein Blut-Bräutigam (*ḥoten*)." [LXX: Still steht das Beschneidungsblut meines Kindes"]. Da ließ *er* ab (*rph* qal; sonst „schlaff werden, zusammenfallen") von *ihm*. 26 Damals sagte sie: „Ein Blutbräutigam im Bezug auf die Beschneidungen." [LXX: Still steht das Beschneidungsblut meines Kindes"].

Die kurze ätiologisch anmutende Notiz in Ex 4,24-26 ist eingebettet in die Erzählung von Moses Rückkehr nach Ägypten. Im Anschluss an seine Kindheitsgeschichte berichtet das Exodusbuch, wie Mose nach dem Mord an einem Ägypter nach Midian floh, wo er bei seinem Schwiegervater Jitro / Jeter, einem JHWH-Priester, lebte. Dort wird er von dessen Gott JHWH aus dem brennenden Dornbusch berufen (Ex 3),

---

24 Zur vermeintlichen Ableitung von *mwl* „beschneiden" vgl. Kommentare wie Houtman (1993), 433f.; Propp (1999), 218; Schmidt (2011), 217; Albertz (2012), 97.

25 Das Verb *pgš* qal findet sich in Gen 32,18; 33,6; Ex 4,27 (Mose und Aaron begegnen sich); 1Sam 25,20; 2Sam 2,13; Jes 34,14; Jer 41,6; im feindlichen Sinne in Ex 4,24; Hos 13,8 mit Akk.; Prv 17,12 mit *b*[e].

26 Das Verb *bqš* „suchen" nimmt gefolgt von der finalen Infinitivkonstruktion (mit *l*[e]-) die – meist negativ konnotierte – „Bedeutung von ,wollen', ,trachten nach', ,beabsichtigen', ,Ziel verfolgen', ,planen' an" und erhält „den Charakter eines Hilfszeitwortes, das der zum Ausdruck gebrachten Tätigkeit eine voluntaristische und finale Färbung gibt" (Wagner [1973], 755f.).

um nach Ägypten zurückzukehren und den Auszug des Gottesvolks anzuführen. Doch scheint V. 24 durch das unterbestimmte rückbezügliche Personalpronomen zu insinuieren, dass JHWH „ihn", nämlich Mose, töten wolle.

Das Szenario der für den modernen Leser irritierenden Notiz kommt der exegetischen Zunft bekannt vor: Schon der Erzvater Jakob musste auf der Reise im nächtlichen Lager an einer Furt mit einer übermenschlichen Gestalt um sein Leben kämpfen. Ähnlich wie in Ex 4,26 endet auch Gen 32 mit einer, für die Tora-Erzählungen so typischen ätiologischen Notiz, die dort in den Bereich der Speisevorschriften verweist.[27]

> Gen 32,33: Deshalb sollen bis auf diesen Tag die Israeliten nicht die Sehne des Hüft-nervs essen (hap.leg. HAL 688), die auf der Hüftpfanne (HAL 468) liegt. Denn er berührte (*ngʿ*) Jakob an der Hüftpfanne, an der Sehne des Hüftnervs.

Wie die Ätiologie in Gen 32,33 betont, dass das Muskelfleisch an den Hüften wegen dieser Begegnung am Jabboq für die JHWH-Anhänger nicht zum Verzehr geeignet ist, so scheint es auch in Ex 4,26 um die Einrichtung einer anderen identitätsstif-tenden Institution, die Beschneidung, zu gehen.[28]

Zum literarischen Befund der Erzählung ist wenigstens kurz Folgendes anzu-merken: Der zitierte Textausschnitt ist literarisch spürbar uneinheitlich.[29] So ist die Motivation zur Rückkehr nach Ägypten zweifach berichtet: in V. 18 als Entschluss des Mose gegenüber dem Schwiegervater (s. die zwei Namensvarianten [vgl. dazu noch Reguel in 2,18]), in V. 19 als Aufforderung durch Gott. Das Mittelstück (4,20b-4,23) verbindet zahlreiche Erzählfäden und Motive der gesamten Exoduserzählung (der Stab in V. 20b erinnert an die Erweiswunder in 4,1-9.17[30], die Verstockung des Herzens in V. 21 oder das Erstlingsopfermotiv in V. 22f. an 13,11-16 – vgl. P) und dürfte deshalb ein spätes redaktionelles Zwischenstück (spätDtr) sein, das die verschiedenen Überlieferungen miteinander verbindet.[31]

---

27 Vgl. dazu Römer (1998), 70-75.

28 Dies scheint Coats (1999), 44ff. vorauszusetzen. S. dazu unten.

29 Zur ausschließlichen Orientierung an der Endgestalt vgl. Jacob (1997), 99-103, und Childs (1974), 93-107 oder auch Kunin (1996), 3-16.

30 Vgl. Bauks (2003), 83-97.

31 Die literarische Zuweisung folgt Albertz (2012), 70-75, der 4,18-22a.24-26 für die älteste Textschicht hält, die in 20b-23 eine spät-deuteronomistische Erweiterung erfahren hat (vgl. Houtman (1993), 447); vgl. differenzierter Schmidt, Exodus, 209f., der in 4,16-17a.19-20a.22f. (J) und 18.20b (E) unterscheidet, die um 4,21 (in priesterschriftlicher Sprache) redaktionell ergänzt sind, während V.24-26 s. E. eine ursprünglich midianitische Tra-

Das archaisch anmutende Textstück in V. 24-26 schließt erzählerisch gut an den in V. 19f. geschilderten Aufbruch an: Mose ist mit seiner Familie auf der Rückreise, um das Rettungswerk Gottes an Israel zu beginnen. Von einem ersten Sohn mit Zippora namens Gerschom[32] ist bereits in Ex 2,(21-)22 die Rede.

Sieht man aber einmal von dem gegebenen Großkontext ab, ist nicht eindeutig, wer in V. 24-26 weiterer Protagonist neben den ausdrücklich Genannten, JHWH und Zippora, ist. Die vorangehenden Verse scheinen zwar nahezulegen, dass es um den künftigen Führer Mose geht. Funktional handelte es sich dann um eine weitere Gefährdungsgeschichte seiner Person, die sich an die Aussetzung als Kind (Ex 2,1-10) bzw. die Nachstellungen nach der Erschlagung eines Ägypters (Ex 2,11-15) anschließt.[33] Ein qualitativer Unterschied besteht darin, dass in Ex 4,24 die Heimsuchung von JHWH selbst ausgeht.[34]

Ein weiterhin sinnstiftendes Motiv ist die Erstgeburt in V. 22f. und die Aussage, dass Israel JHWHs Erstgeburt sei.[35] Nach Rechtstexten wie Ex 22,28 steht die menschliche Erstgeburt Gott zu. Das erforderliche Opfer kann aber substituiert werden.[36] Eine Variante einer Auslösungsgeschichte findet sich in der Erzählung der Aqeda, der Bindung oder Opferung Isaaks.[37] Wie in Gen 22 symbolisches Sterben und Wiedergeburt verhandelt sind, ist Entsprechendes auch für Ex 4,24-26 vorausgesetzt worden.[38] Eine Reprise erfährt das Thema Israel als Erstling JHWHs in der

---

dition war, die in den JE-Erzählzusammenhang eingebettet worden ist; s. auch Propp (1999), 194-197.

32 Auch hier ist eine Namensätiologie angefügt, durch die die midianitische Gastfreundschaft positiv gewürdigt ist: „Ein Schutzbürger bin ich geworden im fremden Land"; vgl. Albertz (2012), 64 – Albertz datiert diesen Strang der Erzählung wegen der biographischen Nähe zu Jerobeam in das Nordreich des 9. Jh. zurück. Zu der ursprünglichen Exoduskomposition mit politischer Ausrichtung gehören auch 1,9-12; 15-22 (Hebammen); 2,1-22 (Kindheit und Flucht); 3,1-4a.5.6b-12aα; 16-17*; 18-22; 4,18-20a.24-26; 5,3-6,1. – Richter (1996), 433ff. postuliert den Anschluss von 4,24-26 an 2,22 (eine „Reguel-Tradition").

33 So begegnet die Wendung „sucht ihn zu töten" ein erstes Mal nach der Tötung des Ägypters in Bezug auf Pharao (*baqaš laharog* in Ex 2,15), der seinen Mord zu ahnden sucht, ein weiteres Mal in dem Rückblick, dass diese Gefahr gebannt ist (4,19 „da alle, die ihm nachstellten [*hamebaqešîm 'æt-naphešækâ*], gestorben waren").

34 Es ist denkbar, dass die Notiz deshalb an dieser Stelle eingelassen worden ist; so z.B. Richter (1996), 435.

35 Die Ansage der zu tötenden ägyptischen Erstgeburt nimmt die letzte Plagenepisode in Ex 12,29-33 vorweg.

36 Vgl. dazu Finsterbusch (2006), 21-45.

37 Vgl. Naumann (2005), 19-50; Bauks (2007), 65-86.

38 Kunin (1996), 4 und Morgenstern (1963), 35-70, hier 36: „The basic principal underlying the removal of a taboo was that the sacrifice, the giving to the spirit or spirits, of a

Auszugsgeschichte, und zwar in der umfassenden Ritualnotiz zur Passafeier (Ex 13,1-2.11-16).[39] Somit ist das Thema des Erstlingsopfers im größeren Erzählkontext präsent. Es ist angewendet auf Mose, auf Israel und auf die Erstgeburt der Ägypter, wobei Mose und Israel dank einer rituellen Handlung Substitution erfahren.

Gerade das Motiv des substituierten Erstlingsopfers fand in kulturgeschichtlichen Diskursen Beachtung. Der Psychoanalytiker F. Maciejewski[40] deutet Ex 4,24-26 sogar als das „biblische Archiv der Beschneidung", das die „Urszene der Beschneidung" mit dem Ziel einer „eigensinnigen Ritualerfindung" (38) erzählt. Dieses Archiv beinhalte s.E. zwei verschiedene Geschichten: einmal die Geschichte der archaisch fundierten göttlichen Tötungsabsicht des symbolischen Sohnes Mose,[41] zum anderen Moses Ansinnen den Sohn als Erstlingsopfer darzubringen, das dann schließlich dank der Mutter durch die Beschneidung substituiert ist.

> Mit dem nunmehr symbolischen Opfer wird aber nicht nur die Kindestötung aufgehoben; auch das andere alte Ritual, die Beschneidung der Jünglinge im Rahmen von Initiationsfeiern, wird radikal transformiert. Die Säuglingsbeschneidung steht als eine Art von Kompromissbildung am Schnittpunkt beider Rituallinien, der Tradition der Kindesopferung sowie derjenigen der Jünglingsbeschneidung.[42]

---

portion, and particularly the first ... redeemed the remainder, freed it from the possession and power of the spirits, and thus rendered it available for participation or use in ordinary, normal, profane existence. ... Or another, equally logical application ... was that the sacrifice of a part of the tabooed object redeemed the remainder. The cutting off of some part of the body of every child, and with this the shedding of some of the child's blood, and the giving of this in proper, ceremonial manner to the spirits or to the particular spirit, which were or was thought to have brought the child into being, redeemed it completely and effectively. Here, quite manifestly, we have the origin of the rite of circumcision."

39 Vgl. Van Weyde (2004), 57-64.

40 Maciejewski (2003), 33-39.

41 Vgl. dazu Bakan (1974), 203-216: Damit eine patriarchal organisierte Gesellschaft ausschließen kann, dass durch Heirat Nachkommen in die Sippe eingebracht werden, die nicht Kinder des Patriarchen sind, wäre die Erstgeburt eigentlich zu töten. Die Beschneidung (wie die Taufe) werde zur symbolischen Alternative des Infantizids und diene der Anerkennung der Lebensberechtigung des Sohnes durch den Vater. – „In the circumcision rite, the ‚children' of Yahweh as Father of the people were accepted into the community and the infanticide impulse was redirected, just as Abraham's intention to sacrifice Isaac was deflected to the ram. In psychoanalytic terms, circumcision was the reverse side of the Oedipus complex, and the incest taboo." (Turner, Body and Society, 106).

42 Maciejewski (2003), 38.

Ich möchte im Folgenden zeigen, dass das „Archiv der Beschneidung" in anderer Weise Auskunft über Körpersymbolik gibt als es die psychoanalytischen Zugänge[43] erkennen lassen, sofern man es diachron liest. M. E. ist Ex 4,24-26 das Ergebnis eines längeren Wachstumsprozesses, der die ursprünglich disparaten Traditionen und Themen vereint und die Notiz zu einer Ritualanweisung in Sachen Beschneidung[44] geformt hat, welche jedoch erst in talmudischer Zeit als bindende Verpflichtung interpretiert wurde.

In synchroner Hinsicht sind folgende Umstände auffällig: Die Rede ist von Moses nicht-israelitischer Frau[45] und (mindestens – s. Ex 1,20) einem Sohn. Die von ihr spontan vorgenommene Beschneidung ist untypisch, denn Beschneidung wird in der Regel nicht durch Frauen durchgeführt.[46] Das Alter des Sohnes ist nicht präzisiert.[47] Doch erhält der durch eine Nichtisraelitin durchgeführte Blutritus Vorbildcharakter, indem durch die Glosse in V. 26b die Verbindung zu einem bedeutenden Ritual Israels hergestellt wird. Eingeschoben ist die Notiz an einer Stelle, die proleptisch auf eine ganz andere Festinstitution anspielt, nämlich das Passafest mitsamt seiner historisierenden Begründung des von Gott rituell vorbereiteten Auszugs Israels aus Ägypten (4,22f.). Wichtig anzumerken ist auch, dass in der Exodus-Großerzählung Zipporas Tat unbeachtet bleibt, an späterer Stelle sogar berichtet wird, dass sie mit ihren zwei Söhnen erst nach dem Auszugsgeschehen

---

43 Vgl. auch Kessler (2001), 204-221 und Wyatt (2009), 405-431.

44 Vgl. Witte (1998), 308: „Ex 4,26, dessen Charakter als Ursprungsnotiz sich daraus ergibt, daß hier eigentlich von der Beschneidung des Mose und damit von der ersten Beschneidung in der Mosezeit überhaupt erzählt wird, dürfte mit Levin (1993), 332, als ein später midraschartiger Nachtrag anzusprechen sein, der verhindern will, daß Mose als Unbeschnittener sein Amt angetreten hätte."

45 Dass es in Ex 4,24-26 auch um ihre Integration in Israel geht (ggf. als eine Frage, die seit der Perserzeit äußerst virulent wird), reflektiert Römer, der den Textausschnitt wegen dieser Thematik für spätdeuteronomistisch hält (Römer [1994], 3-5.11 und [1998], 73-75; vgl. vorsichtiger auch Blum/Blum (1990), 41-54, bes. 53f. mit Anm. 40.

46 Vgl. Gen 17,23; Jos 5,3.7; Ausn. 1Makk 1,60; 2Makk 6,10 – Knauf (1988), 151f. geht von der Erinnerung an kanaanäische Frauenbeschneidung aus, die die Midianiterin Zippora gekannt hat; pTebt. II 293, Z. 10 (TUAT N.F. Bd. 5: Texte zur Heilkunde, Gütersloh 2010, 328) bezeugt einen Antrag auf Beschneidung eines Priestersohnes durch seine Mutter (187 n.Chr.). Morgenstern (1963), 64ff. legt dar, dass der Ritus nur selten, wie in Gen 17 u.ö. dargelegt, Aufgabe des Vaters war, sondern letztlich auf eine Matriarchatsstruktur verweisen könnte, die ein (männliches) Mitglied der mütterlichen Familie den Ritus durchführen lässt.

47 Vgl. z.B. Jacob (1997), 102f. zur Rekonstruktion der Altersangaben und der Identität des Sohnes.

Mose zugeführt worden ist (vgl. Ex 18,2-4) – und ihr so jegliche Mitbeteiligung am Exodusgeschehen abgesprochen ist.[48]

Neben dem rätselhaften Szenario der kleinen Episode sind auch die Unbestimmtheitsstellen (*gaps*)[49] auffällig. So macht der Text z. B. nicht deutlich, ob es bei der Tötungsabsicht um Mose oder um seinen Sohn geht. Während die vorangehenden Textabschnitte die Hauptfigur Mose häufig nennen, lässt dieser Abschnitt die Referenz eigentümlich offen.[50] Auch ist der Akt der Beschneidung nirgends als spontaner Ritus in einer Notlage bekannt. Die Einzeluntersuchung ergibt, dass die Redeeinleitung „Es geschah" (V. 24) einen typischen hebräischen Erzählbeginn darstellt. Eingeführt werden als Erzählfiguren „er", der Sohn, Zippora und JHWH. Es darf angenommen werden, dass Mose erst durch den Großkontext zur Bezugsgröße des rückverweisenden Personalpronomens wurde. V. 26b ist auf den ersten Blick eine recht passende Schlussformulierung, bevor in V. 27 eine neue Episode mit Mose und Aaron als Protagonisten einsetzt. Die Nahtstellen sind also klar nachzuzeichnen: Die typische Erzähleinleitung „Und es geschah …", der knappe Plot und die ätiologisch anmutende Notiz bilden eine Texteinheit.[51]

Problematisch sind drei Dinge, auf die intensiver einzugehen ist: 1) die Ätiologie; 2) die Bezeichnung „Blutbräutigam"; 3) der Ritualverlauf dieser in V. 26b mit einer Beschneidung identifizierten Handlung.

1. Die typischen formalen Marker für eine Ätiologie („Darum nannte man…"; „bis zum heutigen Tage…", „da nannte er …, denn er sprach"[52]) fehlen in Ex 4,26. Vielmehr sind sie durch die kleine Partikel „damals" (*'z*) ersetzt. Gemeinsam ist dem Zeitadverb mit den typisch ätiologischen Formeln, dass ein Bezug zwischen der Begebenheit und dem Leser hergestellt wird und somit der vorgenommenen

---

48 Vgl. Blum/Blum (1990), *47f.*; Römer (1994), 8-10. Vgl. die Darlegung von Albertz (2012), 96-98.

49 Schon die antiken Übersetzungen haben die Unbestimmtheitsstellen registriert und durch Ergänzungen zu präzisieren versucht. In der griechischen Übersetzung wird frei nach dem hermeneutischen Prinzip, dass die Bibel sich selbst erklärt, in Anlehnung an Gen 32,2ff. oder Ex 12,23 ein Engel als potenzielles Ausführungsorgan ergänzt.

50 Hüllstrung (2003), 182-196 weist darauf hin, dass in 4,18-31 die Referenz peinlich genau geklärt ist und der Mosename in V. 18-21 sogar fünfmal wiederholt wird. Mit Ausnahme der syrischen Tradition haben die alten Versionen offen gelassen, ob Mose oder sein Sohn gemeint sei.

51 So die Mehrheit der Exegeten – die literarische Einheitlichkeit bestreitet indes Lescow (1993), 19-26.

52 Vgl. Scherer (2008), § 2.2; kritisch zum ätiologischen Charakter von Ex 4,24ff. Römer (1994), 6f.

Ausdeutung autoritativer Charakter beigemessen wird. Während ätiologische Formeln sonst meist wichtige Angaben zur Herkunft von Ritualbestimmungen geben, die gegenwärtig Gültiges in der Vergangenheit verorten und legitimieren (s. Gen 32,33), handelt es sich hier um eine pseudo-ätiologische Notiz. Es geht nicht etwa um kausale Erklärungen, wie es zu dem Ritus kam, sondern um einen rhetorischen Kunstgriff: Die enigmatische Wendung *ḥatan damîm* wird – wahrscheinlich erst sekundär – mit dem Ritus der Beschneidung identifiziert. Es geht um die Affirmation des Erzählten in seiner eigenen Symbolik.[53] Es ist dabei nicht auszuschließen, dass die rätselhafte Wendung ursprünglich *terminus technicus* eines Rituals A war, um schließlich in semi-ätiologischer Form der Sanktionierung eines anderen Rituals B zu dienen. Doch worum handelt es sich eigentlich bei *ḥatan damîm*?

2. Die rätselhafte Wendung des „Blutbräutigams"[54] wird durch die Wiederholung als Schlüsselbegriff des Abschnitts geradezu markiert. Hebr. *ḥatan*[55] bezeichnet sonst jedes in einem Verschwägerungsverhältnis befindliche, männliche Familienmitglied, wie Schwager, Schwiergervater oder Schwiegersohn.[56] Die Wendung verweist, etwas nachhinkend[57], auf Zipporas Sohn (V. 25a).[58]

Einflussreich für das Verständnis ist die Erklärung von Julius Wellhausen. Seine Untersuchung alt- bzw. früharabischer Einflüsse in der hebräischen Bibel führte zu der These, dass die verstreuten Nachrichten über die in der früharabischen Welt praktizierte Beschneidung Zeugnis geben von einer „Art [...] Reifeprüfung, die der Jüngling bestehen mußte, bevor er heiraten durfte".[59] Der Zusammenhang

---

53 P.J. van Dyk spricht von einer Funktion als „rhetorical device" (Dyk [1990], 26ff.) und expliziert dies an Gen 16,11 (Ismaels Name: Gott hat gehört und seiner Charakterisierung als Wildesel in V. 12, gefolgt von der Namensätiologie des Wüstenbrunnens Lahai Roi in V. 14); vgl. Seebass (1997), 92-94 zu traditionsgeschichtlichen Überlegungen (mit Literatur).

54 Vgl. Propp (1993), 495-518, bes. 501f.; er moniert, dass Bräutigam („bridegroom") die Bezeichnung des Mannes am Tage der Hochzeit ist und somit unpassend ist. Wagner (2010), 458 Anm. 49 übersetzt in V. 26 sowohl *ḥatan* als auch *mwlæt* durch „Beschneider", wobei die zweite Wendung die erste unbekannte erklärt.

55 Vgl. dazu ausführlich Morgenstern (1963), 45-69.

56 In Abgrenzung zu Blum / Blum (1990), die von dieser enigmatischen Wendung her versuchen, die Erzählung zu entschlüsseln, verzichtet Hüllstrung auf eine Übersetzung; zur Begründung Hüllstrung (2003), 189 mit Anm. 30. Evtl. mögliche Übersetzung wäre „Blutsangehöriger" – da Blutsverwandtschaft falsche Assoziationen weckte.

57 Vgl. dazu Blum/Blum (1990), 42f., die den Schluss ziehen, dass es deshalb deutlich um den kontextuell bereits eingeführten Mose als Protagonisten gehen müsse.

58 So z. B. Morgenstern (1963), 44.

59 Wellhausen (1897), 174f.; Wellhausen (1905), 338f.

zwischen der Institution der Beschneidung und dem Status des Bräutigams entspricht der arabischen Wurzel *ḥatana*, die sowohl durch Beschneidung als auch durch Bräutigam übersetzt werden kann (so auch im Mittelhebr. Ges[18], 411). Dieser Befund legt s.E. nahe, dass in Ex 4,24ff. einerseits der Sohn des Mose beschnitten wird, dass dies aber andererseits im Sinne einer stellvertretenden Beschneidung von Mose selbst zu verstehen sei. Erst durch diesen stellvertretenden Ritus werde Mose schließlich zum „Blutbräutigam" und entgehe dem tödlichen Zorn Jahwes. Der Nachsatz in V. 26 hätte dann die Funktion, einen nicht mehr gebräuchlichen *terminus technicus* zu erklären und ihn gleichzeitig in seiner neuen Form, nämlich als Beschneidung von kleineren Kindern, zu sanktionieren.

Über die im Text nicht explizierte Annahme Wellhausens, dass der Sohn ein Kleinkind sei, einmal hinwegsehend, sind folgende Kritikpunkte zu benennen, einmal sein Verständnis von *damîm* „Blut" (Pl.) sowie die Interpretation des deutlich dreiteiligen Ritualakts:

Der Plural *damîm* bezeichnet in der Regel nicht „Blut" als Abstraktum, sondern trägt in den Belegstellen die Konnotation von „viel Blut", z. B. im Sinne einer „Blutschuld".[60] Dieses Verständnis passt insofern in den Großkontext, als Mose wegen seiner Tat, einen Ägypter erschlagen zu haben, Heimsuchung fürchten muss, da die Bluttat mittels einer Sühnehandlung zu kompensieren wäre. Mose würde dann durch einen solchen rituellen Reinigungsakt[61] für seine Führungsaufgabe gewissermaßen erst qualifiziert.[62]

Eine andere Möglichkeit könnte darin bestehen, *damîm* als Anspielung auf einen Blutritus[63] zu verstehen. Überlegenswert ist zudem, ob die Wendung im Anschluss an einen altorientalischen Götternamen zu verstehen ist. Auch in Nordsyrien (Ebla, Emar und Mari) ist eine Gottheit mit dem Namen Damû überliefert. In Südmesopotamien (Isin und Girsu) ist seit der altbabylonischen Zeit eine Gottheit dieses Namens verehrt, die als Sohn Gulas mit heilenden Kompetenzen für

---

60   Zur Bannung der Sphäre der Blutschuld mit reinigender Funktion vgl. Propp (1993), 502 mit Hinweis auf Schneemann (1980), 794; vgl. auch Propp (1999), 236,238.

61   Die Qualifikation eines Menschen für seine Mission durch einen rituellen Reinigungsakt findet sich bereits in Jes 6.

62   Vgl. dazu ausführlich Propp (1993), 504ff. In der Regel wird wenigstens eine vorsätzliche Bluttat durch den Tod des Mörders kompensiert, in diesem Text fände durch den Ritus eine Substitution statt. Warum das möglich ist, bleibt in dem knappen Textstück unausgeführt.

63   Vgl. Christ (1976), 63-65, der lediglich für Ex 4,25f. „Wundblut" voraussetzt und das Pluralnomen im Sinne eines epexegetischen Genetivs deutet, der adjektivisch mit „blutig" (statt „mörderisch" wie z. B. in 2. Sam 16,7f. u.ö.) zu übersetzen ist, während *ḥatan* s.E. einen Beschnittenen bezeichnet (ebd. 58.130).

die Dämonenaustreibung[64] und Krankheiten wie Epilepsie[65] ausgestattet ist.[66] In therapeutischen Texten Mesopotamiens und Ägyptens sind Blutapplikationen im Heilungskontext belegt.[67] Als Bedrohung für Säuglinge und Kinder ist die Dämonin Lamaštu bekannt, die zudem als unablässige Blutfresserin gilt.[68] So erinnert auch Ex 4,24f. in seiner ursprünglichen Form an altorientalische Blutrituale zur Abwehr von Dämonen[69], und die Heimsuchung JHWHs wäre demnach als Erkrankung zu interpretieren. Vergleichbar mit bösen Dämonen aus dem Bereich der Säuglings- und Kindergefährdung (wie z. B. Lamaštu), könnte der identifizierende Satz „du bist *ḥatan damîm*" apotropäische Funktion haben in dem Sinne von „du bist meine Blutschützer".[70] Subjekt des Angriffs wäre ein Kind, welches dank eines besonderen Rituals geheilt wird. Selbst die Unverständlichkeit des Begleitspruchs *ḥatan damîm* wäre kein Unikum. Es begegnet auch im mesopotamischen Kontext, dass Beschwörungsformeln als Arkanwissen unverständlich sind („Abrakadabra-Beschwörung").[71]

---

64 Vgl. die Beschwörung gegen den Dämon „Jegliches Böse" in der 5. Tafel des aus dem 1. Jahrtausend stammenden Handbuchs *Muššu'u* „Einreibung" (BM 46276+; spätbabylonisch); vgl. Böck (2010a), 149; s. einen magisch-medizinischen Text gegen Gallenerkrankung (K 61+; 7. Jh. v. Chr.): Vs. II,49 (ebd., 76) und gegen die Lähmung der sagallu-Krankheit (KA 2453+; 7. Jh. v. Chr.), Rs III,20 (ebd., 104f.).

65 Vgl. Hüllstrung (2003), 192 mit Anm. 39; zu Text und Übersetzung vgl. Schwemer (2010), 127 zu BAM IV 323 (7. Jh.), Rs. 93 „Am 15. Tag (des Monats), dem Tag, da Sîn und Samas zusammen stehen, 94 bekleidest du diesen Mann mit einem Sackgewand. Mit einer Obsidian(klinge) [= ṣurrum] ritzt du seine Schläfen, 95 vergießt sein Blut. Du läßt ihn in einer mit Rohrbündelstandarten umzäunten Einfriedung sitzen ... 95 .... Der Mann spricht wie folgt:...".

66 Vgl. Becking (1999), 175f.; Black/Green (1992), 57; Cagni (1981), 47-86; Dahood (1981), 97-104. Vgl. auch Böck (2013), 21; sie erwähnt Damu als Sohn der Gula/Ningirim, der das Karzillu-Messer trägt (vgl. Köcher [1953], 57-107, bes. col. I:12').101-103 zu BM 98584 + 98589 + K.5461a, Z. 27 (Rekonstruktion der Zeile, die in TUAT.Erg 5, 65 fehlt); vgl. auch Fritz (2003), 299ff.

67 Stol (1993), 105f.

68 Hüllstrung (2003), 192 mit Hinweis auf Farber (1980-83), 439-444 und Farber (2000), 1895-1904. Wiggermann (2000), 217-252, bes. 217ff. und Volk (1999), 1-30, bes. 3ff.

69 Diskutiert wurde auch, ob sich der Ausdruck *ḥatan damîm* als Bezeichnung für einen Dämonen verstehen lässt; vgl. z. B. Christ (1976), Blutvergießen, 130 mit Literatur). Der Name ist z. B. von akk. *ḥatanu* „schützen" abgeleitet worden; so z. B. von J. Blau (1956) – vgl. Houtman (1983), 81-105, hier 100 mit Anm. 43.

70 Die drei letzten Worte in V.26 wurden auch als eine zusammenhängende Wendung gelesen: „Blutbräutigam bzw. Blutschützer für die Beschneidungen" (vgl. die Übersetzung des hebräischen Textes durch Le Boulluec / Sandevoir (1989), 104).

71 Vgl. in der Beschwörung gegen die sagallu-Krankheit (BAM II 130, Vs. II 36-38) in der Übersetzung von Böck (2010b), 104 mit Hinweis auf Prechel/Richter (2001), 333-371. Vgl.

Da der Identifizierungsversuch mit den Kulturen der Umwelt wie z. B. mit dem Gott Damû sehr hypothetisch bleibt[72], ist es methodisch geboten, die Bedeutung der Wendung *ḥatan damîm* offen zu lassen, um sich dem Ritual selbst zuzuwenden.

3. Der geschilderte Ritualverlauf und -kontext ist dreiteilig. Das Ritual besteht aus dem „Abschneiden der Vorhaut", gefolgt vom Berühren der Füße (häufig gedeutet als Euphemismus für Scham) und dem Ritualspruch der Zippora „*ḥatan damîm* bist du für mich!" Durch dessen Performanz wird eine Aura von Faktizität erzeugt, welche die beschriebene Krisis als überwunden darstellt (V. 25a): Denn der Angreifer lässt ab.[73]

Der Umstand, dass eine Frau eine Beschneidung als spontane Reaktion in einer unvorhergesehenen dramatischen Situation vollzieht, widerspricht dem, was wir sonst aus Bibel und Rabbinica über die Beschneidung wissen.[74] Dem Ritus geht zudem sonst eine explizite Aufforderung zur Beschneidung voraus. Der spontan vorgenommenen Beschneidung als Reaktion auf Gefahr fehlt jeder zeremonielle Charakter.

Ein religionsgeschichtliches Beispiel jüngeren Datums lässt sich als Beleg für eine Beschneidung in Notlage anführen, nämlich die Notiz aus Philo von Byblos (Frag. 2,33), die erzählt, wie der Gott Kronos seinen erstgeborenen Sohn dem Vater Ouranos opfert und eine Selbstbeschneidung vornimmt, um mittels dieses Entsühnungsopfers Pest und Tod von seinem Volk abzuwehren.[75]

---

auch Hüllstrung (2003), 192 Anm. 39 mit Hinweis auf Farber (2000), 1901: „Short spells, often in garbled and no longer understandable foreign languages (Hurrain and Elamite elements have occasionally been identified), are hallmarks of many magico-medical rituals."

72 Der in sumerischer Zeit im Ahnen- und Totenkult verortete und mit Fruchtbarkeit befasste Gott Damu wurde in späterer Zeit zusammen mit seiner Mutter Gula als Heilgottheit verehrt (zur Erwähnung Damus in Beschwörungstexten vgl. Fritz (2003), 190-195); seit der altbabylonischen Zeit nahm das Vorkommen an Eigennamen, die mit Damu gebildet wurden, zu (vgl. Fritz (2003), 213-223).

73 Vgl. dazu ausführlich Hüllstrung (2003), 190-193.

74 Es gibt kein wirkliches, ausführlicheres religiöses Gesetz zur Beschneidung (im Vergleich zu anderen religiösen Bräuchen) und scheint sich demnach um etwas Alltägliches zu handeln, das auch durch Frauen vorgenommen werden kann; vgl. 1Makk 1,60 und 2Makk 6,10. S. oben auch Morgenstern in Anm. 45.

75 Überliefert bei Euseb, Praeparatio evangelica I,10,33; vgl. Knauf (1997), 16-19; Quack (2012a), 613.

Das Ritual besteht aus dem Abschneiden der ʿårlâ „Vorhaut".[76] So untypisch das verwendete Verb für „abschneiden" im Kontext der Beschneidung ist, so passend ist das verwendete Werkzeug (*sr* Flintstein; vgl. Jos 5,2; äg. ser[77]), was funktional erklärbar ist. Die Identifizierung mit dem Beschneidungsritus erfolgt erst durch den Deutespruch in V. 26b.[78]

Die Verwendung des Verbums *krt* „abschneiden, trennen" ist ungewöhnlich, da ʿårlâ „Vorhaut" sonst im biblischen Hebräisch in Kombination mit dem Verb *mwl* „beschneiden" verwendet ist.[79] Da das hier bezeugte Verb typisch ist für die priesterschriftliche Tradition des Bundesschlusses zwischen JHWH und Volk[80], lässt sich eine zusätzliche Nähe zu dem thematisch verwandten Text in Gen 17 herstellen. Doch als ursprünglich ist diese Bundesassoziation nicht anzusehen.[81]

Eigentümlich im Kontext der Beschneidung ist der sich anschließende Gestus der Berührung der Füße. Selbst die grundsätzlich mögliche Übersetzung durch „Scham" täuscht nicht darüber hinweg, dass im Vordergrund der Geste die Berührung mit dem ausgetretenen Blut und somit eine Blutapplikation steht. Das Ritual ist also deutlich als Blutritus gezeichnet, wobei die Handlung vor allem der Blutgewinnung dient.[82] Die Blutberührung (hier begleitet von einem performativen Sprechakt) weist apotropäischen Charakter auf, wie er ein zweites Mal begegnet in der Aufforderung zur Bestreichung der Türpfosten und Türschwelle mit dem Blut des Pessach-Lammes (Ex 12,7.13.22). Auch hier dient die Markierung dazu, JHWH bzw. den von ihm gesendeten „Verderber" an der israelitischen Erstgeburt vorbei

---

76 Das Nomen kann metonymisch auch die Unbeschnittenen an sich bezeichnen, denn es geht einher mit kultischer Unreinheit und hat unterscheidende Funktion. Vgl. dazu ausführlicher Mayer (1989), 385-387.

77 Vgl. zum Vokabular der Beschneidung in Ägypten ausführlich Quack (2012a), 580-597, der von neun untersuchten Termini mit z.T. sehr wenigen Belegen, was die genaue Bedeutung sehr diskutiert sein lässt, vor allem *sʿb* als terminus technicus benennt (586).

78 Childs (1974), 100, betont, dass die redaktionelle Ergänzung in V.26b („Beschneidung") die Handlung Zipporas erklären soll und nicht umgekehrt. Die Assoziation ergibt sich durch die Erwähnung der Vorhaut (ʿarlâ) in V. 25.

79 Vgl. Schmidt (2011), 227; Blum/Blum (1990), 42 mit Anm. 7.

80 P – s. den Beitrag von T. Römer in diesem Band.

81 Anders stellt sich der Befund in den rabbinischen Texten seit Mischna und Talmud dar; vgl. dazu ausführlich Blaschke (1998), 108-322; Bauks (2016).

82 Der Hinweis auf *ḥatan* als Beschneidungsterminus ist zwar seit Wellhausen (s.o.) verbreitet, trägt aber m.E. nichts zur Traditionsgeschichte bei, sondern lässt sich vielmehr als Ergebnis der Wirkungsgeschichte verstehen.

gehen zu lassen. Ähnlich dürfte auch Ex 4,24-26 die apotropäische Bedeutung von Beschneidungsblut bezeugen.[83]

III   Auswertung

Das Blut der beschnittenen Vorhaut dürfte ursprünglich als spontan gewonnenes, apotropäisches Schutzmittel gedient haben, um einen göttlichen (oder ursprünglich dämonischen?) Angriff abzuwehren. Problematisch scheint mir allerdings die Deutung im Sinne einer nachgeholten Beschneidung als Reaktion auf die göttliche Tötungsabsicht.[84] Denn es gibt weder im Alten Testament noch außerhalb des Alten Testaments Hinweise darauf, dass auf unterlassene Beschneidung der Tod stünde, und zwar weder der Tod des Unbeschnittenen noch der Tod dessen, der die Beschneidung vorzunehmen hätte, also des Vaters.[85] Das Beschnittensein wird als Voraussetzung geschildert, um zur israelitischen Kultusgemeinde dazuzugehören (vgl. Ex 12,44.48). Der Akt trägt demnach vor allem inkludierenden Charakter.

W. Propp setzt für den Wachstumsprozess des Textes vier Stadien voraus, die jeweils neue Sinnebenen gestiftet haben: Ursprünglich dürfte es um einen Blutritus gegangen sein, der erst im Nachhinein an die Beschneidung rückgebunden wurde. Da die Beschneidung als *rite de passage* mit der Heiratsfähigkeit des Initianten einherging, konnte die Reinterpretation der in ihrem ursprünglichen Sinn nicht mehr geläufigen Wendung *ḥatan damîm* gelingen. Erst in einem zweiten Schritt erhielt der Initiationsritus explizit religiöse Bedeutung, indem er zur Voraussetzung für die Teilnahme am Kult (z. B. im Pessachkontext) wurde. Im Anschluss wurde der Zeitpunkt der Durchführung von der Adoleszenz in die frühe Kindheit

---

83   Vgl. Hartenstein (2007), 119-137, bes. 133, in Anlehnung an Propp, Exodus, 236.238 z. B. zur Bannung der Sphäre der Blutschuld mit reinigender Funktion. – Auf das Verhältnis von Beschneidung und Blutritus – Erstgeburt und Pessach geht Sarna (1991), 25f. ausführlicher ein.

84   Im Grunde handelt es sich um eine logische Konsequenz aus V.26a; vgl. Hüllstrung (2003), 188. Diese Deutung findet sich zuerst in den Targumim (s. unten).

85   Gen 17,14 lässt als Straffolge nicht etwa die Todesstrafe erkennen, *krt* nif. bezeichnet vielmehr „den Ausschluß aus ..." (vgl. Seebass [1997], 107; Wagner [2010], 459). In Jos 5,5 wird sogar ausdrücklich festgestellt, dass alle Israeliten, die während der 40jährigen Wüstenwanderung geboren wurden, unbeschnitten waren. Darauf hat Jahwe aber nicht mit der Tötung reagiert, sondern mit der schlichten Aufforderung, die Beschneidung nachzuholen. Wagner (2010), 455 sieht in dem Text die Erinnerung an eine Zeit, in der noch nicht beschnitten wurde und in 4,26 die Erinnerung, dass der Beschneidungsritus ursprünglich aus Ägypten kam.

transferiert „when the bloodied child became a symbol of the paschal night, when the endangered first-born of Israel were saved by the blood of the lamb".[86] Und dieses theologisch veränderte Verständnis lieferte schließlich die Voraussetzung, dass das Thema in die Mosebiographie Eingang finden konnte und als ein Sühneakt aus familiärer Solidarität interpretiert wurde. Das bedeutet aber, dass der Titel *hatan damîm* ursprünglich mit der Moseerzählung nichts zu tun hatte. Die geschaffene Ätiologie für die Kinderbeschneidung bildet die neuere Fassung eines alten apotropäischen Ritus, die verschiedene Motive in sich vereint.[87] Offensichtlich ging es auch um die Integration der midianitischen Frau des Mose, Zippora, in die JHWH-Gemeinschaft durch symbolische Blutsverwandtschaft.[88] Ursprünglich war es wohl auch nicht JHWH, der attackiert hat, sondern ein Dämon oder ein anderes zwischenmenschliches Wesen, welches mittels eines apotropäischen Ritus – vielleicht midianitischen Ursprungs – von Zippora besänftigt wurde.[89]

Die im Ritualspruch rezitierte und im Deutewort interpretierte Wendung des „Blutbräutigams" bzw. „Blutangehörigen" ist seltsam und schon von der frühen Rezeptionsgeschichte als seltsam wahrgenommen worden.

Interessant ist in diesem Kontext die griechische Übersetzung der LXX (4./3. Jh. v. Chr.):

> 24 Es geschah aber unterwegs, in der Herberge, da überfiel ihn der Bote des Herrn und suchte ihn zu töten. 25 Und Sepphora nahm einen Kieselstein, schnitt die Vorhaut ihres Sohnes ab, fiel (ihm) zu Füßen und sagte: Still steht das Beschneidungsblut meines Kindes! 26 Da ging er von ihr fort, weil sie gesagt hatte: Still steht das Beschneidungsblut meines Kindes (LXX Deutsch).

LXX gibt im Vergleich mit dem hebräischen Text einen stringenter komponierten Handlungsverlauf wieder: An die Stelle Gottes tritt ein Bote, die Frau schreitet ein und beschneidet den Sohn, fällt vor dem Boten nieder, äußert einen Ritualspruch, der entweder den Vollzug der Handlung bestätigt oder lediglich der Blutstillung dient.[90] Der Ritualspruch, der nun des Deuteworts nicht mehr bedarf, wird zu der Begründung, warum es gelang, die Gefahr auszusetzen. Die rätselhafte ätiologische Notiz *hatan damîm* „ein Blut-Bräutigam im Bezug auf die Beschneidung" wird in

---

86 Propp (1993), 515.

87 Vgl. Albertz (2012), 96f.

88 Blum/Blum (1990), *47f.*

89 Schmidt (2011), 230f.

90 Kosmala (1962), 14-28, bes. 28 sieht hier die Anwesenheit von Blut deklariert, nicht die Feststellung, dass die Blutung gestillt wird und übersetzt: „The blood of the circumcision of my son was or is there (the result of the circumcision being still visible)."

eine Erfüllungsnotiz transformiert. Hier zeigt sich, dass schon die antiken Übersetzungen mit der Wendung Schwierigkeiten gehabt und sie in eine Positivaussage im Sinne des Handlungserfolgs umgemünzt haben, der sich aus der Beschneidung ergeben hat.[91]

Jüdisches Rewriting im Jubiläenbuch verzichtet nicht nur weitgehend auf die Details der Szene (z. B. Jub 48,2f[92]; vgl. 4Q Gen-Ex[a] frag. 24-25i), sondern entzieht zudem Gott die Verantwortung, indem es einen dämonischen Übergriff beschreibt: Dieser Variante nach ist der Dämon Mastema am Werk, der durch die Tötung des Mose versuchte, die Ägypter zu retten, was durch Gottes Eingreifen unterbunden werden musste, um den Auszug Israels aus Ägypten zu ermöglichen. Anders argumentieren Targum Pseudo-Jonatan und Targum Jerusalem z.St., wenn sie den Angriff damit begründen, dass Mose seinen Sohn nicht beschnitten hatte. In diesem Sinne explizieren VitMos (sl) 598 und Chron-Jerahm 47,1.2: „Sie ruhten sich an einem bestimmten Platz aus, und ein Engel kam herab und griff ihn an wegen der Übertretung des Bundes, den Gott mit seinem Sklaven Abraham gemacht hatte, weil er seinen ältesten Sohn nicht beschnitten hatte, und er wollte ihn erschlagen.“[93] Während in den Targumim der Vernichter bzw. der Engel des Todes handelt, ist es nach dem Talmud bNedarim 32a Satan (vgl. MidrWajoscha 96).[94] Einig sind sich alle diese Versionen darin, dass der Text von der Tötung des Mose und nicht der seines Sohnes ausgeht.

Die Integration des ursprünglichen Blutritus in den größeren Kontext des Exodus als dem grundlegenden „jüdischen Nationalepos“ ermöglicht eine Reihe von Assoziationen, die zu seiner Ausdeutung beitragen: die Integration apotropäischer Riten, die Integration von deren nichtisraelitischen „Trägern“, die Gefährdung des Helden Mose, Bestimmungen zur Erstgeburt oder aber Bestimmungen zur Auslösung von Blutschuld bieten einige dem Kontext geschuldete Lesehilfen, die dem rätselhaften Text erst Sinn verleihen. Wichtig erscheint mir, dass keine der

---

91 Vgl. ausführlich Wyatt (2009), 412f.; Wenigstens in ExodB (Vaticanus) ist die Ätiologie am Schluss ausgelassen; vgl. Gurtner (2013), 40f.; anders Wevers (1990), 55, der von einem Schreiberfehler ausgeht.

92 Berger (1981)1, 542f.: 2 Und du weißt, was er mit dir geredet hat auf dem Berge Sinai und was der Fürst Mastema mit dir tun wollte, als du nach Ägypten zurückkehrtest, auf dem Weg bei der Tanne am schattigen Ort. 3 Wollte er dich nicht mit aller seiner Macht töten und die Ägypter retten aus deiner Hand, als er dich sah, daß du gesund warst, daß du Gericht wirken solltest und Rache gegen die Ägypter? 4 Und ich befreite dich aus seiner Hand […].

93 Zitiert nach Berger (1981), 543.

94 Vgl. Berger (1981), 543 ad 2; Blaschke (1998), 477; zu frühchristlichen Interpretationen vgl. Blaschke (1998), 469-490; Jacob (2008), 311-323.

Perspektivierungen exklusiven Anspruch erheben kann, sondern Vieldeutigkeit im Textausschnitt absichtsvoll belassen ist. Dazu tragen die vielen *gaps* bei. Dank des Deuteworts, welches das apotropäische Blutritual *ḥatan damîm* „im Bezug auf die Beschneidung(en)" verortet, wird der Beschneidung schützende Wirkung zugewiesen. Dass Beschneidung die rituelle Substitution des Erstlingsopfers sei, hat auf der endgestaltlichen Textebene (im Kontext von Ex 4,22f.) zwar durchaus Anhalt, lässt sich aber aus Ex 4,24-26 weder ableiten noch begründen.

Trotz des Unverständnisses des ursprünglichen Blutrituals und seiner Bezeichnung hat sich dieser Text allmählich zu einer Ätiologie der Beschneidung entwickelt. So zog insbesondere die rabbinische Auslegung aus dieser Notiz das Fazit, dass die Beschneidung ein zentrales Merkmal jüdischer Identität ist:

> „R. Jehosua B. Qorcha sagte: Bedeutend ist die Beschneidung, dass ihrethalben dem gerechten Mosche nicht einmal eine Stunde Aufschub gewährt wurde." (Ned. 31b)[95]

Dem rabbinischen Diktum folgt eine Reihe von Begründungen, warum Mose auf der Reise den Sohn nicht beschnitt und sich bzw. seinen Sohn überhaupt gefährdet habe. Einigkeit zeigen die Rabbinen in ihrer durchaus kontroversen Diskussion über die Stelle lediglich bezüglich des *Imperativs der Beschneidung*, obwohl er sich aus der rätselhaften Begebenheit selbst nicht weiter ableiten lässt.[96] Historische Untersuchungen zum rituellen Umgang mit der Beschneidung in griechisch-römischer Zeit[97] lassen erkennen, dass bis weit in die rabbinische Zeit Beschneidung keinesfalls ausreichte, um jüdische Identität[98] zu markieren. Identitätsbildung erwies sich schon in der Antike mit ihren multikulturellen Bezügen als ein sehr komplizierter Vorgang. Dringlich ist zudem die Frage, inwieweit es sich überhaupt um ein Zeichen *religiöser* Zugehörigkeit handelte. Wie bereits das Fehlen eigensprachlicher

---

95　Zitiert nach Goldschmidt (1996), 429; vgl. Jacob (1997), 100; Sarna (1991), 25. Es gehe um die Gefährdung des unbeschnittenen Sohnes, weil Mose diese versäumt habe, vgl. auch Banks (2016).

96　Vgl. Sarna (1991), 25: „the brief narrative in verses 24-26 underscores the paramount importance of the institution of circumcision and the surpassing seriousness of its neglect." Jacob (1997), 102f. bezieht die Notiz auch auf Mose, den göttlichen Zorn aber auf die Tatsache, dass Mose seine Familie zu dieser wichtigen Mission mitnimmt (mit Verweis auf die Kommentare von A. Ibn Esra und S.D. Luzzato).

97　Vgl. Spann (2012), 225-242.

98　Für Collins (1997), 211f. stellt sie bestenfalls ein Erkennungszeichen der Juden durch Nichtjuden dar. Zu der Frage, was eigentlich jüdisch sein bedeutet, vgl. Mason (2007), 457-512. Er unterscheidet in Jehudîm, Judaios, Judaismus/Judaisierung (zuerst in 2Makk belegt, wird aber erst im 3. Jh. bei den Kirchenvätern zu einem Glaubenssystem, das vom Christianismus abgesetzt wird, 471ff.).

Begrifflichkeit für Religion in der Antike anzeigt,[99] hat die Körpersymbolik lange
Zeit erst einmal auf das Ethnos anstelle des Ethos verwiesen.[100]

Spannend bleibt die Frage nach den Gründen für die Aufnahme dieser archaisch
wirkenden Tradition in den Mosezyklus, die von den exegetischen Kommentaren
weitgehend ausgeblendet wird.[101] Die Rezeptionsgeschichte verdeutlicht nicht nur
die Mühe, den Text in seinem (neuen) literarischen Kontext zu plausibilisieren,
sondern zeigt auch an, dass ein eindeutiges Verständnis nicht zu erreichen ist.
Stattdessen präsentiert sie Variationen der Auslegung, die sich aus dem größeren
Zusammenhang ergeben.

Ich komme zurück zu S. Freud und seiner Beobachtung:

> Nur als absichtlichen Widerspruch gegen den verräterischen Sachverhalt kann man
> die rätselhafte, unverständlich stilisierte Stelle in Exodus auffassen, daß Jahve einst
> dem Moses gezürnt, weil er die Beschneidung vernachlässigt hatte, [...] Und hier findet
> sich der Anlaß zu einem entscheidenden *Streich* gegen die ägyptische Herkunft der
> Beschneidungssitte. *Jahve hat sie bereits von Abraham verlangt, hat sie als Zeichen des*
> *Bundes zwischen sich und Abrahams Nachkommen eingesetzt.* (Der Mann Moses, 144f.)

Für Freud war die Beschneidung als Bundeszeichen schlecht gewählt, da schon in
der Antike bekannt war, dass sie auch in Ägypten weit verbreitet war. Er wertet die
Einführung der Beschneidung als Legitimierung und Kompensation Israels, um die
Nachfolge des inzwischen in Ägypten ausgemerzten Monotheismus in gebührender
Form anzutreten. Nach Moses Tod sei der Ritus in die Abrahamzeit vorverlegt (Gen
17) und mit der Notiz in Ex 4 die absolute Wertigkeit betont worden. Erst so wurde
Beschneidung in einen genuin israelitischen Brauch umgedeutet.

Im Anschluss an die psychoanalytischen Deutungen des Sinns der Beschneidung als symbolische Unterwerfung bei Freud oder aber als symbolisches Opfer
bei Morgenstern und Maciejewski interpretiert der Alttestamentler R. Kessler Ex
4,24-26 als eine Fehlleistung, die „zum Einen ein überraschendes und befremden-
des Element enthält, das doch zum Andern mehr oder weniger geschickt in den
herrschenden Diskurs der Exoduserzählung integriert ist."[102] Doch ist diese Heran-

---

99 Vgl. Mason (2007), 378-481; zum römischen Kontext vgl. Rüpke (2006), 11-18.42f.

100 Vgl. Mason (2007), 480-488, der Ethnos in Verbindung mit Ethos, Kultus, bestimmte
    Orte, Familientraditionen und Magie u. a. stellt und Konversion als einen Vorgang
    beschreibt, der dem Wechsel von einer ethnischen Gruppe in eine andere gleicht (491).

101 Vgl. aber Kessler (2001), 206ff. zur Midianiterhypothese; Morgenstern (1963), 38f. zur
    Keniterhypothese.

102 Vgl. Kessler (2001), 208; er reduziert die Narration auf den hilflosen Helden (Mose), den
    zürnenden Gott und die rettende Frau als Spuren eines kryptischen Gedächtnisses, auf
    das der im kulturellen Gedächtnis verankerte Ritus der Beschneidung „reagiert".

gehensweise modern, indem Traditionsliteratur an das Individuelle des Menschen rückgebunden wird, über das der antike literarische Text an sich keine Aussagen zulässt. Im Vergleich dazu ist die kulturanthropologische oder -soziologische Herangehensweise weniger hypothetisch und mit den geläufigen historisch-kritischen Zugängen weitaus kompatibler. Sie verzichtet auf eine universal anthropologische, geradezu gattungsgeschichtlich vorgegebene Voraussetzung, die auch die individuelle Biographie prägt, nämlich, dass Beschneidung eine Variante darstellt, um auf eine Urerfahrung, wie die Erinnerung an den „Vatermord", zu reagieren und dies in Form eines symbolischen Akts zu kompensieren. Der Rückschluss von der Phylo- auf die Ontogenese ist nicht unvermittelt zu denken. Beschneidung als natürliches Symbol kann nicht universell definiert werden. Stattdessen ist ihre strukturelle Bedeutung im jeweiligen gesellschaftlichen Rahmen zu berücksichtigen. Beschneidung wurde durch die auf den Bund Gottes mit Israel verweisende Begründung zu einem Symbol der Gruppenidentität[103], das im Falle eines Verbots an Bedeutung gewinnt, da es die Gruppenloyalität verstärkt und einfordert. Somit bringt konkret vorgenommene Verkörperung soziale Strukturen symbolisch zum Ausdruck, die ihre Berechtigung aus einer Kombination von Tradition und Narration bezieht und daraus argumentative Gründe ableitet. Damit dürfte sich M. Jungs aufgestellte These bewähren, dass verkörperte Erfahrung („embodied experience"), eingebettete Wertkonzeptionen („embedded social practices") und freistehende Gründe („discursive, freestanding reasoning") in einem reziproken und nicht etwa hierarchischen Verhältnis zueinander stehen.[104]

Literatur

Abusch (2003): Ra'anan Abusch, „Negotiating Difference: Genital Mutilation in Roman Slave Law and the History of the Bar Kokhba Revolt". In: P. Schäfer (Hg.), *The Bar Kokhba War Reconsidered. New perspectives on the second Jewish revolt against Rome*, Tübingen, 71-91.
Albertz (2012): Rainer Albertz, *Exodus 1-18 (ZBK)*, Zürich.
Assmann (2001): Jan Assmann, *Mose, der Ägypter. Entzifferung einer Gedächtnisspur*, 3. Auflage, Frankfurt/M.

---

103 Vgl. M. Jungs Ausführungen zur indexalischen Bedeutung von Dtn 4 in diesem Band sowie Ackermann, der den Körper mit M. Douglas als das „mikrokosmische Abbild der Gesellschaft" umschreibt, der das Sozialsystem „systematisch zum Ausdruck bringt".
104 S. M. Jung, in diesem Band.

Bakan (1974): David Bakan, "Paternity in Judeo-Christian Tradition". In: Allan W. Eister (Hg.), *Changing Perspectives in the Scientific Study of Religion*, New York, 203-216.

Bauks (2003): Michaela Bauks, „Das Dämonische im Menschen. Einige Anmerkungen zur priesterschriftlichen Theologie (Ex 7-14)". In: A. Lange, H. Lichtenberger (Hg.), *Dämonen – Demons. Die Dämonologie der alttestamentlich-jüdischen und frühchristlichen Literatur im Kontext ihrer Umwelt*, Tübingen, 83-97.

Bauks (2007): Michaela Bauks, „The Theological Implications of Child Sacrifice in and beyond the Biblical Context in Relation to Genesis 22 and Judges 11". In: K. Finsterbusch / A. Lange / K.F. Römheld / L. Lazar (Hg.), *Human Sacrifice in Jewish and Christian Tradition, (SHR 112)*, Leiden, 65-86.

Bauks (2016): Michaela Bauks, "Exod 4:24-26 – The Genesis of the 'Torah' of Circumcision in Post-Exilic and Rabbinic Discourses", in: S. Gillmayr-Bucher/M. Häusl (Hg.), Sedayah and Torah in Post-exilic Discourse, Sheffield (in Vorbereitung)

Bazzana (2010): Giovanni Battista Bazzana, "The Bar Kokhba Revolt and Hadrian's Religious Policy". In: M. Rizzi (Hg.), *Hadrian and the Christians*, Berlin, 85-109.

Becking (1999): Bob Becking, "Art. Blood". In: *Dictionary of Deities and Demons in the Bible*, 2. Auflage, 175-176.

Berger (1981): Klaus Berger, *Das Jubiläenbuch* (JSHRZ II/3), Gütersloh.

Black / Green (1992): Jeremy Black / Anthony Green, *Gods, Demons and Symbols of Ancient Mesopotamia. An illustrated Dictionary*, London.

Blaschke (1998): Andreas Blaschke, *Beschneidung. Zeugnisse der Bibel und verwandte Texte*, Tübingen.

Blau (1956): Joshua Blau, „The Ḥatan damim (Ex. IV, 24-26)". In: Tarbiz 26, 1-3.

Blum / Blum (1990): Erhard Blum / Ruth. Blum, „Zippora und ihr *chatan-dāmîm*". *In: E. Blum / C. Macholz / E.W. Stegemann (Hg.), Die hebräische Bibel und ihre zweifache Nachgeschichte (FS R. Rendtorff), Neukirchen-Vluyn, 41-54.*

Böck (2010a): Barbara Böck, „Muššu'u-Beschwörungen". In: *Texte aus der Umwelt des Alten Testaments, Neue Folge (TUAT.NF), Bd. 5: Texte zur Heilkunde*, Gütersloh, 147-152.

Böck (2010b) Barbara Böck, „Krankheiten der Extremitäten und unteren Körperhälfte". In: *Texte aus der Umwelt des Alten Testaments, Neue Folge (TUAT.NF), Bd. 5: Texte zur Heilkunde*, Gütersloh, 69-106.

Böck (2013): Barbara Böck, *The Healing Goddess Gula. Towards an Understanding of Ancient Babylonian*, Leiden.

Cagni (1981): Luigi Cagni, „Il sangue nella letteratura assiro-babilonese". In: *Sangue e Antropologia Biblica*, Rom, 47-86.

Childs (1974): Brevard Childs, *The Book of Exodus (OTL)*, Philadelphia.

Christ (1976): Hieronymus Christ, Blutvergießen im Alten Testament. Der gewaltsame Tod des Menschen untersucht am hebräischen Wort dam (Theologische Dissertationen 12), Basel.

Coats (1999): George W. Coats, Exodus 1-18 [FOTL IIA], Grand Rapids, MI.

Collins (1997): John J. Collins, „A Symbol of Otherness. Circumcision and Salvation in the First Century". In: J.J. Collins, Seer, Sybils and Sages in Hellenistic-Roman Judaism (SJSJ 54), Leiden u. a., 211-235.

Dahood (1981): Mitchell J. Dahood, „Il dio Damu nelle tavolette do Ebla". In: Sangue et Antropologia Biblica, Rom, 97-104.

Douglas (1986): Mary Douglas, *Ritual, Tabu und Körpersymbolik. Sozialanthropologische Studien in Industriegesellschaft und Stammeskultur*, Frankfurt.

Dyk (1990): Peet J. van Dyk, „The Function of So-Called Etiological Elements in Narratives". In: *Zeitschrift für die alttestamentliche Wissenschaft*. 102, 19-33.

Eckhardt (2012): Benedikt Eckhardt, „Introduction". In: ders. (Hg.) Groups, *Normativity, and Rituals. Jewish Identity and Politics between the Maccabees and Bar Kokhba*, Leiden/ Boston, 2-5.

Farber (1980-83): Walter Farber, „Art. Lamaštu". In: Reallexikon der Assyrologie und vorderasiatischen Archäologie. Bd. 6, 439-466.

Farber (2000): Walter Farber, "Art. Witchcraft, Magic, and Divination in Ancient Mesopotamia". In: Civilizations of the Ancient Near East (CANE) III+IV (2000) 1895-1909.

Feucht (2003): Erika Feucht, Beschneidung in Ägypten, in: S. Meyer (Hg.), Temple of the Whole World. Studies in Honour of Jan Assmann, Leiden/Boston, 81-94

Finsterbusch (2006): Karin Finsterbusch, *Vom Opfer zur Auslösung. Analyse ausgewählter Texte zum Thema Erstgeburt im Alten Testament, Vetus Testamentum* 66, 21-45.

Freud (1950): Sigmund Freud, „Der Mann Moses und die monotheistische Religion". In: ders., *Gesammelte Werke, Bd. XVI: Werke aus den Jahren 1932-1939*, hg. von A. Freud, Frankfurt.

Freud (1999a): Sigmund Freud, „Das Tabu der Virginität". In: ders., *Gesammelte Werke,* , *Bd. XII: Werke aus den Jahren 1919-1920*, hg. v. A. Freud u. a., Frankfurt.

Freud (1999b): Sigmund Freud, „Aus der Geschichte einer infantilen Neurose". In: ders., *Gesammelte Werke, Bd. XII: Werke aus den Jahren 1919-1920*, hg. v. A. Freud u. a., Frankfurt.

Freud (1999c): Sigmund Freud, „Totem und Tabu". In: ders., *Gesammelte Werke, Bd. IX:* hg. v. A. Freud u. a., Frankfurt.

Fritz (2003): Michael M. Fritz, *„Und weinten um Tammuz". Die Götter Dumuzi-Ama'ushumgal'anna und Damu* (AOAT 307), Münster.

Hartenstein (2007): Friedhelm Hartenstein: „Zur symbolischen Bedeutung des Blutes im Alten Testament" . In: J. Frey / J. Schröter (Hg.), *Deutungen des Todes Jesu im Neuen Testament* (utb 2953), Tübingen.

Goldschmidt (1996): Lazarus Goldschmidt, *Der Babylonische Talmud, Bd. 5*, Darmstadt.

Grunert (2002): Stefan Grunert, „ ‚Nicht nur sauber, sondern rein' – Rituelle Reinigungsanweisungen aus dem Grab des Anchmahor in Saqqara". In: Studien zur altägyptischen Kultur 30, 137-151.

Gugutzer (2004): Robert Gugutzer, *Zur Soziologie des Körpers*, Bielefeld.

Gurtner (2013): Daniel M. Gurtner, *Exodus. A Commentary on the Greek Text of Codex Vaticanus*, Leiden.

Hannig / Witthuhn (2010): Rainer Hannig / Orell Witthuhn, „Ägyptische medizinische Texte des 2. Jt. v. Chr." , In: *Texte aus der Umwelt des Alten Testaments, Neue Folge (TUAT. NF), Bd. 5: Texte zur Heilkunde,* Gütersloh, 217-273.

Hendel (2008a): Ronald Hendel, "Mary Douglas and Anthropological Modernism", *Journal of Hebrew Scriptures* 8, Art. 8.

Hendel (2008b): Ronald Hendel, "Remembering Mary Douglas: *Kashrut,* Cultures, and Thought-Style", Jewish Studies 45, 3-15.

Houtman (1983): Cornelis Houtman, "Exodus 4:24-26 and Its Interpretation". In: *Journal of Northwest Semitic Languages*. Vol. 11, 81-105.

Houtman (1993): Cornelis Houtman, *Exodus.* Vol. 1 *(HCOT)*, Kampen.

Hüllstrung (2003): Wolfgang Hüllstrung, „Wer versuchte wen zu töten? Ein Beitrag zum Verständnis von Exodus 4,24-26". In: A. Lange / H. Lichtenberger (Hg.), *Dämonen – Demons. Die Dämonologie der alttestamentlich-jüdischen und frühchristlichen Literatur im Kontext ihrer Umwelt*, Tübingen, 182-196.

Jacob (1997): Benno Jacob, Das Buch Exodus, hg. v. Sh. Mayer, Stuttgart.

Jacob (2008): Andrew S. Jacob, Blood will out. Jesus' Circumcision and Christian Readings of Exod 4:24-24, Henoch 30, 310-332.

Jördens (2010): Andrea Jördens, „Griechische Texte aus Ägypten: Texte zur Beschneidung". In: *Texte aus der Umwelt des Alten Testaments, Neue Folge (TUAT.NF), Bd. 5: Texte zur Heilkunde*, Gütersloh, 326-330.

Kessler (2001): Rainer Kessler, „Psychoanalytische Lektüre biblischer Texte – das Beispiel von Ex 4,24-26". In: *Evangelische Theologie* 61, 204-221.

Knauf (1988): Ernst A. Knauf, *Midian. Untersuchungen zur Geschichte Palästinas und Nordarabiens am Ende des 2. Jahrtausend v. Chr.*, Wiesbaden.

Knauf (1997): Ernst A. Knauf, „Supplementa Ismaelitica". In: *Biblische Notizen 40*, 16-19.

Köcher (1953): Franz Köcher, „Der babylonische Göttertypentext". In: *Mitteilungen des Instituts für Orientforschung 1*, 57-107.

Kosmala (1962): Hans Kosmala, „That Bloody Husband": In: *Vetus Testamentum* 12, 14-28.

Kuhlmann (2002): Peter Kuhlmann, *Religion und Erinnerung*, Göttingen.

Kunin (1996): Seth D. Kunin, „The Bridegroom of Blood: A Structural Analysis". In: *Journal for the Study of Old Testament*. Vol. 21, 3-16.

Le Boulluec / Sandevoir (1989): Alain le Boulluec / Pierre Sandevoir, *Exodus [BdA 2]*, Paris.

Lescow (1993): Theodor Lescow, „Ex 4,24-26: Ein archaischer Bundesschlußritus". In: *Zeitschrift für die alttestamentliche Wissenschaft*. 105, 19-26.

Levin (1993): Christoph Levin, Der Jahwist (FRLANT 157), Göttingen.

Levinson (1997): Bernard M. Levinson, *Deuteronomy and the Hermeneutics of Legal Innovation*, Oxford.

Maciejewski (2003): Franz Maciejewski, „Das biblische Archiv der Beschneidung". In: *Biblische Notizen 117*, 33-39.

Mason (2007): Steve Mason, "Jews, Judaeans, Judaizing, Judaism. Problems of categorization in Ancient Judaism". In: *Journal for the Study of Judaism* 38, 457-512.

Mayer (1989): G. Mayer, "Art. ʿarel, ʿårlâ". In: *Theologische Wörterbuch zum Alten Testament*. Bd. 6, 385-387.

Meyer (1906): Eduard Meyer, *Die Israeliten und ihre Nachbarstämme*, Halle.

Morgenstern (1963): Julian Morgenstern, "The 'Bloody Husband' (?) (Exod. 4:24-26) once again". In: *Hebrew Union College Annual 34*, 35-70.

Naumann (2005): Thomas Naumann, „Die Preisgabe Isaaks (Gen 22) im Kontext der biblischen Abraham-Sara-Geschichte". In: B. Janowski / N. Greiner (Hg.), *Genesis 22 in Judentum, Christentum und Islam*, Tübingen, 19-50.

Olyan (2000): Saul Olyan, Rites and Rank: Hierarchy in Biblical Representations of Cult, Princeton/N.J.

Oppenheimer (2003): Aharon Oppenheimer, „The Ban on Circumcision as a Cause of the Revolt: A Reconsideration". In: P. Schäfer (Hg.), *The Bar Kokhba War Reconsidered. New perspectives on the second Jewish revolt against Rome*, Tübingen, 55-69.

Prechel / Richter (2001): Doris Prechel / Thomas Richter, „Abrakadabra oder Althurritisch. Betrachtungen zu einigen altbabylonischen Beschwörungstexten". In: dies. / J. Klinger (Hg.), *Kulturgeschichten. Altorientalische Studien (FS V. Haas)*, Saarbrücken, 333-371.

Propp (1993): William H. C. Propp, "That Bloody Bridegroom (Exodus IV 24-6)". In: *Vetus Testamentum 43*, 495-518.

Propp (1999): William H. C. Propp, *Exodus 1-18 (AB 2)*, New York.

Quack (2012a): Joachim Friedrich Quack, „Zur Beschneidung im Alten Ägypten". In: A. Berlejung / J. Dietrich / J.F. Quack (Hg.), *Menschenbild und Körperkonzepte*, Tübingen, 561-651.

Quack (2012b): Joachim Friedrich Quack, „Die traditionelle Beschneidung, ihr Verbot und ihre Sondergenehmigung in Ägypten". In: J. Heil / S.J. Kramer (Hg.), *Beschneidung: Das Zeichen des Bundes in der Kritik. Zur Debatte um das Kölner Urteil*, Berlin, 17-22.

Richter (1996): Hans-Friedemann Richter, „Gab es einen ‚Blutbräutigam'? Erwägungen zu Ex 4,24-26". In: M. Vervenne (Hg.), *Studies in the Book of Exodus. Redaction – Reception – Interpretation (BEThL 126)*, Leuven, 433-441.

Römer (1994): Thomas Römer, *De l'archaïque au subversif: le cas d'Exode 4/24-26*, Etudes Théologiques et Religieuses 69, 1-12.

Römer (1998): Thomas Römer, *Dieu obscur. Le sexe, la cruauté et la violence dans l'Ancien Testament (Essais bibliques 27)*, Genève.

Rüpke (2006): Jörg Rüpke, *Die Religion der Römer. Eine Einführung*, 2. Auflage, München.

Sarna (1991): Nahum Sarna, *Exodus (JPS Torah-Commentary)*, Bd. 2, Philadelphia.

Sasson (1966): Jack Sasson, „Circumcision in the Ancient Near East". In: *Journal of Biblical Literatur 85*, 473-476.

Schäfer (1981): Peter Schäfer, *Der Bar Kokhba-Aufstand. Studien zum zweiten jüdischen Krieg gegen Rom*, Tübingen.

Scherer (2008): Andreas Scherer, „Art. Ätiologie". In: *Das Wissenschaftliche Bibellexikon im Internet* (www.wibilex.de, Zugriffsdatum: 12.10.2013).

Schmidt (2011): Werner H. Schmidt, *Exodus 1,1-6,30 (BKAT 2/1)*, 2. Auflage, Neukirchen-Vluyn.

Schneemann (1980): Gisela Schneemann, „Deutung und Bedeutung der Beschneidung nach Ex. 4,24-26". In: *Theologische Literaturzeitung 105*, 794.

Schneemann (1989): Gisela Schneemann, Die Deutung und Bedeutung der Beschneidung nach Exodus 4,24-26, Communio viatorum 32, 21-37 und 193-200.

Schwemer (2010): Daniel Schwemer, „Therapien gegen von Geistern oder von Hexerei verursachten Leiden". In: *Texte aus der Umwelt des Alten Testaments, Neue Folge (TUAT. NF), Bd. 5: Texte zur Heilkunde*, Gütersloh, 123-135.

Seebass (1997): Horst Seebass, Genesis II/I (Vätergeschichte I), Neukirchen-Vluyn.

Sellin (1992): Ernst Sellin, *Mose und seine Bedeutung für die israelitisch-jüdische Religionsgeschichte*, Leipzig.

Spann (2012): Korbinian Spann, „The Meaning of Circumcision for Strangers in Rabbinic Literature". In: Benedikt Eckhardt (Hg.) *Jewish Identity and Politics between the Maccabees and Bar Kokhba. Groups, Normativity, and Rituals*, Leiden/Boston, 225-242.

Stol (1993): Marten Stol, *Epilepsy in Mesopotamia (CM 2)*, Groningen.

Turner (2008): Bryan S. Turner, *Body and Society: Explorations in Social Theory*, 3. Auflage, London.

Van Ruiten / Vandermeersch (1994): Jacques van Ruiten / Patrick Vandermeersch, „Psychoanalyse en historisch-kritische exegese. De actualiteit van Freud's boek over Mozes en het monotheïsme: op zoek naar het 'echte feit'". In: *Tijdschrift voor Theologie 34*, 269-291.

Van Weyde (2004): Karl W. van Weyde, *The Appointed Festivals of YHWH: The Festival Calandar in Leviticus 23 and the sukkot Festival in Other Biblical Texts* (FAT II/4), Tübingen.

Volk (1999): Konrad Volk, „Kinderkrankheiten nach der Darstellung babylonisch-assyrischer Keilschrifttexte". In: *Orientalia N.S. 68*, 1-30.

Wagner (1973): Siegfried Wagner, Art. *bāqaš*, ThWAT 1, 754-769.

Wagner (2010): Volker Wagner, „Profanisierung und Säkularisierung der Beschneidung im Alten Testament". In: *Vetus Testamentum 60*, 447-464.

Wellhausen (1897): Julius Wellhausen, *Reste arabischen Heidentums. Skizzen und Vorarbeiten Heft III*, 2. Auflage Berlin.

Wellhausen (1905): Julius Wellhausen, *Prolegomena zur Geschichte Israels*, 6. Auflage, Berlin.

Westendorf (2010): Wolfhart Westendorf, „Einleitung". In: *Texte aus der Umwelt des Alten Testaments, Neue Folge (TUAT.NF), Bd. 5: Texte zur Heilkunde*, Gütersloh, 209.

Wevers (1990): John W. Wevers, *Notes on the Greek Text of Exodus (SCS 30)*, Atlanta/GA.

Wiggermann (2000): Frans Wiggerman, „Lamaštu, daughter of Anu. A Profile". In: M. Stol / F. Wiggermann (Hg.), *Birth in Babylonia and the Bible. Its Mediterranean Setting (CM)*, Groningen, 217-252.

Willis (2010): John T. Willis, *Yahweh and Moses in Conflict: The Role of Exodus 4:24-26 in the Book of Exodus*, Bern.

Witte (1998): Markus Witte, Die biblische Urgeschichte. Redaktions- und theologiegeschichtliche Beobachtungen zu Gen 1,1-11,26 (BZAW 265), Berlin/New York.

Wyatt (2009): Nicolas Wyatt, "Circumcision and Circumstance: Male Genital Mutilation in Ancient Israel and Ugarit". In: *Journal for the Study of the Old Testament 33*, 405-431.

Register

Kursiv: Textstelle ist zitiert oder ausführlich besprochen

Bibelstellen:

Hebräische Bibel

Weitere Literatur

Mesopotamische Literatur

Ägyptische Literatur

Graeco-romanische Literatur:

Herodot

Antike christliche Literatur

Clemens von Alexandria,

Eusebius

Sachregister

Namensregister / Autoren

Autorinnen und Autoren

Andreas Ackermann hat Ethnologie, Kunstgeschichte und Volkskunde in Frankfurt am Main studiert. Promotion in Soziologie über Multikulturalität in Frankfurt am Main und Singapur (Bielefeld). Nach Stationen am Kulturwissenschaftlichen Institut (Essen) und dem Frobenius-Institut (Frankfurt am Main) wurde er als Professor für Kulturwissenschaft mit Schwerpunkt Ethnologie an die Universität Koblenz-Landau berufen. Zu seinen Forschungsschwerpunkten gehören Ästhetische Ethnologie, Multikulturalität und Visuelle Ethnologie.

Michaela Bauks, Studium der Romanistik und Ev. Theologie in Bochum Liège, Wien, Hamburg; 1995 Promotion im Alten Testament an der Universität Heidelberg (*Die Welt am Anfang. Zum Verhältnis von Vorwelt und Weltentstehung in Gen 1 und in der altorientalischen Literatur*, WMANT 74, 1997); 2002 Habilitation an der Universität Strasbourg (*Die Feinde des Psalmisten und die Freunde Ijobs. Untersuchungen zur Freund-Klage im Alten Testament am Beispiel von Psalm 22* (SBS 203), 2004); 1995 2005 Professorin für Altes Testament an der Faculé de Théologie Protestante (IPT) Montpellier, Frankreich; seit 2005 Professorin für Bibelwissenschaft (Altes Testament) und Religionsgeschichte an der Universität Koblenz-Landau, Campus Koblenz.

Thorsten Gieser, Studium der Ethnologie, Religionswissenschaft und Umweltwissenschaften an der Universität Heidelberg. M.A. in Ethnologie und Religionswissenschaft (2004). Promotion in Ethnologie (PhD in Social Anthropology, 2009) an der University of Aberdeen, Großbritannien. Research Consultant an der Middlesex University. Seit 2011 Akademischer Mitarbeiter im Seminar Ethnologie des Instituts für Kulturwissenschaft der Universität Koblenz-Landau. Forschungsgebiete sind

Kulturphänomenologie, Ethnologie der Sinne/des Körpers, Umweltethnologie (insb. Landschaftsphänomenologie), Psychologische Ethnologie (insb. Dialogical Self, Emotion/Affekt, Empathie) und Religionsethnologie.

Judith Hartenstein, Studium der Evangelischen Theologie in Bonn und Berlin, Promotion an der Humboldt-Universität zu Berlin (1997). Vikariat und Zweites Theologisches Examen. Wissenschaftliche Assistentin an der Kirchlichen Hochschule Wuppertal und am Fachbereich Evangelische Theologie in Marburg, Habilitation 2006. Lehrtätigkeit und Gastprofessuren in Mainz, Berlin und Koblenz. Seit 2013 Professorin für Evangelische Theologie mit dem Schwerpunkt Neues Testament und Religionspädagogik, Universität Koblenz-Landau, Campus Landau. Forschungsgebiete sind apokryphe Literatur zum Neuen Testament, insbesondere koptisch-gnostische Evangelien wie das Mariaevangelium, Johannesevangelium, hermeneutische Fragen der Schriftauslegung (Gender).

Matthias Jung , Studium der Philosophie und kath. Theologie; Promotion 1989, Habilitation 2007; danach Lehr- und Forschungstätigkeit sowie Gastprofessuren in Chemnitz, Jena, Erfurt, Berlin, Bochum, Atlanta und St. Louis/USA. Seit 20010 Professor für Moral- und Rechtsphilosophie an der Universität Koblenz-Landau, Campus Koblenz. Monographien (Auswahl): *Erfahrung und Religion*, Freiburg 1999; *Dilthey zur Einführung*, Hamburg 1996, ²2014; *Hermeneutik zur Einführung*, Hamburg 2001, ⁴2012; *Der bewusste Ausdruck, Anthropologie der Artikulation*, Berlin 2009; *Gewöhnliche Erfahrung*, Tübingen 2014.

PD Dr. Martin F. Meyer lehrt seit 1997 als wissenschaftlicher Mitarbeiter Philosophie an der Universität Koblenz-Landau, Campus Koblenz. Zu seinen wichtigsten Veröffentlichungen zählen die Monographien *Philosophie als Meßkunst. Platons epistemologische Handlungstheorie* (1994) sowie *Aristoteles und die Geburt der biologischen Wissenschaft* (2015) und die von ihm herausgegebenen Bücher *Zur Geschichte des Dialogs. Philosophische Positionen von Sokrates bis Habermas* (2006), *Zur Kulturgeschichte der Scham* (zus. mit M. Bauks 2010) und *Zur Kulturgeschichte der Botanik* (zus. mit M. Bauks 2013).

Käte Meyer-Drawe, 1. und 2. Staatsexamen für das Lehramt an Grund- und Hauptschulen. Promotion in Bielefeld (1977); Habilitation in Bochum (1983). Seit 1984

Professorin für Allgemeine Pädagogik am Institut für Erziehungswissenschaft der Ruhr-Universität Bochum. Wissenschaftliche Schwerpunkte in Forschung und Lehre: Selbst- und Weltdeutungen unter dem Einfluss moderner Technologien; Maschinenbilder in der Pädagogik; die Bedeutung der Leiblichkeit für Bildung, Erziehung und Lernen; Comeniologische Studien; Erarbeitung eines zeitgemäßen Subjektkonzepts; eine pädagogische Theorie des Lernens.
Publikationen (Auswahl): *Illusionen von Autonomie. Diesseits von Ohnmacht und Allmacht des Ich.* München 2000. *Leiblichkeit und Sozialität. Phänomenologische Beiträge zu einer pädagogischen Theorie der Inter-Subjektivität.* München 2001. *Menschen im Spiegel ihrer Maschinen.* München. 2007. *Diskurse des Lernens.* München 2012.

Saul M. Olyan ist Samuel Ungerleider Jr. Professor am Institut für Judaistik und Professor am Institut für Religionswissenschaft an der Brown Universität, Providence, USA. Seine wichtigsten Veröffentlichungen zu Anthropologie und Ritualforschung sind *Social Inequality in the World of the Text: The Significance of Ritual and Social Distinctions in the Hebrew Bible* (Göttingen: Vandenhoeck & Ruprecht, 2011); *Disability in the Hebrew Bible: Interpreting Mental and Physical Differences* (Cambridge University Press, 2008); *Biblical Mourning: Ritual and Social Dimensions* (Oxford University Press, 2004); *Rites and Rank: Hierarchy in Biblical Representations of Cult* (Princeton University Press, 2000).

Thomas Reinhardt hat in Frankfurt und Basel Ethnologie, Germanistik und Romanistik studiert. Promotion (1999) und Habilitation (2005) im Fach Historische Ethnologie in Frankfurt/Main. Lehrtätigkeiten an den Universitäten Paderborn, Frankfurt/Main, Köln und München. Seit 2012 Professor am Institut für Ethnologie der LMU München. 2014 Chaire Alfred Grosser an der Sciences Po Paris. Forschungsschwerpunkte: Anthropologie der Medien und Visuelle Anthropologie, Strukturalismus und Poststrukturalismus, Semiotik, Gedächtnispolitiken und Ethnologie der Verwandtschaft. Regionale Schwerpunkte: Afrika, Europa und Afroamerika.

Thomas Römer, Studium der Theologie und Religionswissenschaften in Heidelberg, Tübingen und Paris (1974-1981). Assistent an der theologischen Fakultät der Universität Genf mit Lehrauftrag für biblisches Hebräisch und Ugaritisch (1984–1989). Promotion zum Dr. Theol. der Universität Genf (1988): *Israels Väter* (veröffentlicht

1990). Maître d'enseignement et de recherche, Universität Genf (1989-1991). Dort Assistenzprofessor (1991-1993). Seit 1993 Professor für die Hebräische Bibel an der Theologischen und Religionswissenschaftlichen Fakultät der Universität Lausanne. Dekan dieser Fakultät (1999-2003). Diverse Gastprofessuren (Paris, Montpellier, Zürich, Managua, Mexico). Berufung an das Collège de France, Paris, Lehrstuhl « Milieux Bibliques » (2007). Außerordentlicher Professor der Theologischen Fakultät der Universität Pretoria (2013). Doktor h.c. der Universität von Tel Aviv (2015).

Magnus Schlette, Referent für Philosophie an der Forschungsstätte der Evangelischen Studiengemeinschaft und Privatdozent für Philosophie an der Universität Erfurt; Forschungsschwerpunkte: Anthropologie, Hermeneutik, Ästhetik, Religionsphilosophie. Publikationen: *Die Idee der Selbstverwirklichung. Zur Grammatik des modernen Individualismus*, Campus: Frankfurt/M. 2013.

Prof. Dr. Rüdiger Schmitt, ist Nachwuchsgruppenleiter am Exzellenzcluster Religion und Politik in den Kulturen der Vormoderne und der Moderne an der Westfälischen Wilhelms-Universität Münster. Aktuelle Forschungsprojekte befassen sich mit der Mantik im Alten Israel und der Rezeptionsgeschichte der Lehre von den vier Weltreichen in Daniel 2.

Susan A. J. Stuart ist Senior Lecturer für Kritische Philosophie an der University of Glasgow. Ihr Schwerpunkt liegt auf dem Gebiet der Phänomenologischen und Hermeneutischen Philosophie, wobei sie sich auf die Ausarbeitung des Konzeptes der *Enkinaesthesia* (eines dynamischen, physisch-affektiven Wirkungs- und Erfahrungsfelds) konzentriert. Buchpublikation (neben zahlreichen Aufsätzen): (zus. mit G. Dogic-Crnkovic), *Computation, Information, Cognition : The Nexus and The Liminal*. Newcastle. 2007.